Theory of Electromagnetic Waves

THEORY OF ELECTROMAGNETIC WAVES

Jin Au Kong

Department of Electrical Engineering
and Research Laboratory of Electronics
Massachusetts Institute of Technology

A Wiley-Interscience Publication

JOHN WILEY & SONS

New York · London · Sydney · Toronto

phys

Library of Congress Cataloging in Publication Data:

Kong, Jin Au, 1942–
Theory of electromagnetic waves.

"A Wiley-Interscience publication."
Bibliography: p.
Includes index.
1. Electromagnetic waves. I. Title.

QC661.K65 537 74-34258
ISBN 0-471-50190-5

Printed in the United States of America

10 9 8 7 6 5 4 3 2 1

AB 6/1/04

TO MY PARENTS

Preface

This book presents a unified theory of macroscopic electrodynamics in accordance with the principle of special relativity from the point of view of the form invariance of Maxwell's equations and the constitutive relations. Electromagnetic waves are studied, with attention given to their behavior inside media that are linear but not isotropic. Topics essential to the understanding of electromagnetic waves are so selected and presented as to make the book a useful graduate text.

In describing physical laws of nature, equations that are form invariant with respect to uniformly moving observers are called Lorentz covariant. Written in tensor notation, both Maxwell's equations and the constitutive relations are manifestly covariant. The Lorentz covariant description of the constitutive relations is called bianisotropic. Chapter 1 outlines the macroscopic Maxwell's theory, discussing the fundamental equations and the constitutive relations. Chapter 2 deals with electromagnetic fields and special relativity. The transformation of field quantities and the constitutive relations from one frame to another is derived from the Lorentz transformation for space and time. Isotropic or anisotropic media, when viewed from a moving frame, become bianisotropic. Problems involving one moving medium can usually be solved in the rest frame of the medium, with the results tranformed back to the observer's frame of reference. When more than two moving media are involved, the concept of bianisotropic media is indispensable because in the rest frame of one medium all others are in motion. In Chapter 7, four-dimensional notation is introduced and macroscopic Maxwell's equations and energy momentum conservation laws are derived from variational principles.

Throughout the book electromagnetic waves are the primary concern. For instance, when an electromagnetic wave encounters a medium, we are concerned with how the wave is affected by the medium, rather than the medium's reaction to the wave field. Because of the increasing interest in media that are not isotropic, the formulation and solution of problems here are so oriented that they can be conveniently generalized to treat such media. Simple cases of anisotropic and bianisotropic media are illustrated whenever the algebra, which may become very complicated, does not overshadow the main results and their physical interpretation. Chapter 3 develops the kDB system to treat plane waves in homogeneous anisotropic and bianisotropic media. Reflection and transmission of plane waves are considered in Chapter 4. In Chapter 5 we study guided waves and generalize the formulation to cover moving gyrotropic media. Chapter 6 examines radiation due to point sources, scattering of dipole fields by stratified uniaxial media, and scattering of plane waves by corrugated surfaces, cylinders, and spheres. To provide the student with proper mathematical background, we introduce the saddle point method and make extensive use of the Hankel functions. After such exposure the student should have sufficient confidence to face more complicated techniques and special functions. In Chapter 7 theorems and concepts useful in solving electromagnetic wave problems are illustrated. Quantization of electromagnetic waves in macroscopic media is also considered.

We use boldface letters to denote vectors and boldface letters with an overbar to denote matrices and dyadics. For time-harmonic fields we choose the time-dependent factor $\exp(-i\omega t)$, which leads to familiar equations in quantum theory. It also leads to the definition of an impedance which is the complex conjugate of that used in circuit theory, but circuit concepts are not emphasized in this book. For four-dimension tensors, the contravariant and covariant convention is adopted. The use of potential functions in treating electromagnetic wave problems is not emphasized because they do not appear to have direct physical interpretation and because they lead to algebraic complication in dealing with media that are not isotropic. Potential functions are introduced in Chapter 7.

This book is the outgrowth of a graduate course that I have been teaching at Massachusetts Institute of Technology. The development of the various concepts relies heavily on published work. I have not attempted the task of referring to all relevant publications. The list of books and journal articles in the Reference section is at best representative and by no means exhaustive. Some of the results contained in the book were obtained while I was a graduate student and postdoctoral research engineer at Syracuse University and a Vinton Hayes postdoctoral fellow at M.I.T. I am grateful to Professor David K. Cheng of Syracuse University, and Professors Paul L. Penfield, Jr., Louis D. Smullin, and Henry J. Zimmermann of M.I.T. for the opportunity, the support, and the encouragement to pursue research in electromagnetic fields. I am also indebted to my teaching assistants Leung Tsang and Nathaniel J. Fisch for their comments and suggestions, and to many of my students for their enthusiastic response which gives me joy and satisfaction in teaching. And I thank Dr. Helen L. Thomas for reading and editing the manuscript, my wife Wen Yuan for her assistance in typing, and Ms. Cynthia Kopf for preparing the final version.

J. A. KONG

Cambridge, Massachusetts
November 1974

Contents

1

Macroscopic Maxwell's Theory

In the study of electromagnetic wave propagation and radiation, our primary interest is in the behavior of the wave as a result of its interaction with the environment. The environment is defined electromagnetically by specifying the constitutive relations and the geometrical configurations of material media. In principle, any macroscopic medium can be treated as a collection of microscopic elementary particles. These particles constitute sources for the electromagnetic fields and interact with them. A macroscopic description of the medium can be obtained from microscopic theory by a proper averaging proccess, and for many years this formidable task has attracted research interest. In our studies, we do not resort to microscopic theory. Our approach is phenomenological, and the foundation is macroscopic Maxwell's theory.

The fundamental relations are Maxwell's equations that govern the behavior of electromagnetic fields. The properties of media are described phenomenologically by constitutive relations. Maxwell's equations, together with the constitutive relations, form a self-consistent set of equations. For boundaries and sources where a differential equation description is not applicable, Maxwell's equations are supplemented by boundary conditions. In principle, with properly prescribed source distributions, geometrical configurations, and medium properties, solutions of electromagnetic fields in a given problem can be obtained. We shall first present the fundamental laws of electromagnetism and derive the boundary conditions. The concept of bianisotropic media is then introduced to provide a general description of the constitutive relations.

1.1 FUNDAMENTAL EQUATIONS

1.1a Maxwell's Equations

The behavior of electromagnetic fields in the presence of material media is governed by Maxwell's equations of macroscopic electrodynamics:

$$\nabla \times \mathbf{E} + \frac{\partial \mathbf{B}}{\partial t} = 0 \quad \text{(Faraday's induction law)} \tag{1.1a}$$

$$\nabla \times \mathbf{H} - \frac{\partial \mathbf{D}}{\partial t} = \mathbf{J} \quad \text{(generalized Ampère's law)} \tag{1.1b}$$

$$\nabla \cdot \mathbf{B} = 0 \quad \text{(Gauss' magnetic field law)} \tag{1.1c}$$

$$\nabla \cdot \mathbf{D} = \rho \quad \text{(Gauss' electric field law)} \tag{1.1d}$$

where $\mathbf{E}$ = electric field strength (volts/meter),
$\mathbf{H}$ = magnetic field strength (amperes/meter),
$\mathbf{B}$ = magnetic flux density (webers/meter^2),
$\mathbf{D}$ = electric displacement (coulombs/meter^2),
$\mathbf{J}$ = electric current density (amperes/meter^2),
ρ = electric charge density (coulombs/meter^3).

The celebrated displacement current term $\partial\mathbf{D}/\partial t$ in Ampère's law is due to Maxwell. The continuity equation follows from (1.1b) and (1.1d):

$$\nabla\cdot\mathbf{J}+\frac{\partial\rho}{\partial t}=0. \tag{1.1e}$$

Equation 1.1e expresses the conservation of electric charge and is a fundamental law of physics.

We are usually interested in field quantities whose time dependence is monochromatic, that is, in functions of a single angular frequency ω. We write

$$\mathbf{E}(t)=\mathrm{Re}\left[\mathbf{E}(\omega)e^{-i\omega t}\right]$$

and similarly for other field quantities $\mathbf{H}$, $\mathbf{B}$, $\mathbf{D}$, $\mathbf{J}$, and ρ. The field vector $\mathbf{E}(\omega)$ in the frequency domain is a complex quantity, in contrast to $\mathbf{E}(t)$ in the time domain, which is always real. In the frequency domain, Maxwell's equations become

$$\nabla\times\mathbf{E}=i\omega\mathbf{B}, \tag{1.2a}$$

$$\nabla\times\mathbf{H}=-i\omega\mathbf{D}+\mathbf{J}, \tag{1.2b}$$

$$\nabla\cdot\mathbf{B}=0, \tag{1.2c}$$

$$\nabla\cdot\mathbf{D}=\rho. \tag{1.2d}$$

Note that now all field quantities in (1.2) are complex. We shall not use different symbols to distinguish complex quantities in the frequency domain and real quantities in the time domain; their meanings should be clear from the context. In (1.2), we have suppressed the time-dependence factor $e^{-i\omega t}$.

1.1b Boundary Conditions

A boundary condition is associated with each of Maxwell's equations. Consider a surface boundary joining two different media (Fig. 1.1a), and let $\hat{n}$ be the surface normal to the boundary pointing into medium 1. The boundary conditions corresponding to Maxwell's equations are as follows.

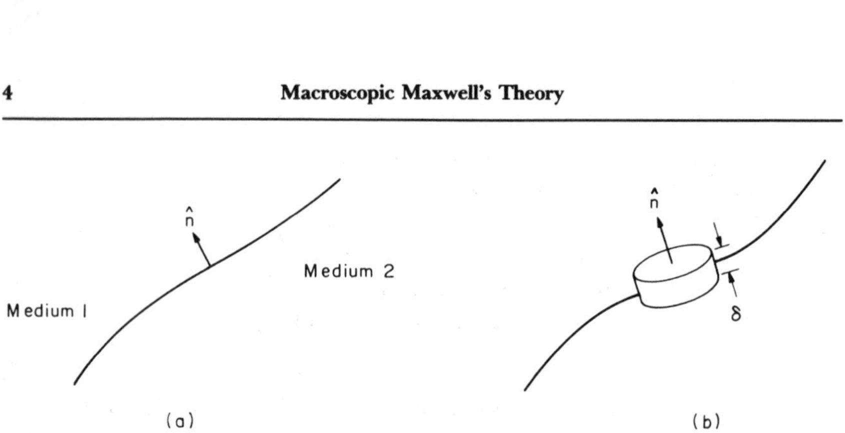

Figure 1.1 **(*a*) A surface boundary between two different media. (*b*) A small pillbox for deriving boundary conditions.**

Tangential **E** continuous:

$$\hat{n}\times(\mathbf{E}_1-\mathbf{E}_2)=0. \tag{1.3a}$$

Discontinuity in tangential **H** equals surface current density $\mathbf{J}_s$:

$$\hat{n}\times(\mathbf{H}_1-\mathbf{H}_2)=\mathbf{J}_s. \tag{1.3b}$$

Normal **B** continuous:

$$\hat{n}\cdot(\mathbf{B}_1-\mathbf{B}_2)=0. \tag{1.3c}$$

Discontinuity in normal **D** equals surface charge density ρ_s:

$$\hat{n}\cdot(\mathbf{D}_1-\mathbf{D}_2)=\rho_s. \tag{1.3d}$$

We see that the boundary conditions take the same form as Maxwell's equations with the curl operator replaced by $\hat{n}$, the time derivative by zero, fields by their differences across the boundary, and **J** and ρ by surface current and charge densities.

To derive the boundary conditions, we need to introduce vector identities. Consider the integration of a vector field **A** in a volume V enclosed by a surface S with surface normals $\hat{s}$. The following formulas are useful:

$$\iiint_V dV\,\nabla\cdot\mathbf{A}=\oiint_S dS\,\hat{s}\cdot\mathbf{A}, \tag{1.4a}$$

$$\iiint_V dV\,\nabla\times\mathbf{A}=\oiint_S dS\,\hat{s}\times\mathbf{A}. \tag{1.4b}$$

Note that (1.4a) is Gauss' theorem that turns an integration of the divergence of a vector field into an integration of the field over the enclosed surface. Equation 1.4b is derivable from (1.4a) by noting that $\nabla \cdot (\mathbf{C} \times \mathbf{A}) = -\mathbf{C} \cdot (\nabla \times \mathbf{A})$ if $\mathbf{C}$ is a constant vector. Applying (1.4a) to $\nabla \cdot (\mathbf{C} \times \mathbf{A})$, we find (1.4b) dot-multiplied by $\mathbf{C}$ on both sides. Assuming that $\mathbf{C}$ is an arbitrary constant vector, we obtain (1.4b).

At the boundary surface, differential description 1.1 fails to apply. We assume that Maxwell's equations in integral form are valid in all regions, including the boundary surfaces. By integrating (1.1) over a small pillbox volume (Fig. 1.1b) across the boundary between two media and applying (1.4a) and (1.4b) to the integrals, we have

$$\oint\!\!\oint dS(\hat{s} \times \mathbf{E}) + \int\!\!\int\!\!\int dV \frac{\partial \mathbf{B}}{\partial t} = 0, \tag{1.5a}$$

$$\oint\!\!\oint dS(\hat{s} \times \mathbf{H}) - \int\!\!\int\!\!\int dV \frac{\partial \mathbf{D}}{\partial t} = \int\!\!\int\!\!\int dV \mathbf{J}, \tag{1.5b}$$

$$\oint\!\!\oint dS(\hat{s} \cdot \mathbf{B}) = 0, \tag{1.5c}$$

$$\oint\!\!\oint dS(\hat{s} \cdot \mathbf{D}) = \int\!\!\int\!\!\int dV \rho. \tag{1.5d}$$

Although they may be discontinuous, field vectors $\mathbf{B}$ and $\mathbf{D}$ are assumed to be finite across the boundary. Let the volume of the pillbox approach zero by letting δ go to zero and then letting the circular contour shrink to a point. The partial derivatives can be taken out of the integrals when the boundary is stationary. We dispose of terms proportional to δ; thus terms involving derivatives with respect to time will be dropped. Terms involving cross products and dot products by $\hat{s}$ will also be dropped except when $\hat{s}$ is in the direction of $\hat{n}$ or $-\hat{n}$. For the term involving integration over $\mathbf{J}$, we assume that the boundary may support a surface current $\mathbf{J}_s$ such that $\lim_{\delta \to 0} \delta \mathbf{J} = \mathbf{J}_s$. Therefore $\mathbf{J}$ can be infinite along the boundary surface, and the right-hand side of (1.5b) will remain finite. Thus we obtain boundary conditions 1.3a and 1.3b from (1.5a) and (1.5b). Assuming a surface charge layer $\rho_s = \lim_{\delta \to 0} \delta \rho$, we similarly obtain boundary conditions 1.3c and 1.3d from (1.5c) and (1.5d).

Boundary conditions 1.3 are good only for stationary boundaries. When the boundary surface is moving, we can apply the same procedure as before, but we must note that the partial derivatives with respect to time can no longer be taken out of the integrals. In accordance with kinematic theory, when the boundary is moving,

$$\frac{d}{dt} \int\!\!\int\!\!\int dV \mathbf{A} = \int\!\!\int\!\!\int dV \frac{\partial \mathbf{A}}{\partial t} + \oint\!\!\oint dS(\hat{s} \cdot \mathbf{v})\mathbf{A}, \tag{1.6}$$

where $\mathbf{v}$ is the velocity of the boundary. A surface term involving motion must be added to the time derivative of the vector field $\mathbf{A}$ over the volume enclosed by the surface. Introducing (1.6) into (1.5), we obtain the following boundary conditions.[24]

$$\hat{n}\times(\mathbf{E}_1-\mathbf{E}_2)-(\hat{n}\cdot\mathbf{v})(\mathbf{B}_1-\mathbf{B}_2)=0, \tag{1.7a}$$

$$\hat{n}\times(\mathbf{H}_1-\mathbf{H}_2)+(\hat{n}\cdot\mathbf{v})(\mathbf{D}_1-\mathbf{D}_2)=\mathbf{J}_s, \tag{1.7b}$$

$$\hat{n}\cdot(\mathbf{B}_1-\mathbf{B}_2)=0, \tag{1.7c}$$

$$\hat{n}\cdot(\mathbf{D}_1-\mathbf{D}_2)=\rho_s, \tag{1.7d}$$

instead of (1.3). The changes are in (1.7a) and (1.7b). Instead of the continuity of tangential $\mathbf{E}$ fields across the boundary, we now have the discontinuity of tangential $\mathbf{E}$ fields equal to $(\hat{n}\cdot\mathbf{v})(\mathbf{B}_1-\mathbf{B}_2)$. Similarly, the surface current $\mathbf{J}_s$ is not due to discontinuity in $\mathbf{H}$ alone; normal $\mathbf{D}$ components also contribute. In the case of $\hat{n}\cdot\mathbf{v}=0$, that is, the velocity is parallel to the surface, (1.7) reduces to (1.3).

1.1c Poynting's Theorem

Energy conservation follows immediately from Maxwell's equations. We dot-multiply (1.1a) by $\mathbf{H}$, (1.1b) by $\mathbf{E}$, and subtract. By making use of the vector identity $\nabla\cdot(\mathbf{E}\times\mathbf{H})=\mathbf{H}\cdot\nabla\times\mathbf{E}-\mathbf{E}\cdot\nabla\times\mathbf{H}$, we obtain Poynting's theorem:

$$\nabla\cdot(\mathbf{E}\times\mathbf{H})+\mathbf{H}\cdot\frac{\partial\mathbf{B}}{\partial t}+\mathbf{E}\cdot\frac{\partial\mathbf{D}}{\partial t}=-\mathbf{E}\cdot\mathbf{J}. \tag{1.8}$$

Poynting's vector,

$$\mathbf{S}=\mathbf{E}\times\mathbf{H}, \tag{1.9}$$

is interpreted as the power flow density in dimensions of watts per square meter. On the right-hand side of (1.8), $-\mathbf{E}\cdot\mathbf{J}$ is the power supplied by the current $\mathbf{J}$, and $\mathbf{H}\cdot\partial\mathbf{B}/\partial t+\mathbf{E}\cdot\partial\mathbf{D}/\partial t$ represents the time rate of change of the stored electric and magnetic energy. Consider an infinitesimal volume surrounding a point. Equation 1.8 states that the power supplied by the source inside the volume is equal to the sum of the increase in electromagnetic energy and the Poynting's power flowing out of the volume.

Poynting's theorem can be similarly derived in complex notation from (1.2a) and (1.2b). We have

$$\nabla\cdot(\mathbf{E}\times\mathbf{H}^*)=i\omega(\mathbf{H}^*\cdot\mathbf{B}-\mathbf{E}\cdot\mathbf{D}^*)-\mathbf{E}\cdot\mathbf{J}^*. \tag{1.10}$$

The complex Poynting's vector is defined to be

$$\mathbf{S}=\mathbf{E}\times\mathbf{H}^{*}\cdot \tag{1.11}$$

Note that, although we use the same symbol **S** in (1.9) and (1.11), the two are different: one is real; the other, complex. We repeat that we shall not introduce different symbols to distinguish real and complex quantities. The last term in (1.10), $-\mathbf{E}\cdot\mathbf{J}^{*}$, is interpreted as the complex power density delivered to the field by the source **J**.

The instantaneous values of the products of field quantities such as power flow or energy stored can be determined only after the instantaneous field quantities have been determined. Since the time dependence is harmonic and everything oscillates periodically with angular frequency ω, the power and energy of interest are those that are time-averaged. The time average of the product of any two complex fields can be easily obtained. For instance, the time-average Poynting's vector, denoted $\langle\mathbf{S}\rangle$, is given by

$$\langle\mathbf{S}\rangle=\frac{1}{2}\,\mathrm{Re}(\mathbf{E}\times\mathbf{H}^{*}). \tag{1.12}$$

Let us verify for the time average of S_x resulting from $\mathrm{Re}(E_y)$ and $\mathrm{Re}(H_z)$. Contributions from $\mathrm{Re}(E_z)$ and $\mathrm{Re}(H_y)$ are not considered. We write

$$\langle S_x\rangle=\frac{1}{T}\int_0^T dt\,\mathrm{Re}(E_y)\,\mathrm{Re}(H_z). \tag{1.13}$$

where T is one full period of time variation and

$$\mathrm{Re}(E_y)=\mathrm{Re}\left\{E_0(r)e^{-i[\omega t+\phi_1(r)]}\right\}=E_0\cos[\omega t+\phi_1(r)],$$

$$\mathrm{Re}(H_z)=\mathrm{Re}\left\{H_0(r)e^{-i[\omega t+\phi_2(r)]}\right\}=H_0\cos[\omega t+\phi_2(r)],$$

We assume that $E_0(r)$ and $H_0(r)$ are both real functions of space and that ϕ_1 and ϕ_2 contain all information about the phase dependence. After integration, we find

$$\langle S_x\rangle=\frac{1}{2}E_0(r)H_0(r)\cos(\phi_1-\phi_2)$$

$$=\frac{1}{2}\,\mathrm{Re}(E_yH_z^{*}). \tag{1.14}$$

Relation 1.14 is true for products of any two complex field components. Thus in general we can state that the time average of two time-harmonic complex quantities is equal to half of the real part of the product of one quantity and the complex conjugate of the other.

1.2 CONSTITUTIVE RELATIONS

There are eight scalar equations in Maxwell's equations. Gauss' electric field law is a consequence of Ampère's law and the equation of continuity, Gauss' magnetic field law is a consequence of Faraday's law, based on the physical fact that nowhere has a single magnetic monopole been found. Consequently, among the eight scalar equations only six are independent. With a given source distribution **J** and ρ, we wish to solve for the electromagnetic field vectors. The given source must satisfy the equation of continuity. The electromagnetic field vectors are **E**, **B**, **D**, and **H**; they have 12 scalar components. Therefore Maxwell's six independent scalar equations are not sufficient to determine the field vectors—six more scalar equations must be supplied. Those are constitutive relations. This observation is supported by simple physical considerations. Maxwell's equations do not take into account the properties of media, which must play a role in the final solutions. The constitutive relations provide a description of media. In mathematical terms, constitutive relations describe functional dependence among field vectors.

1.2a Constitutive Matrices

Constitutive relations in the most general form can be written as

$$c\mathbf{D}=\overline{\mathbf{P}}\cdot\mathbf{E}+\overline{\mathbf{L}}\cdot c\mathbf{B}, \tag{1.15a}$$

$$\mathbf{H}=\overline{\mathbf{M}}\cdot\mathbf{E}+\overline{\mathbf{Q}}\cdot c\mathbf{B}, \tag{1.15b}$$

where $c=3\times10^8$ m/s is the velocity of light in vacuum, and $\overline{\mathbf{P}}$, $\overline{\mathbf{Q}}$, $\overline{\mathbf{L}}$, and $\overline{\mathbf{M}}$ are all 3×3 matrices. Their elements are called *constitutive parameters.* The reason that we write constitutive relations in the present form is based on relativistic considerations. First, the fields **E** and $c\mathbf{B}$ form a single tensor in four-dimensional space, and so do $c\mathbf{D}$ and **H**. Second, constitutive relations written in the form (1.15) are Lorentz-covariant. These aspects will be discussed in Chapter 2.

Equation 1.15 can be rewritten in the form

$$\begin{bmatrix} c\mathrm{D} \\ \mathrm{H} \end{bmatrix}=\overline{\mathrm{C}}\cdot\begin{bmatrix} \mathrm{E} \\ c\mathrm{B} \end{bmatrix}, \tag{1.16a}$$

and $\overline{C}$ is a 6×6 constitutive matrix:

$$\overline{C}=\begin{bmatrix} \overline{P} & \overline{L} \\ \overline{M} & \overline{Q} \end{bmatrix}, \tag{1.16b}$$

which has the dimension of admittance.

The constitutive matrix $\overline{C}$ may be functions of space-time coordinates, thermodynamical and continuum-mechanical variables, or electromagnetic field strengths. According to the functional dependence of $\overline{C}$, we can classify the various media as (i) inhomogeneous if $\overline{C}$ is a function of space coordinates, (ii) nonstationary if $\overline{C}$ is a function of time, (iii) time-dispersive if $\overline{C}$ is a function of time derivatives, (iv) spatial-dispersive if $\overline{C}$ is a function of spatial derivatives, (v) nonlinear if $\overline{C}$ is a function of the electromagnetic field, and so forth. In the general case $\overline{C}$ may be a function of integral-differential operators and coupled to fundamental equations of other physical disciplines. In this book we are concerned with linear media.

We have defined constitutive relations by expressing **D** and **H** in terms of **E** and **B**. We may also express constitutive relations in the form of **D** and **B** as a function of **E** and **H**:

$$\begin{bmatrix} \mathbf{D} \\ \mathbf{B} \end{bmatrix}=\overline{C}_{EH}\cdot\begin{bmatrix} \mathbf{E} \\ \mathbf{H} \end{bmatrix}, \tag{1.17a}$$

where

$$\overline{C}_{EH}=\begin{bmatrix} \overline{\epsilon} & \overline{\xi} \\ \overline{\zeta} & \overline{\mu} \end{bmatrix}=\frac{1}{c}\begin{bmatrix} \overline{P}-\overline{L}\cdot\overline{Q}^{-1}\cdot\overline{M} & \overline{L}\cdot\overline{Q}^{-1} \\ -\overline{Q}^{-1}\cdot\overline{M} & \overline{Q}^{-1} \end{bmatrix}. \tag{1.17b}$$

Here $\overline{C}_{EH}$ is the constitutive matrix under **EH** representation. To express **E** and **H** in terms of **B** and **D**, we write

$$\begin{bmatrix} \mathbf{E} \\ \mathbf{H} \end{bmatrix}=\overline{C}_{DB}\cdot\begin{bmatrix} \mathbf{D} \\ \mathbf{B} \end{bmatrix}, \tag{1.18a}$$

where

$$\overline{C}_{DB}=\begin{bmatrix} \overline{\kappa} & \overline{\chi} \\ \overline{\gamma} & \overline{\nu} \end{bmatrix}=c\begin{bmatrix} \overline{P}^{-1} & -\overline{P}^{-1}\cdot\overline{L} \\ \overline{M}\cdot\overline{P}^{-1} & \overline{Q}-\overline{M}\cdot\overline{P}^{-1}\cdot\overline{L} \end{bmatrix}. \tag{1.18b}$$

Here $\overline{\mathbf{C}}_{DB}$ is the constitutive matrix under **DB** representation. The other possible construction for expressing **E** and **B** in terms of **H** and **D** is not shown because it will not be needed in later developments.

1.2b Symmetry Conditions

Under time-harmonic excitations, the constitutive matrix $\overline{\mathbf{C}}$ is usually complex and frequency-dependent. In source-free regions where $\mathbf{J}=0$, the time average of the divergence of Poynting's vector, in view of Poynting's theorem (1.10), is

$$\begin{aligned}\langle \nabla\cdot \mathrm{S}\rangle &= \frac{1}{2}\,\mathrm{Re}[i\omega(\mathrm{B}\cdot\mathrm{H}^* - \mathrm{E}\cdot\mathrm{D}^*)]\\ &= \frac{1}{4}\{[i\omega(\mathrm{B}\cdot\mathrm{H}^* - \mathrm{E}\cdot\mathrm{D}^*)] + [i\omega(\mathrm{B}\cdot\mathrm{H}^* - \mathrm{E}\cdot\mathrm{D}^*)]^*\}\\ &= \frac{1}{4} i\omega\Big[\frac{1}{c}\mathrm{E}^*\cdot(\overline{\mathrm{P}} - \overline{\mathrm{P}}^+)\cdot\mathrm{E} + c\mathrm{B}^*\cdot(\overline{\mathrm{Q}}^+ - \overline{\mathrm{Q}})\cdot\mathrm{B}\\ &\qquad + \mathrm{E}^*\cdot(\overline{\mathrm{L}} + \overline{\mathrm{M}}^+)\cdot\mathrm{B} - \mathrm{B}^*\cdot(\overline{\mathrm{L}}^+ + \overline{\mathrm{M}})\cdot\mathrm{E}\Big].\end{aligned} \tag{1.19}$$

The medium can be characterized as passive if $\langle \nabla\cdot\mathbf{S}\rangle<0$, active if $\langle\nabla\cdot\mathbf{S}\rangle>0$, and lossless if $\langle\nabla\cdot\mathbf{S}\rangle=0$. For lossless media, we obtain from (1.19) the *lossless conditions*

$$\overline{\mathrm{P}} = \overline{\mathrm{P}}^+, \tag{1.20a}$$

$$\overline{\mathrm{Q}} = \overline{\mathrm{Q}}^+, \tag{1.20b}$$

$$\overline{\mathrm{M}} = -\overline{\mathrm{L}}^+, \tag{1.20c}$$

where the superscript + denotes a transpose and complex conjugate.

Under **EH** representation, (1.20) can be transformed into

$$\bar{\boldsymbol{\epsilon}} = \bar{\boldsymbol{\epsilon}}^+, \tag{1.21a}$$

$$\bar{\boldsymbol{\mu}} = \bar{\boldsymbol{\mu}}^+, \tag{1.21b}$$

$$\bar{\zeta} = \bar{\xi}^+. \tag{1.21c}$$

Under **DB** representation, the lossless conditions become

$$\bar{\kappa}=\bar{\kappa}^{+}, \tag{1.22a}$$

$$\bar{\nu}=\bar{\nu}^{+}, \tag{1.22b}$$

$$\bar{\chi}=\bar{\gamma}^{+}. \tag{1.22c}$$

For a nondispersive medium where all constitutive parameters are real,

$$\bar{\mathrm{P}}=\bar{\mathrm{P}}^{t}, \tag{1.23a}$$

$$\bar{\mathrm{Q}}=\bar{\mathrm{Q}}^{t}, \tag{1.23b}$$

$$\bar{\mathrm{M}}=-\bar{\mathrm{L}}^{t}, \tag{1.23c}$$

where the superscript t denotes the transpose of the matrices. It can be seen from (1.20) and (1.23) that $\bar{\mathbf{M}}$ is completely determined by $\bar{\mathbf{L}}$. The matrices $\bar{\mathbf{P}}$ and $\bar{\mathbf{Q}}$ are hermitian; when they are real, they are symmetrical. The matrix $\bar{\mathbf{P}}$ has 6 independent elements, and so does $\bar{\mathbf{Q}}$. In general, $\bar{\mathbf{L}}$ has 9 elements. Therefore the constitutive matrix $\bar{\mathbf{C}}$ contains 21 independent elements.

1.2c Isotropic, Anisotropic, and Bianisotropic Media

In the definition of the constitutive relations, the constitutive matrices $\bar{\mathbf{L}}$ and $\bar{\mathbf{M}}$ relate electric and magnetic fields. When $\bar{\mathbf{L}}$ and $\bar{\mathbf{M}}$ are not identically zero, the medium is *bianisotropic*. When there is no coupling between electric and magnetic fields, $\bar{\mathbf{L}}=\bar{\mathbf{M}}=0$ and the medium is *anisotropic*. For an anisotropic medium, if $\bar{\mathbf{P}}=c\epsilon\bar{\mathbf{I}}$ and $\bar{\mathbf{Q}}=(1/c\mu)\bar{\mathbf{I}}$ with $\bar{\mathbf{I}}$ denoting the 3×3 unit matrix, the medium is *isotropic*.

Isotropic Media. The constitutive relations for an isotropic medium can be written simply as

$$\mathrm{D}=\epsilon\mathrm{E}, \qquad \text{where } \epsilon=\text{permittivity}, \tag{1.24a}$$

$$\mathrm{B}=\mu\mathrm{H}, \qquad \text{where } \mu=\text{permeability}. \tag{1.24b}$$

Thus the field vector **E** is parallel to **D** and the field vector **H** is parallel to **B**. In free space void of any matter, $\mu=\mu_0$ and $\epsilon=\epsilon_0$,

$$\mu_0=4\pi\times 10^{-7}\ \text{henry/meter} \tag{1.25a}$$

$$\epsilon_0\approx 8.85\times 10^{-12}\ \text{farad/meter} \tag{1.25b}$$

Inside a material medium, the permittivity ϵ is determined by the electrical properties of the medium and the permeability μ by the magnetic properties of the medium.

A dielectric material can be described by a free-space part and a part that is due to the material medium alone. The material part can be characterized by a polarization vector $\mathbf{P}$ such that $\mathbf{D}=\epsilon_0\mathbf{E}+\mathbf{P}$. The polarization $\mathbf{P}$ symbolizes the electric dipole moment per unit volume of the dielectric material. In the presence of an external electric field, the polarization vector may be caused by induced dipole moments, alignment of the permanent dipole moments of the medium, or migration of ionic charges.

A magnetic material can also be described by a free-space part and a part characterized by a magnetization vector $\mathbf{M}$ such that $\mathbf{B}=\mu_0\mathbf{H}+\mu_0\mathbf{M}$. A medium is diamagnetic if $\mu \leqslant \mu_0$, paramagnetic if $\mu \geqslant \mu_0$. Diamagnetism is caused by induced magnetic moments that tend to oppose the externally applied magnetic field. Paramagnetism is due to alignment of magnetic moments. When placed in an inhomogeneous magnetic field, a diamagnetic material tends to move toward regions of weaker magnetic field, and a paramagnetic material toward regions of stronger magnetic field. Ferromagnetism and antiferromagnetism are highly nonlinear effects. Ferromagnetic substances are characterized by spontaneous magnetization below the Curie temperature. The medium also depends on the history of applied fields, and in many instances the magnetization curve forms a hysteresis loop. In an antiferromagnetic material, the spins form sublattices that become spontaneously magnetized in an antiparallel arrangement below the Néel temperature.

Time dispersion is a common phenomenon for most media in the presence of time-variant fields. As an example, the permittivity of water drops from $80\epsilon_0$ to approximately $1.8\epsilon_0$ as the frequency increases from static to optical ranges. The reason for this decrease is that the alignment of water molecules, which possess permanent dipole moments, is much more ineffective at optical frequencies than in slowly varying fields. As another example, consider the permittivity of an electron plasma:

$$\epsilon(\omega)=\epsilon_0\left(1-\frac{\omega_p^2}{\omega^2}\right), \tag{1.26}$$

where the plasma frequency ω_p is defined to be

$$\omega_p=\sqrt{\frac{Ne^2}{m\epsilon_0}} \approx 56.4\sqrt{N}\ ,$$

with e = electron charge $= -1.6\times 10^{-19}$ C,
m = electron mass $= 9.1\times 10^{-31}$ kg,
N = electron density.

Equation 1.26 explicitly displays the frequency dependence of the permittivity ϵ. It is noted that ϵ is always less than ϵ_0. Equation 1.26 is derived in Problem 1.4 under the assumption of a collisionless process.

The collision process in an electron plasma consumes energy and causes absorption. Let the collision frequency ω_{eff} measure the number of effective collisions an electron makes per unit time. The permittivity becomes (see problem 1.5)

$$\epsilon(\omega)=\epsilon_0\left[\left(1-\frac{\omega_p^2}{\omega^2+\omega_{\text{eff}}^2}\right)+i\frac{\omega_{\text{eff}}\omega_p^2}{\omega(\omega^2+\omega_{\text{eff}}^2)}\right]. \tag{1.27}$$

The imaginary part can also be regarded as a conductive term, since conductivity also accounts for loss. Because our interest is in the various forms of the permittivity, we treat conductivity as an imaginary part of ϵ. We can be easily convinced from Ampère's law that a medium with conductivity σ is equivalent to a dielectric medium with an imaginary part σ/ω added to the permittivity. The ratio $\sigma/\omega\epsilon$ usually is called the *loss tangent* of the medium. The permittivity of a lossy medium is seen to violate the lossless condition (1.21a). The real and imaginary parts of the complex permittivity for a time-dispersive medium are related by the Kramers-Kronig relation, which is a statement of causality and is derived in Problem 1.3.

Anisotropic Media. The constitutive relations for anisotropic media are usually written in the **EH** representation as

$$\mathbf{D}=\bar{\epsilon}\cdot\mathbf{E},\qquad \text{where } \bar{\epsilon}=\text{permittivity tensor}, \tag{1.28a}$$

$$\mathbf{B}=\bar{\mu}\cdot\mathbf{H},\qquad \text{where } \bar{\mu}=\text{permeability tensor}. \tag{1.28b}$$

The field vector **E** is no longer parallel to **D**, and the field vector **H** is no longer parallel to **B**. A medium is *electrically anisotropic* if it is described by the permittivity tensor $\bar{\epsilon}$ and *magnetically anisotropic* if it is described by the permeability tensor $\bar{\mu}$. Note that a medium can be both electrically and magnetically anisotropic.

Crystals are described in general by symmetric permittivity tensors. There always exists a coordinate transformation that transforms a symmetric matrix into a diagonal matrix. In this coordinate system, called the *principal system*,

$$\bar{\epsilon} = \begin{bmatrix} \epsilon_x & 0 & 0 \\ 0 & \epsilon_y & 0 \\ 0 & 0 & \epsilon_z \end{bmatrix}. \tag{1.29}$$

The three coordinate axes are referred to as the principal axes of the crystal. For cubic crystals, $\epsilon_x = \epsilon_y = \epsilon_z$ and they are isotropic. In tetragonal, hexagonal, and rhombohedral crystals, two of the three parameters are equal. Such crystals are *uniaxial*. Here there is a two-dimensional degeneracy; the principal axis that exhibits this anisotropy is called the *optic axis*. For a uniaxial crystal with

$$\bar{\epsilon} = \begin{bmatrix} \epsilon & 0 & 0 \\ 0 & \epsilon & 0 \\ 0 & 0 & \epsilon_z \end{bmatrix}, \tag{1.30}$$

the z axis is the optic axis. The crystal is *positive uniaxial* if $\epsilon_z > \epsilon$; it is *negative uniaxial* if $\epsilon_z < \epsilon$. In orthorhombic, monoclinic, and triclinic crystals, all three crystallographic axes are unequal. We have $\epsilon_x \neq \epsilon_y \neq \epsilon_z$, and the medium is *biaxial*.

An electron plasma as described in (1.26) becomes anisotropic when an external dc magnetic field $\mathbf{B}_0$ is applied. The permittivity tensor becomes

$$\bar{\epsilon} = \begin{bmatrix} \epsilon & -i\epsilon_g & 0 \\ i\epsilon_g & \epsilon & 0 \\ 0 & 0 & \epsilon_z \end{bmatrix} \tag{1.31}$$

under the assumption that $\mathbf{B}_0$ is in the $\hat{z}$ direction. The magnetic field enters the constitutive parameters via the cyclotron frequency ω_c:

$$\omega_c = \frac{eB_0}{m}.$$

The parameters in (1.31) are defined by

$$\epsilon = \epsilon_0 \left(1 - \frac{\omega_p^2}{\omega^2 - \omega_c^2} \right),$$

$$\epsilon_g = \epsilon_0 \left(\frac{\omega_p^2 \omega_c}{\omega(\omega^2 - \omega_c^2)} \right),$$

$$\epsilon_z = \epsilon_0 \left(1 - \frac{\omega_p^2}{\omega^2} \right).$$

In the case of an infinitely strong magnetic field, we have

$$\epsilon = \epsilon_0,$$

$$\epsilon_g = 0,$$

$$\epsilon_z = \epsilon_0 \left(1 - \frac{\omega_p^2}{\omega^2} \right),$$

and the plasma becomes a uniaxial medium.

An anisotropic medium characterized by a hermitian permittivity tensor such as (1.31) is called *gyroelectric* or *electrically gyrotropic*. An anisotropic medium characterized by a hermitian permeability tensor $\bar{\mu}$ such as

$$\bar{\mu} = \begin{bmatrix} \mu & -i\mu_g & 0 \\ i\mu_g & \mu & 0 \\ 0 & 0 & \mu_z \end{bmatrix} \tag{1.32}$$

is called *gyromagnetic* or *magnetically gyrotropic*. An example is a ferrite subjected to a dc magnetic field in the $\hat{z}$ direction. Note that the gyrotropic elements in a gyrotropic medium, although imaginary, do not account for any loss. As we have shown, the lossless conditions require that the permittivity and the permeability tensors be hermitian. When absorptive effects are introduced, however, the permittivity and the permeability tensors will no longer be hermitian.

Bianisotropic Media. For isotropic or anisotropic media, the constitutive relations relate the two electric field vectors and the two magnetic field vectors by either a scalar or a tensor. Such media become polarized when placed in an electric field and become magnetized when placed in a magnetic field. A bianisotropic medium provides the cross coupling between the electric and magnetic fields. When placed in an electric or a magnetic field, a bianisotropic medium becomes both polarized and magnetized.

Magnetoelectric materials, theoretically predicted by Dzyaloshinskii[32] and by Landau and Lifshitz,[65] were observed experimentally in 1960 by Astrov[4] in antiferromagnetic chromium oxide. The constitutive relations that Dzyaloshinskii proposed for chromium oxide have the following form:

$$\mathbf{D}=\begin{bmatrix}\epsilon & 0 & 0\\ 0 & \epsilon & 0\\ 0 & 0 & \epsilon_z\end{bmatrix}\cdot\mathbf{E}+\begin{bmatrix}\xi & 0 & 0\\ 0 & \xi & 0\\ 0 & 0 & \xi_z\end{bmatrix}\cdot\mathbf{H}, \tag{1.33a}$$

$$\mathbf{B}=\begin{bmatrix}\xi & 0 & 0\\ 0 & \xi & 0\\ 0 & 0 & \xi_z\end{bmatrix}\cdot\mathbf{E}+\begin{bmatrix}\mu & 0 & 0\\ 0 & \mu & 0\\ 0 & 0 & \mu_z\end{bmatrix}\cdot\mathbf{H}. \tag{1.33b}$$

It was then shown by Indenbom[49] and by Birss[9] that 58 magnetic crystal classes can exhibit the magnetoelectric effect. Rado[85] proved that the effect is not restricted to antiferromagnetics; ferromagnetic gallium iron oxide is also magnetoelectric.

In 1948, the gyrator was introduced by Tellegen[102] as a new element, in addition to the resistor, the capacitor, the inductor, and the ideal transformer, for describing a network. To realize his new network element, Tellegen conceived of a medium possessing constitutive relations of the form

$$\mathbf{D}=\epsilon\mathbf{E}+\xi\mathbf{H}, \tag{1.34a}$$

$$\mathbf{B}=\xi\mathbf{E}+\mu\mathbf{H}, \tag{1.34b}$$

where $\xi^2/\mu\epsilon$ is nearly equal to 1. Tellegen considered that the model of the medium had elements possessing permanent electric and magnetic dipoles parallel or antiparallel to each other, so that an applied electric field that aligns the electric dipoles simultaneously aligns the magnetic dipoles; and a magnetic field that aligns the magnetic dipoles simultaneously aligns the electric dipoles. Tellegen also wrote general constitutive relations 1.17 and examined the symmetry properties by energy conservation.

In (1.34), ϵ, μ, and ξ are all real scalars. Since the prefix "bi" describes the dependence of **D** or **B** on both **E** and **H**, the medium may be called *biisotropic*. The constitutive parameter ξ may be real or complex when the medium is lossless. The constitutive relation that satisfies the lossless condition is

$$\mathbf{D} = \epsilon\mathbf{E} + \xi\mathbf{H}, \tag{1.35a}$$

$$\mathbf{B} = \xi^*\mathbf{E} + \mu\mathbf{H}. \tag{1.35b}$$

When $\xi = 0$, the medium becomes isotropic. As we shall prove in Chapter 7, if ξ is pure imaginary, the biisotropic medium is reciprocal; otherwise the medium is nonreciprocal.

Media in motion were the first bianisotropic media to receive attention in electromagnetic theory. In 1888, Roentgen[87] discovered that a moving dielectric becomes magnetized when it is placed in an electric field. In 1905, Wilson[110] showed that a moving dielectric in a uniform magentic field becomes electrically polarized. Almost any medium becomes bianisotropic when it is in motion. In Chapter 2, we shall derive constitutive relations for uniformly moving media using the Lorentz transformation of field vectors.

PROBLEMS

1.1. Show that for lossless nondispersive bianisotropic media Poynting's theorem becomes

$$\frac{\partial W}{\partial t} + \nabla \cdot \mathbf{S} = -p,$$

where

$$W = \tfrac{1}{2}(\mathbf{D}\cdot\mathbf{E} + \mathbf{B}\cdot\mathbf{H}) = \text{total stored energy density},$$

$$\mathbf{S} = \mathbf{E}\times\mathbf{H} = \text{Poynting's vector power density},$$

$$-p = -\mathbf{J}\cdot\mathbf{E} = \text{power density supplied by an external agent}.$$

This is in a general form expressing conservation of energy.

1.2. The Lorentz force law relates electromagnetism to mechanics:

$$\mathbf{f} = \rho\mathbf{E} + \mathbf{J}\times\mathbf{B}.$$

It can be used to define the fields **E** and **B**. Substitute Maxwell's equations for ρ and **J** in the right-hand side and show that for lossless nondispersive bianisotropic media

$$\mathbf{f} = -\frac{\partial}{\partial t}(\mathbf{D}\times\mathbf{B}) - \nabla\cdot\left[\tfrac{1}{2}(\mathbf{D}\cdot\mathbf{E}+\mathbf{B}\cdot\mathbf{H})\mathbf{I} - \mathbf{DE} - \mathbf{BH}\right].$$

The interpretation of the terms is

$$\mathbf{G} = \mathbf{D}\times\mathbf{B} = \text{momentum density},$$

$$\overline{\mathbf{T}} = \tfrac{1}{2}(\mathbf{D}\cdot\mathbf{E}+\mathbf{B}\cdot\mathbf{H})\bar{\mathbf{I}} - \mathbf{DE} - \mathbf{BH} = \text{Maxwell stress tensor}.$$

Thus we have the theorem

$$\frac{\partial \mathbf{G}}{\partial t} + \nabla\cdot\overline{\mathbf{T}} = -\mathbf{f};$$

which expresses conservation of momentum. Compare this with Poynting's theorem, which expresses conservation of energy. [*Hint*: Using Maxwell's equations, we obtain

$$\mathbf{f} = -\frac{\partial}{\partial t}(\mathbf{D}\times\mathbf{B}) + \mathbf{E}\nabla\cdot\mathbf{D} + \mathbf{H}\nabla\cdot\mathbf{B} - \mathbf{B}\times(\nabla\times\mathbf{H}) - \mathbf{D}\times(\nabla\times\mathbf{E}).$$

It is easier to use index notation and write the ith component of $[\mathbf{B}\times(\nabla\times\mathbf{H})]$ as

$$[\mathbf{B}\times(\nabla\times\mathbf{H})]_i = \varepsilon_{ijk}B_j\varepsilon_{klm}\frac{\partial}{\partial x_l}H_m = B_j\frac{\partial}{\partial x_i}H_j - B_j\frac{\partial}{\partial x_j}H_i,$$

where we use the identity $\varepsilon_{ijk}\varepsilon_{klm} = \delta_{il}\delta_{jm} - \delta_{im}\delta_{jl}$. The Levi-Cività symbol ε_{ijk} is defined as

$$\varepsilon_{ijk} = \begin{cases} 1 & \text{if } ijk \text{ is an even permutation of } 123, \\ -1 & \text{if } ijk \text{ is an odd permutation of } 123, \\ 0 & \text{if any two of } ijk \text{ are equal.} \end{cases}$$

The ith component of $\mathbf{f}$ is then

$$f_i = -\varepsilon_{ijk}\frac{\partial}{\partial t}D_jB_k + \frac{\partial}{\partial x_j}(D_jE_i + B_jH_i) - D_j\frac{\partial}{\partial x_i}E_j - B_j\frac{\partial}{\partial x_i}H_j.$$

The proof is complete when we apply the constitutive relations and lossless conditions for a nondispersive bianisotropic medium to the last two terms. Note that the last two terms are easier to write with index notation than with vector notation, where new operational definitions must be introduced. The reader should be able to use either notation interchangeably].

1.3. Consider a linear, isotropic, time-dispersive medium with permittivity $\epsilon(\omega)=\epsilon'(\omega)+i\epsilon''(\omega)$. The real and the imaginary parts of the complex permittivity are related by the causality conditions (*Kramers-Kronig relation*):

$$\epsilon'(\omega)-\epsilon_\infty=\frac{1}{\pi}P\int_{-\infty}^{\infty}dz\,\frac{\epsilon''(z)}{z-\omega},$$

$$\epsilon''(\omega)=-\frac{1}{\pi}P\int_{-\infty}^{\infty}dz\,\frac{\epsilon'(z)-\epsilon_\infty}{z-\omega},$$

where P denotes the principal value of the integral.

To prove this, note that the linear relationship between fields **D** and **E** can be written as

$$\mathbf{D}(t)=\epsilon_0\mathbf{E}(t)+\int_{-\infty}^{t}d\tau\,\epsilon_0\chi_e(t-\tau)\mathbf{E}(\tau)$$

$$=\epsilon_0\mathbf{E}(t)+\epsilon_0\int_0^{\infty}d\tau\,\chi_e(\tau)\mathbf{E}(t-\tau).$$

The convolution integral term signifies that $\mathbf{D}(t)$ is determined by **E** at all previous times, because of causality. With time dependence $e^{-i\omega t}$ for the fields, show that the permittivity is

$$\epsilon(\omega)=\epsilon_0\left[1+\int_0^{\infty}d\tau\,\chi_e(\tau)e^{i\omega\tau}\right].$$

Regarding this as a complex function of the complex variable ω, show that $\epsilon(\omega)$ is a single-valued regular function over the upper half of the complex ω plane and that $\epsilon(-\omega^*)=\epsilon^*(\omega)$. Use Cauchy's theorem, and integrate $\int_c dz[\epsilon(z)-\epsilon_\infty]/(z-\omega)$ over a semicircle of infinite radius with the straight side along the real axis, but indented around the pole $z=\omega$. Show that

$$P\int_{-\infty}^{\infty}dz\,\frac{\epsilon(z)-\epsilon_\infty}{z-\omega}-i\pi[\epsilon(\omega)-\epsilon_\infty]=0.$$

Separating this expression into real and imaginary parts yields the Kramers-Kronig relation. Show that, if the conductivity σ is involved, there will be an additional term σ/ω, on the right-hand side of $\epsilon''(\omega)$, because of the additional pole at $z=0$.

1.4. As an example of dispersive media, derive the constitutive relations for a plasma. Consider an electron plasma with ions in the background so that the total charge is neutral. Show that, upon application of an electric field $\mathbf{E}$, the polarization vector $\mathbf{P}$ obeys the equation

$$\frac{d^2\mathbf{P}}{dt^2}=\frac{Ne^2}{m}\mathbf{E}.$$

Under time-harmonic excitation, derive the permittivity $\epsilon(\omega)$ in (1.26). When collision is involved, add a damping term to this equation to obtain

$$\left(\frac{d^2}{dt^2}+\omega_{\text{eff}}\frac{d}{dt}\right)\mathbf{P}=\frac{Ne^2}{m}\mathbf{E}$$

and derive the permittivity in (1.27).

1.5. When an electron plasma is in a dc magnetic field $\mathbf{B}_0$, show that the equation for the polarization vector $\mathbf{P}$ becomes

$$\frac{d^2\mathbf{P}}{dt^2}=\frac{e}{m}\left(Ne\mathbf{E}+\frac{d\mathbf{P}}{dt}\times\mathbf{B}_0\right).$$

Define a vector $\boldsymbol{\omega}_c=e\mathbf{B}_0/m$. Show that, for time dependence $e^{-i\omega t}$,

$$\mathbf{P}=\epsilon_0\left[\frac{\omega_p^2}{\omega_c^2-\omega^2}\mathbf{E}-\frac{\omega_p^2\boldsymbol{\omega}_c}{\omega^2(\omega_c^2-\omega^2)}\boldsymbol{\omega}_c\cdot\mathbf{E}-i\frac{\omega_p^2}{\omega(\omega_c^2-\omega^2)}\boldsymbol{\omega}_c\times\mathbf{E}\right].$$

Under the assumption that $\mathbf{B}_0$ is in the direction of the z axis, derive the permittivity tensor $\bar{\epsilon}(\omega)$ in (1.31).

1.6. In this problem we examine dispersion in the vicinity of a resonant frequency. Show that, if an electron is originally bound to an ion as in an atom,

$$\left(\frac{d^2}{dt^2}+g\omega_0\frac{d}{dt}+\omega_0^2\right)\mathbf{P}=\frac{Ne^2}{m}\mathbf{E},$$

where g is a damping constant, and ω_0 is the characteristic frequency of the electron which accounts for the restoring force. Derive the complex permittivity

$$\epsilon(\omega)=\epsilon_0\left(1+\frac{\omega_p^2}{\omega_0^2-ig\omega\omega_0-\omega^2}\right)$$

$$=\epsilon'(\omega)+i\epsilon''(\omega)$$

and plot the real and imaginary parts of $\epsilon(\omega)$. Identify the region of normal dispersion, where ϵ' increases with the frequency, and the region of anomalous dispersion, where ϵ' decreases with the frequency. Show that ϵ'' is maximum at the resonant frequency ω_0.

1.7. In the case of induced dipole moments, the polarization **P** is proportional to the polarizability per unit volume, $N\alpha$:

$$\mathbf{P}=N\alpha\mathbf{E}_{\text{loc}}.$$

The local electric field at the place of the induced dipole comprises the externally applied field and the field caused by the surrounding dipoles. If a spherical cavity is assumed to surround the particle, show that the local field is

$$\mathbf{E}_{\text{loc}}=\mathbf{E}+\frac{\mathbf{P}}{3\epsilon_0}.$$

From these two equations, show that

$$\frac{\epsilon}{\epsilon_0}=\frac{1+(2N\alpha/3\epsilon_0)}{1-(N\alpha/3\epsilon_0)}.$$

This is the well-known Clausius-Mossoti (Lorentz-Lorenz) equation.

1.8. In the case of media possessing permanent moments, the polarization **P** and the magnetization **M** are given classically by the Langevin equation, which explicitly displays the temperature dependence. The Langevin function is defined as

$$L(x)=\coth x-\frac{1}{x}.$$

For a paramagnetic material with magnetic moments Nm,

$$M=NmL\left(\frac{mH}{kT}\right),$$

where $k=1.38\times10^{-23}$ J/K is Boltzmann's constant, and T is the absolute temperature in degrees Kelvin. Show that in the low-field limit, $mH\ll kT$, the medium is linear.

1.9. The ionosphere extends from approximately 50 km to 1000 km above the ground. The ionization is attributed, for the most part, to ultraviolet radiation from the sun. On the whole, the ion density first increases with altitude and then decreases, but it shows ledges where it varies more slowly with altitude. These ledges are the D, E, F_1, and F_2 layers. These maxima arise because the solar radiation and the composition of the atmosphere both change with altitude. The heights and the intensities of ionization of these layers change with the hour of the day, the season of the year, the sunspot cycle, and so on. The electron density varies from $\sim 10^{10}/\text{m}^3$ to $10^{12}/\text{m}^3$ in going from the lowest to the highest layer. Calculate the plasma frequency and the corresponding permittivity at 5-MHz frequency. How does the permittivity change with frequency? How does the earth's magnetic field ($\approx 6\times 10^{-5}$ Wb/m^2) change the permittivity?

1.10. Describe what is meant, mathematically and physically, by saying that a medium is (*a*) homogeneous versus inhomogeneous, (*b*) linear versus nonlinear, (*c*) time or spatial dispersive versus nondispersive, (*d*) cold versus warm, (*e*) compressible versus incompressible, (*f*) passive versus active, and (*g*) isotropic versus anisotropic versus bianisotropic. Give an example for each case.

1.11. How do you categorize the following constitutive relations?

(*a*) The phenomenon of natural optical activity can be explained with the use of the constitutive relation (Landau and Lifshitz[65])

$$D_i = \epsilon_{ij} E_j + \gamma_{ijk} \frac{\partial E_j}{\partial x_k},$$

where ϵ_{ij} and γ_{ijk} are functions of frequency and $\gamma_{ijk} = -\gamma_{jik}$.

(*b*) In view of the optical activities in quartz crystals, the constitutive relation for a quartz crystal is phenomenologically described as[74]

$$E_j = \kappa_{ij} D_i + \frac{g_{ij}}{\mu_0 \epsilon_0} \frac{\partial B_i}{\partial t},$$

$$H_j = \frac{B_j}{\mu_0} - \frac{g_{ji}}{\mu_0 \epsilon_0} \frac{\partial D_i}{\partial t},$$

where g_{ij} is a gyration tensor.

(*c*) Cholesteric liquid crystals can be modeled[27] by a spiral structure with constitutive relations given by

$$\mathbf{D}=\begin{bmatrix} \epsilon(1+\delta\cos Kz) & \epsilon\delta\sin Kz & 0 \\ \epsilon\delta\sin Kz & \epsilon(1-\delta\cos Kz) & 0 \\ 0 & 0 & \epsilon_z \end{bmatrix}\cdot\mathbf{E},$$

where the spiral direction is along the z axis.

(*d*) When a magnetic field is applied to a conductor carrying a current, an electric field is developed. This is called the *Hall effect.* Using an atomic model, show that the constitutive relation that accounts for the Hall effect is given by

$$\mathbf{J}=\sigma(\mathbf{E}-R\sigma\mathbf{E}\times\mathbf{B}),$$

where σ is the conductivity, and R is the Hall coefficient. For copper, $\sigma\approx 6.7\times 10^{7}$ mho/m and $R\approx -5.5\times 10^{-11}$ m^3/Coul.

(*e*) The phenomenon of pyroelectricity in a crystal is observed when it is heated. The constitutive relation for a *pyroelectric material* can be written as

$$\mathbf{D}=\mathbf{D}_0+\bar{\boldsymbol{\epsilon}}\cdot\mathbf{E},$$

where a spontaneous term $\mathbf{D}_0$ exists even in the absence of an external field.

(*f*) The phenomenon in which dipole moments are induced in a crystal by mechanical stress is called *piezoelectricity*. A piezoelectric material is characterized by a piezoelectric tensor $\gamma_{i,kl}=\gamma_{i,lk}$ such that

$$D_i=D_{0i}+\epsilon_{ik}E_k+\gamma_{i,kl}s_{kl},$$

where s_{kl} is the stress tensor to second order in electric fields. All pyroelectric media are also piezoelectric.

(*g*) An isotropic dielectric can exhibit the *Kerr effect* when placed in an electric field. In this case the permittivity can be written as

$$\epsilon_{ij}=\epsilon\delta_{ij}+\alpha E_iE_j,$$

where ϵ is the unperturbed permittivity. The principal axis of ϵ_{ij} coincides with the electric field.

(*h*) In an electro-optical material that exhibits *Pockel's effect*, the constitutive relation can be written as

$$D_i = \epsilon_{ij} E_j + \alpha_{ijk} E_j E_k,$$

where $\alpha_{ijk} = \alpha_{jik}$ is a third-rank tensor symmetrical in i and j, and therefore has 18 independent elements.

2

Electromagnetic Theory and Special Relativity

Electromagnetic theory was the most important stimulus that prompted Einstein's theory of special relativity. The principle of relativity, which requires that all physical laws be form-invariant as described by all observers, is basic in formulating the laws of nature. Space and time constitute the coordinates for descriptions of physical phenomena. Galilean transformation of space and time was used to provide a basis for deriving transformation laws between observers in relative motion. The principle of relativity that is based on the Galilean transformation is referred to as *Galilean relativity*. Under the Galilean transformation, the laws of Newtonian mechanics are form-invariant, but the laws of electromagnetism change their form. In 1904, Lorentz examined the conditions for form invariance of Maxwell's equations in vacuum between moving observers. In 1905, Einstein deduced Lorentz transformation laws from the single postulate that the velocity of light in vacuum is a universal constant and the assumption that vacuum is linear, isotropic, and homogeneous. Einstein's principle of relativity that is based on the Lorentz transformation is *special relativity*. The laws of Newtonian mechanics, since they were not form-invariant under the Lorentz transformation, have been revised. Physical laws that are form-invariant under the Lorentz transformation are *Lorentz-covariant*.

In this chapter, we shall state the two postulates of special relativity: (i) *Among uniformly moving observers space and time coordinates obey Lorentz transformation laws*; and (ii) *physical laws are form-invariant under the Lorentz transformation laws for space and time*. On the basis of these two postulates, the constancy of the velocity of light is a direct consequence of the Lorentz covariance of Maxwell's equations in vacuum. In 1908, Minkowski postulated that the macroscopic Maxwell's equations in material media are Lorentz-covariant. With this postulate and the Lorentz transformation for space and time coordinates, we can obtain transformation formulas for electromagnetic field vectors. Then the constitutive relations for various moving media can be derived.

2.1 SPACE-TIME TRANSFORMATION

2.1a Lorentz Transformation

Consider the simple case in which the coordinate axes of observers S and S' are parallel with their origins coinciding at time $t=0$. Observer S' moves uniformly with velocity $\mathbf{v}$ relative to S (Fig. 2.1). The Lorentz transformation of space-time coordinates between these two moving observers, with the use of dyadic notation, is given by

$$\text{LT}\begin{cases} ct' = \gamma(ct - \boldsymbol{\beta}\cdot\mathbf{r}), & (2.1\text{a}) \\ \mathbf{r}' = \bar{\boldsymbol{\alpha}}\cdot\mathbf{r} - \gamma\boldsymbol{\beta}ct, & (2.1\text{b}) \end{cases}$$

where

$$\bar{\boldsymbol{\alpha}} = \bar{\mathbf{I}} + (\gamma - 1)\frac{\boldsymbol{\beta}\boldsymbol{\beta}}{\beta^2}, \tag{2.2}$$

$$\gamma = \frac{1}{\sqrt{1-\beta^2}}, \tag{2.3}$$

$$\boldsymbol{\beta} = \frac{\mathbf{v}}{c}, \qquad \beta^2 = \boldsymbol{\beta}\cdot\boldsymbol{\beta}. \tag{2.4}$$

Here, $c = 3\times10^8$ m/s is the velocity of light in vacuum, and $\bar{\mathbf{I}}$ stands for a 3×3 unit dyad. If $\mathbf{v}$ is along the z axis, $\boldsymbol{\beta} = \hat{z}\beta$, and (2.1) becomes

$$ct' = \gamma(ct - \beta z), \tag{2.5a}$$

$$x' = x, \tag{2.5b}$$

$$y' = y, \tag{2.5c}$$

$$z' = \gamma(z - \beta ct). \tag{2.5d}$$

We observe that the time coordinate is not a universal constant; two physical events that are simultaneous in S' will no longer be simultaneous in S.

Note that the 3×3 dyad $\bar{\boldsymbol{\alpha}}$ is symmetrical, $\bar{\boldsymbol{\alpha}}^t = \bar{\boldsymbol{\alpha}}$, where the superscript t denotes transpose. We can deduce from this matrix some useful identities

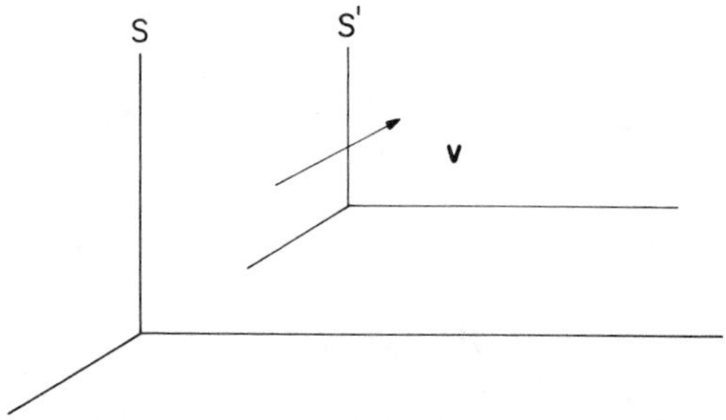

Figure 2.1 Observer S' is moving with velocity v with respect to observer S.

such as

$$\bar{\boldsymbol{\alpha}}^{-1}=\bar{\mathbf{I}}+\left(\frac{1}{\gamma}-1\right)\frac{\boldsymbol{\beta}\boldsymbol{\beta}}{\beta^2} \tag{2.6a}$$

$$=\bar{\boldsymbol{\alpha}}-\gamma\boldsymbol{\beta}\boldsymbol{\beta}, \tag{2.6b}$$

$$\bar{\boldsymbol{\alpha}}^2=\bar{\mathbf{I}}+\gamma^2\boldsymbol{\beta}\boldsymbol{\beta}, \tag{2.6c}$$

$$(\bar{\boldsymbol{\alpha}}^{-1})^2=\bar{\mathbf{I}}-\boldsymbol{\beta}\boldsymbol{\beta}, \tag{2.6d}$$

$$\boldsymbol{\beta}\cdot\bar{\boldsymbol{\alpha}}=\bar{\boldsymbol{\alpha}}\cdot\boldsymbol{\beta}=\gamma\boldsymbol{\beta}, \tag{2.6e}$$

$$\boldsymbol{\beta}\cdot\bar{\boldsymbol{\alpha}}^{-1}=\bar{\boldsymbol{\alpha}}^{-1}\cdot\boldsymbol{\beta}=\frac{1}{\gamma}\boldsymbol{\beta}. \tag{2.6f}$$

For any vector **A**,

$$|\bar{\boldsymbol{\alpha}}\cdot\mathbf{A}|^2=|\mathbf{A}|^2+\gamma^2(\boldsymbol{\beta}\cdot\mathbf{A})^2. \tag{2.6g}$$

We can also define a 3×3 dyad $\bar{\boldsymbol{\beta}}$ such that

$$\bar{\boldsymbol{\beta}}\cdot\mathbf{A}\equiv\boldsymbol{\beta}\times\mathbf{A}. \tag{2.6h}$$

In explicit matrix form

$$\bar{\boldsymbol{\beta}}=\begin{bmatrix} 0 & -\beta_z & \beta_y \\ \beta_z & 0 & -\beta_x \\ -\beta_y & \beta_x & 0 \end{bmatrix} \tag{2.6i}$$

and

$$\bar{\boldsymbol{\alpha}}\cdot\bar{\boldsymbol{\beta}}=\bar{\boldsymbol{\beta}}\cdot\bar{\boldsymbol{\alpha}}=\bar{\boldsymbol{\alpha}}^{-1}\cdot\bar{\boldsymbol{\beta}}=\bar{\boldsymbol{\beta}}\cdot\bar{\boldsymbol{\alpha}}^{-1}=\bar{\boldsymbol{\beta}}. \tag{2.6j}$$

From $(\bar{\boldsymbol{\beta}})^2\cdot\mathbf{A}=\boldsymbol{\beta}\times(\boldsymbol{\beta}\times\mathbf{A})=\boldsymbol{\beta}\boldsymbol{\beta}\cdot\mathbf{A}-\beta^2\mathbf{A}$ it follows that

$$\bar{\boldsymbol{\beta}}^2=\boldsymbol{\beta}\boldsymbol{\beta}-\beta^2\bar{\mathbf{I}}. \tag{2.6k}$$

Although both $\bar{\boldsymbol{\alpha}}$ and $\bar{\boldsymbol{\alpha}}^{-1}$ are symmetric, $\bar{\boldsymbol{\beta}}$ is skew-symmetric.

An important identity can be derived from the LT (2.1). Forming the difference of magnitudes squared of $\mathbf{r}'$ and ct', and using (2.6e) and (2.6g), we find

$$|\mathbf{r}'|^2 - |ct'|^2 = |\mathbf{r}|^2 - |ct|^2. \tag{2.7}$$

Equation 2.7 is important because it is independent of the relative velocity $\mathbf{v}$ between S and S'. It is a numerical constant that is invariant under the Lorentz transformation. Its square root can be regarded as expressing the length of a four-dimensional vector representing the space and time coordinates of a physical event. Evidently, in this four-dimensional space, called *Minkowski space*, the length of a vector can be imaginary as well as real.

2.1b First-Order Lorentz Transformation

When $\mathbf{v}$ is small so that only terms of the order of $\mathbf{v}/c$ are significant, we have $\bar{\alpha} = \bar{\mathbf{I}}$, $\gamma = 1$, and LT 2.1 becomes

$$\text{FOLT}\begin{cases} ct' = ct - \boldsymbol{\beta} \cdot \mathbf{r}, & (2.8\text{a}) \\ \mathbf{r}' = \mathbf{r} - \boldsymbol{\beta} ct. & (2.8\text{b}) \end{cases}$$

We call transformations 2.8 *first-order Lorentz transformations* (*FOLT*). As is seen from FOLT, the space term $\boldsymbol{\beta} \cdot \mathbf{r}$ in the time transformation may not be negligible, since $\mathbf{r}$ can be large, while $\boldsymbol{\beta}$ is small.

2.1c Galilean Transformation

Before 1905, time was regarded as a universal quantity. For two observers in relative uniform motion, the space coordinate changed because of motion, but the time coordinate remained the same:

$$\text{GT}\begin{cases} t' = t, \\ \mathbf{r}' = \mathbf{r} - \mathbf{v}t. \end{cases}$$

This transformation law of space and time is the *Galilean transformation* (*GT*). We note also that the Galilean transformation is not a limiting case of the LT when velocity $\mathbf{v}$ is small. Rather, LT reduces to GT when $\mathbf{v}$ is small and when $\mathbf{r}$ is small compared with ct/β, as is seen from FOLT 2.8a. Mathematically, LT reduces to GT when $c \to \infty$.

The transformation laws for space and time under GT, LT, and FOLT have been given from S to S'. The transformations from S' to S can be obtained by finding their inverses. For the LT, we have

$$\text{LT}\begin{cases} ct = \gamma(ct' + \boldsymbol{\beta} \cdot \mathbf{r}'), & (2.9\text{a}) \\ \mathbf{r} = \bar{\alpha} \cdot \mathbf{r}' + \gamma \boldsymbol{\beta} ct'. & (2.9\text{b}) \end{cases}$$

The FOLT follows from (2.9) by letting $\gamma=1$, and the GT follows from (2.9) by letting $c\to\infty$. Physically, the formulas can be obtained by changing $\boldsymbol{\beta}$ in the transformation laws from S to S' to $-\boldsymbol{\beta}$, because, as S' is moving with $\mathbf{v}$ relative to S, S is moving with $-\mathbf{v}$ relative to S'.

2.2 TRANSFORMATION OF FIELD VECTORS

2.2a Maxwell-Minkowski Theory

According to Minkowski's postulate, macroscopic Maxwell's equations are Lorentz-covariant. Suppose that, from the point of view of an observer S, macroscopic electrodynamics is described by Maxwell's equations:

$$\nabla\times\mathbf{E}+\frac{\partial\mathbf{B}}{\partial t}=0, \tag{2.10a}$$

$$\nabla\cdot\mathbf{B}=0, \tag{2.10b}$$

$$\nabla\times\mathbf{H}-\frac{\partial\mathbf{D}}{\partial t}=\mathbf{J}, \tag{2.10c}$$

$$\nabla\cdot\mathbf{D}=\rho, \tag{2.10d}$$

with the charge conservation law

$$\nabla\cdot\mathbf{J}+\frac{\partial\rho}{\partial t}=0. \tag{2.10e}$$

Then, from the point of view of an observer S' moving with respect to S, Maxwell's equations take the same forms:

$$\nabla'\times\mathbf{E}'+\frac{\partial\mathbf{B}'}{\partial t'}=0, \tag{2.11a}$$

$$\nabla'\cdot\mathbf{B}'=0, \tag{2.11b}$$

$$\nabla'\times\mathbf{H}'-\frac{\partial\mathbf{D}'}{\partial t'}=\mathbf{J}', \tag{2.11c}$$

$$\nabla'\cdot\mathbf{D}'=\rho', \tag{2.11d}$$

with the charge conservation law

$$\nabla'\cdot\mathbf{J}'+\frac{\partial\rho'}{\partial t'}=0, \tag{2.11e}$$

where primes denote quantities associated with S'. The fundamental field quantities are **E**, **B**, **D**, and **H**. If **E** and **B** are regarded as pure field quantities, then **D** and **H** contain information about the material media. Following Sommerfeld,[96] we refer to **E** and **B** as the entities of intensity, and **D** and **H** as the entities of quantity. In four-dimensional Minkowski space the entities of intensity form a field tensor of second rank, and the entities of quantity form an excitation tensor of second rank. By Minkowski's postulate, we can find the transformation laws for all field variables from the Lorentz transformation of space and time.

The formulation that we have just described is called the *Minkowski formulation*. Following the concept that all material media can be regarded as source terms in Maxwell's equations and that only two electromagnetic vectors are fundamental quantities has led to alternative formulations for macroscopic electromagnetic theory, of which the Amperian and the Chu formulations are best known. With each formulation models of the constituents of media are elaborated. In the Amperian formulation (Panofsky and Phillips[82]), constituents of a dipolar medium are visualized as two basic elements, an electric dipole and a current loop. The Amperian model is closely related to the atomic structure, where electrons circulating the nuclei act as a current loop. In the Chu formulation (Fano, Chu, and Adler;[34] Penfield and Haus[83]), a dipolar medium is visualized as containing electric and magnetic dipoles. The Chu formulation is useful because there are no inherently moving parts in a magnetic dipole, as opposed to a current loop. When higher-order moments than the dipole moment are significant in a medium, the task of modeling becomes more involved. No clear model for the Minkowski formulation is proposed.

We must appreciate that a model serves as a guide for visualizing the behavior of a medium in the presence of an electromagnetic field. A model is indispensable when the interest lies in the reaction of a medium under the action of a field. From the point of view of electromagnetic wave propagation and radiation, we are interested not in the medium but in the way the electromagnetic waves behave. Hence there is little need for a detailed model of the medium. The medium properties are prespecified by the constitutive relations. It has been shown by Tai[100] that, with the constitutive relations given, all formulations are equivalent. We favor the Maxwell-Minkowski theory not only because of its simplicity and elegance but also because of its practical applicability.

2.2b Derivation of the Transformation Formulas

Transformation formulas for field vectors are direct consequences of the Lorentz transformation for space and time and Minkowski's postulate of the Lorentz covariance of Maxwell's equations. From the LT given in (2.1) we obtain Lorentz transformation formulas for space-time derivatives:

$$\frac{\partial}{\partial ct}=\gamma\left(\frac{\partial}{\partial ct'}-\boldsymbol{\beta}\cdot\nabla'\right), \tag{2.12a}$$

$$\nabla=\bar{\boldsymbol{\alpha}}\cdot\nabla'-\gamma\boldsymbol{\beta}\frac{\partial}{\partial ct'}. \tag{2.12b}$$

To derive transformation laws for all field vectors, we substitute (2.12) in Maxwell's equations in the S frame and require them to have the same forms in the S' frame. First, consider the charge conservation equation. Transformation from S to S' gives

$$\left(\bar{\boldsymbol{\alpha}}\cdot\nabla'-\gamma\boldsymbol{\beta}\frac{\partial}{\partial ct'}\right)\cdot\mathbf{J}+\gamma\left(\frac{\partial}{\partial ct'}-\boldsymbol{\beta}\cdot\nabla'\right)c\rho=0.$$

Thus the charge conservation equation is Lorentz-covariant if

$$c\rho'=\gamma(c\rho-\boldsymbol{\beta}\cdot\mathbf{J}), \tag{2.13a}$$

$$\mathbf{J}'=\bar{\boldsymbol{\alpha}}\cdot\mathbf{J}-\gamma\boldsymbol{\beta}c\rho. \tag{2.13b}$$

Note that a charge distribution stationary in S certainly produces a current in S', but from (2.13a) a uniform current element in S also generates a charge distribution in S', which is a relativistic effect and cannot be seen under GT.

Next, we introduce (2.12) into Ampère's law and Gauss' electric field law:

$$\left(\bar{\boldsymbol{\alpha}}\cdot\nabla'-\gamma\frac{\partial}{\partial ct'}\boldsymbol{\beta}\right)\times\mathbf{H}-\gamma\left(\frac{\partial}{\partial ct'}-\boldsymbol{\beta}\cdot\nabla'\right)c\mathbf{D}=\mathbf{J}, \tag{2.14a}$$

$$\left(\bar{\boldsymbol{\alpha}}\cdot\nabla'-\gamma\frac{\partial}{\partial ct'}\boldsymbol{\beta}\right)\cdot c\mathbf{D}=c\rho. \tag{2.14b}$$

To find transformation laws for $\mathbf{D}'$ and $\mathbf{H}'$, we wish to cast (2.14) into the form

$$\nabla'\cdot\mathbf{D}'=\rho',$$

$$\nabla'\times\mathbf{H}'-\frac{\partial}{\partial t'}\mathbf{D}'=\mathbf{J}'.$$

In view of (2.13a), we have $\gamma(2.14\text{b})-\gamma\boldsymbol{\beta}\cdot(2.14\text{a})=c\rho'$, which gives

$$\gamma[(\bar{\boldsymbol{\alpha}}\cdot\nabla')\cdot c\mathbf{D}-\boldsymbol{\beta}\cdot(\bar{\boldsymbol{\alpha}}\cdot\nabla')\times\mathbf{H}-\gamma(\boldsymbol{\beta}\cdot\nabla')(\boldsymbol{\beta}\cdot c\mathbf{D})]=c\rho'.$$

Using

$$\boldsymbol{\beta}\cdot[(\bar{\boldsymbol{\alpha}}\cdot\nabla')\times\mathbf{H}]=\boldsymbol{\beta}\cdot\left[\left(\nabla'+(\gamma-1)\frac{(\boldsymbol{\beta}\cdot\nabla')}{\beta^2}\boldsymbol{\beta}\right)\times\mathbf{H}\right]=\boldsymbol{\beta}\cdot\nabla'\times\mathbf{H}$$

and the identity $\bar{\boldsymbol{\alpha}}^{-1}=\bar{\boldsymbol{\alpha}}-\gamma\boldsymbol{\beta}\boldsymbol{\beta}$, we have

$$\nabla'\cdot[(\gamma\bar{\boldsymbol{\alpha}}^{-1}\cdot c\mathbf{D}+\gamma\boldsymbol{\beta}\times\mathbf{H})]=c\rho'. \tag{2.15a}$$

By the same token, we use (2.13b) to calculate $\bar{\boldsymbol{\alpha}}\cdot$(2.14a)$-\gamma\boldsymbol{\beta}$(2.14b)$=\mathbf{J}'$, which gives

$$\bar{\boldsymbol{\alpha}}\cdot[(\bar{\boldsymbol{\alpha}}\cdot\nabla')\times\mathbf{H}]+\gamma\bar{\boldsymbol{\alpha}}\cdot(\boldsymbol{\beta}\cdot\nabla')c\mathbf{D}-\boldsymbol{\beta}[(\bar{\boldsymbol{\alpha}}\cdot\nabla')\cdot c\mathbf{D}]$$
$$+\frac{\partial}{\partial ct'}[-\gamma\bar{\boldsymbol{\alpha}}\cdot(\boldsymbol{\beta}\times\mathbf{H})-\gamma(\bar{\boldsymbol{\alpha}}\cdot c\mathbf{D})+\gamma^2\boldsymbol{\beta}\boldsymbol{\beta}\cdot c\mathbf{D}]=\mathbf{J}'.$$

Using the fact that

$$\begin{aligned}\bar{\boldsymbol{\alpha}}\cdot[(\bar{\boldsymbol{\alpha}}\cdot\nabla')\times\mathbf{H}]&=\nabla'\times\mathbf{H}+\frac{\gamma-1}{\beta^2}[(\boldsymbol{\beta}\cdot\nabla')\boldsymbol{\beta}\times\mathbf{H}-\boldsymbol{\beta}\nabla'\cdot(\boldsymbol{\beta}\times\mathbf{H})]\\&=\nabla'\times\left\{\mathbf{H}-\frac{\gamma-1}{\beta^2}\;[\boldsymbol{\beta}\times(\boldsymbol{\beta}\times\mathbf{H})]\right\}\\&=\nabla'\times(\gamma\bar{\boldsymbol{\alpha}}^{-1}\cdot\mathbf{H})\end{aligned}$$

and

$$\begin{aligned}\bar{\boldsymbol{\alpha}}\cdot(\boldsymbol{\beta}\cdot\nabla')c\mathbf{D}-\boldsymbol{\beta}[(\bar{\boldsymbol{\alpha}}\cdot\nabla')\cdot c\mathbf{D}]&=(\boldsymbol{\beta}\cdot\nabla')c\mathbf{D}-\boldsymbol{\beta}(\nabla'\cdot c\mathbf{D})\\&=-\nabla'\times(\boldsymbol{\beta}\times c\mathbf{D}),\end{aligned}$$

we have

$$\nabla'\times(\gamma\bar{\boldsymbol{\alpha}}^{-1}\cdot\mathbf{H}-\gamma\boldsymbol{\beta}\times c\mathbf{D})-\frac{\partial}{\partial ct'}(\gamma\bar{\boldsymbol{\alpha}}^{-1}\cdot c\mathbf{D}+\gamma\boldsymbol{\beta}\times\mathbf{H})=\mathbf{J}'. \tag{2.15b}$$

In view of (2.15), we obtain the transformation formulas for **D** and **H**:

$$\text{LT}\qquad\begin{bmatrix}c\mathbf{D}'\\\mathbf{H}'\end{bmatrix}=\gamma\begin{bmatrix}\bar{\boldsymbol{\alpha}}^{-1}&\bar{\boldsymbol{\beta}}\\-\bar{\boldsymbol{\beta}}&\bar{\boldsymbol{\alpha}}^{-1}\end{bmatrix}\cdot\begin{bmatrix}c\mathbf{D}\\\mathbf{H}\end{bmatrix}, \tag{2.16a}$$

where, as before, $\bar{\boldsymbol{\beta}}$ is defined operationally by $\bar{\boldsymbol{\beta}} \cdot \mathbf{A} = \boldsymbol{\beta} \times \mathbf{A}$ on any vector $\mathbf{A}$.

In a similar way, we can show that the form of Faraday's induction law and Gauss' magnetic field law is preserved under LT, provided that **E** and **B** transform as

$$\text{LT} \qquad \begin{bmatrix} \mathbf{E}' \\ c\mathbf{B}' \end{bmatrix} = \gamma \begin{bmatrix} \bar{\alpha}^{-1} & \bar{\boldsymbol{\beta}} \\ -\bar{\boldsymbol{\beta}} & \bar{\alpha}^{-1} \end{bmatrix} \cdot \begin{bmatrix} \mathbf{E} \\ c\mathbf{B} \end{bmatrix}, \tag{2.16b}$$

which is identical in form to (2.16a). Transformation formulas 2.16 express the fact that **H** and **D** fields transform as an entity (entity of quantity), and so do **E** and **B** fields (entity of intensity).

It is interesting to see that field components parallel to the velocity are left unchanged:

$$\left.\begin{aligned} \mathbf{E}'_{\|} &= \mathbf{E}_{\|}, \\ \mathbf{B}'_{\|} &= \mathbf{B}_{\|}, \\ \mathbf{D}'_{\|} &= \mathbf{D}_{\|}, \\ \mathbf{H}'_{\|} &= \mathbf{H}_{\|}, \end{aligned}\right\} \tag{2.17a}$$

and the perpendicular components transform as

$$\left.\begin{aligned} \mathbf{E}'_{\perp} &= \gamma(\mathbf{E}_{\perp} + \boldsymbol{\beta} \times c\mathbf{B}_{\perp}), \\ c\mathbf{B}'_{\perp} &= \gamma(c\mathbf{B}_{\perp} - \boldsymbol{\beta} \times \mathbf{E}_{\perp}), \\ c\mathbf{D}'_{\perp} &= \gamma(c\mathbf{D}_{\perp} + \boldsymbol{\beta} \times \mathbf{H}_{\perp}), \\ \mathbf{H}'_{\perp} &= \gamma(\mathbf{H}_{\perp} - \boldsymbol{\beta} \times c\mathbf{D}_{\perp}). \end{aligned}\right\} \tag{2.17b}$$

This is in contrast to the transformation of space coordinates, where the perpendicular components are left unchanged.

The FOLT and GT of the field quantities can be easily deduced from the LT. In the case of FOLT, let $\gamma = 1$, and we have

$$\text{FOLT} \qquad \begin{bmatrix} \mathbf{E}' \\ c\mathbf{B}' \end{bmatrix} = \begin{bmatrix} \bar{\mathrm{I}} & \bar{\boldsymbol{\beta}} \\ -\bar{\boldsymbol{\beta}} & \bar{\mathrm{I}} \end{bmatrix} \cdot \begin{bmatrix} \mathbf{E} \\ c\mathbf{B} \end{bmatrix},$$

$$\text{FOLT} \qquad \begin{bmatrix} c\mathbf{D}' \\ \mathbf{H}' \end{bmatrix} = \begin{bmatrix} \bar{\mathrm{I}} & \bar{\boldsymbol{\beta}} \\ -\bar{\boldsymbol{\beta}} & \bar{\mathrm{I}} \end{bmatrix} \cdot \begin{bmatrix} c\mathbf{D} \\ \mathbf{H} \end{bmatrix}.$$

For GT, let $c \to \infty$:

$$\text{GT}\begin{cases} \mathbf{E}' = \mathbf{E} + \mathbf{v} \times \mathbf{B}, \\ \mathbf{B}' = \mathbf{B}, \\ \mathbf{D}' = \mathbf{D}, \\ \mathbf{H}' = \mathbf{H} - \mathbf{v} \times \mathbf{D}. \end{cases}$$

The inverses of all of these transformation formulas can be written with $\boldsymbol{\beta}$ replaced by $-\boldsymbol{\beta}$.

According to both LT and GT, a pure $\mathbf{B}$ field in S produces an electric field $\mathbf{E}' = \mathbf{v} \times \mathbf{B}$ in S'. Thus a voltage is induced in a moving conductor when its velocity has a component perpendicular to the $\mathbf{B}$-field lines. According to LT, for a pure $\mathbf{E}$ field in S, a magnetic field is witnessed from a moving frame. Thus a stationary electron, when viewed from a moving frame, exhibits a magnetic field. But, according to GT, $\mathbf{B}'$ is equal to $\mathbf{B}$.

We denote the 6×6 matrix in (2.16) by $\bar{\mathbf{L}}_6$:

$$\bar{\mathbf{L}}_6(\boldsymbol{\beta}) = \gamma \begin{bmatrix} \bar{\boldsymbol{\alpha}}^{-1} & \bar{\boldsymbol{\beta}} \\ -\bar{\boldsymbol{\beta}} & \bar{\boldsymbol{\alpha}}^{-1} \end{bmatrix}. \tag{2.18a}$$

When the velocity is along the z axis, the LT matrix $\mathbf{L}_6$ becomes

$$\bar{\mathbf{L}}_6 = \gamma \left[\begin{array}{ccc|ccc} 1 & 0 & 0 & 0 & -\beta & 0 \\ 0 & 1 & 0 & \beta & 0 & 0 \\ 0 & 0 & 1/\gamma & 0 & 0 & 0 \\ \hline 0 & \beta & 0 & 1 & 0 & 0 \\ -\beta & 0 & 0 & 0 & 1 & 0 \\ 0 & 0 & 0 & 0 & 0 & 1/\gamma \end{array}\right]. \tag{2.18b}$$

The inverse transformation of (2.16) is determined by the inverse of $\bar{\mathbf{L}}_6(\boldsymbol{\beta})$. It can be verified that

$$\begin{aligned} \bar{\mathbf{L}}_6^{-1}(\boldsymbol{\beta}) &= \bar{\mathbf{L}}_6(-\boldsymbol{\beta}) \\ &= \gamma \begin{bmatrix} \bar{\boldsymbol{\alpha}}^{-1} & -\bar{\boldsymbol{\beta}} \\ \bar{\boldsymbol{\beta}} & \bar{\boldsymbol{\alpha}}^{-1} \end{bmatrix}. \end{aligned} \tag{2.19}$$

By physical reasoning, the inverse of a pure Lorentz transformation is equivalent to changing the direction of velocity.

Let us explore further some properties of the $\bar{\mathbf{L}}_6$ matrix. Since $\bar{\boldsymbol{\alpha}}$ is symmetric and $\bar{\boldsymbol{\beta}}$ is skew-symmetric, we have

$$\bar{\mathbf{L}}_6^t = \gamma \begin{bmatrix} (\bar{\boldsymbol{\alpha}}^{-1})^t & (-\bar{\boldsymbol{\beta}})^t \\ \bar{\boldsymbol{\beta}}^t & (\bar{\boldsymbol{\alpha}}^{-1})^t \end{bmatrix} = \bar{\mathbf{L}}_6, \tag{2.20}$$

where the superscript t denotes the transpose of the matrix. Thus $\bar{\mathbf{L}}_6$ is a symmetric 6×6 matrix. We can also show that

$$\bar{\mathbf{L}}_6^t \cdot \begin{bmatrix} \bar{\mathbf{I}} & \bar{0} \\ \bar{0} & -\bar{\mathbf{I}} \end{bmatrix} \cdot \bar{\mathbf{L}}_6 = \begin{bmatrix} \bar{\mathbf{I}} & \bar{0} \\ \bar{0} & -\bar{\mathbf{I}} \end{bmatrix}, \tag{2.21a}$$

$$\bar{\mathbf{L}}_6^t \cdot \begin{bmatrix} \bar{0} & \bar{\mathbf{I}} \\ \bar{\mathbf{I}} & \bar{0} \end{bmatrix} \cdot \bar{\mathbf{L}}_6 = \begin{bmatrix} \bar{0} & \bar{\mathbf{I}} \\ \bar{\mathbf{I}} & \bar{0} \end{bmatrix}. \tag{2.21b}$$

Numerous other identities can also be derived for the $\bar{\mathbf{L}}_6$ and $\bar{\boldsymbol{\alpha}}$ matrices.

2.2c Classification of Electromagnetic Fields

Equation 2.21 can be used to find relations that are invariant under the Lorentz transformation. Using LT for the entity of intensity (2.16b), we have

$$[\mathbf{E}' \; c\mathbf{B}'] \cdot \begin{bmatrix} \bar{\mathbf{I}} & \bar{0} \\ \bar{0} & -\bar{\mathbf{I}} \end{bmatrix} \cdot \begin{bmatrix} \mathbf{E}' \\ c\mathbf{B}' \end{bmatrix} = [\mathbf{E} \; c\mathbf{B}] \cdot \bar{\mathbf{L}}_6^t \cdot \begin{bmatrix} \bar{\mathbf{I}} & \bar{0} \\ \bar{0} & -\bar{\mathbf{I}} \end{bmatrix} \cdot \bar{\mathbf{L}}_6 \cdot \begin{bmatrix} \mathbf{E} \\ c\mathbf{B} \end{bmatrix}. \tag{2.22}$$

In view of (2.21a), (2.22) gives

$$|\mathbf{E}'|^2 - |c\mathbf{B}'|^2 = |\mathbf{E}|^2 - |c\mathbf{B}|^2. \tag{2.23}$$

It can be seen that the velocity between observers S and S' does not appear in (2.23). The difference between the magnitude squared of $\mathbf{E}$ and the magnitude squared of $c\mathbf{B}$ is therefore a numerical constant independent of motion. Any quantity that is invariant under LT is a Lorentz invariant. Note the difference between Lorentz invariance and Lorentz covariance: the former refers to a scalar number, the latter to a physical law. Another Lorentz invariant is obtained by replacing $\begin{bmatrix} \bar{\mathrm{I}} & \bar{0} \\ \bar{0} & -\bar{\mathrm{I}} \end{bmatrix}$ in (2.22) with $\begin{bmatrix} \bar{0} & \bar{\mathrm{I}} \\ \bar{\mathrm{I}} & \bar{0} \end{bmatrix}$ and using (2.21b). The result is

$$\mathbf{E}' \cdot \mathbf{B}' = \mathbf{E} \cdot \mathbf{B}. \tag{2.24}$$

According to (2.23) and (2.24), we can classify the electromagnetic field as follows.

Free-Space Wave Fields Characterized by $\mathbf{E} \cdot \mathbf{B} = 0$ *and* $|\mathbf{E}| = |c\mathbf{B}|$. The field vectors $\mathbf{E}$ and $c\mathbf{B}$ are always perpendicular in direction and equal in magnitude. The magnitude changes from observer to observer, as is characteristic of a plane wave in free space. Consider the case $\mathbf{E} = \hat{x}E$ and $\mathbf{B} = \hat{y}B$, and assume that an observer S' is moving along the z axis of S. Then by (2.17b) we immediately obtain

$$E' = E\left(\frac{1-\beta}{1+\beta}\right)^{1/2},$$

$$B' = B\left(\frac{1-\beta}{1+\beta}\right)^{1/2}.$$

The amplitude of the wave field decreases as the velocity along the $\hat{z}$ direction increases. When it reaches c, namely, $\beta = 1$, the amplitude is zero. Thus an observer moving in the $\mathbf{E} \times \mathbf{B}$ direction with the velocity of light sees no field at all. An observer moving in a direction opposite to that of $\mathbf{E} \times \mathbf{B}$ with $\mathbf{v} \to c$ sees the amplitude approaching infinity. We can consider another simple case in which S' moves along the $\mathbf{E}$-field direction; the situation is shown in Fig. 2.2. As the velocity becomes larger, the $\mathbf{E}'$-field vector tilts its direction so that $(-\mathbf{E}' \times c\mathbf{B}')$ tends to be parallel to $\mathbf{v}$. The angle of tilt of $\mathbf{E}$ is $\theta = \sin^{-1}\beta$, which increases as β increases.

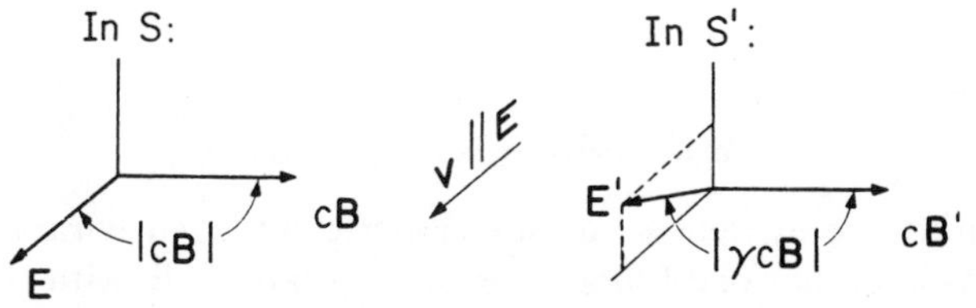

Figure 2.2 Observer S' is moving with velocity v in the direction of electric field E as observed by S.

Electric Fields Characterized by $\mathbf{E} \cdot \mathbf{B} = 0$ *and* $|\mathbf{E}| > |c\mathbf{B}|$. From (2.17b) we see that there exists an observer moving in the $\mathbf{E} \times \mathbf{B}$ direction who sees only an electric field and no magnetic field. The velocity of this observer is given by $\beta = |c\mathbf{B}|/|\mathbf{E}|$. Apparently $|\mathbf{v}|$ is smaller than c or $\beta < 1$. It is interesting to note that this observer, called S', is not the only one who does not see a magnetic field. All observers moving along the $\mathbf{E}'$-field vector relative to S' do not experience a magnetic field!

Magnetic Fields Characterized by $\mathbf{E} \cdot \mathbf{B} = 0$ *and* $|\mathbf{E}| < |c\mathbf{B}|$. This is the dual of the second case: S' and all observers moving in the $\mathbf{B}'$ direction with respect to the observer see only a magnetic field, where S' moves with a speed $\beta = |\mathbf{E}|/|c\mathbf{B}|$ along the $\mathbf{B} \times \mathbf{E}$ direction relative to S.

Wrench Fields Characterized by $\mathbf{E} \cdot \mathbf{B} \neq 0$. There are six cases in this class. It is clear that there are frames where $\mathbf{E}$ and $\mathbf{B}$ fields are parallel or antiparallel. When they are parallel and $\mathbf{E} \cdot \mathbf{B} > 0$, we have a positive wrench field. When they are antiparallel and $\mathbf{E} \cdot \mathbf{B} < 0$, we have a negative wrench field. The wrench field is electric if $|\mathbf{E}| > |c\mathbf{B}|$ because here the electric field dominates. The wrench field is magnetic if $|\mathbf{E}| < |c\mathbf{B}|$.

2.3 TRANSFORMATION OF CONSTITUTIVE RELATIONS

The Lorentz transformation formulas for the electromagnetic field vectors derived in Section 2.2 can be used to derive transformation laws for the constitutive relations. A medium at rest in one frame becomes a medium in motion when viewed from another frame. The derivation of equivalent constitutive relations for a moving medium in the laboratory frame is useful conceptually and practically. It is indeed true that a problem involving one moving medium can always be solved in its rest frame and the results can be transformed back to the laboratory frame. In practice, the Lorentz transformation method cannot be applied when more than two relatively moving media are involved because, in the rest frame of one medium, all others are in motion. Thus constitutive relations for moving media have to be determined.

2.3a Moving Isotropic Media

To treat media in motion, we must consider two reference frames in relative motion. Assume that one frame S' is moving uniformly with respect to the laboratory frame S. Assume also that all quantities associated with S' are denoted by primes. Consider the simplest case, when an isotropic medium possessing constitutive relations

$$\mathbf{D}' = \epsilon' \mathbf{E}' \tag{2.25a}$$

and

$$\mathbf{H}' = \frac{\mathbf{B}'}{\mu'} \tag{2.25b}$$

is moving uniformly with velocity $\mathbf{v}$. Employing transformation formulas 2.16, we obtain the constitutive matrix for the moving medium:

$$\overline{\mathbf{C}} = \overline{\mathbf{L}}_6^{-1} \cdot \begin{bmatrix} c\epsilon' \overline{\mathbf{I}} & \overline{0} \\ \overline{0} & \dfrac{1}{c\mu'} \overline{\mathbf{I}} \end{bmatrix} \cdot \overline{\mathbf{L}}_6$$

$$= \frac{\gamma^2}{c\mu'} \begin{bmatrix} (n^2 - \beta^2)\overline{\mathbf{I}} - (n^2 - 1)\boldsymbol{\beta}\boldsymbol{\beta} & (n^2 - 1)\overline{\boldsymbol{\beta}} \\ (n^2 - 1)\overline{\boldsymbol{\beta}} & (1 - n^2\beta^2)\overline{\mathbf{I}} + (n^2 - 1)\boldsymbol{\beta}\boldsymbol{\beta} \end{bmatrix} \tag{2.26}$$

where

$$n^2 = c^2 \mu' \epsilon'. \tag{2.27}$$

We see that, in vacuum with $n = 1$, $\overline{\mathbf{C}}$ reduces to $(1/c\mu')\,\overline{\mathbf{I}}$ and the medium is isotropic. When $\beta = 0$, (2.26) reduces to (2.25).

When the velocity is along the $\hat{z}$ direction, (2.26) becomes

$$\overline{\mathbf{C}} = \frac{\gamma^2}{c\mu'} \begin{bmatrix} n^2 - \beta^2 & 0 & 0 & 0 & -(n^2-1)\beta & 0 \\ 0 & n^2 - \beta^2 & 0 & (n^2-1)\beta & 0 & 0 \\ 0 & 0 & n^2(1-\beta^2) & 0 & 0 & 0 \\ 0 & -(n^2-1)\beta & 0 & 1 - n^2\beta^2 & 0 & 0 \\ (n^2-1)\beta & 0 & 0 & 0 & 1 - n^2\beta^2 & 0 \\ 0 & 0 & 0 & 0 & 0 & 1/\gamma^2 \end{bmatrix} \tag{2.28}$$

Note that, although we assumed μ' and ϵ' to be scalar numbers, the derivation remains valid even if they are not. We have to remember that μ' and ϵ' are measured in the rest frame of the medium. If they are dependent on parameters pertaining to their rest frame, these parameters must be properly transformed. For instance, when the medium in its rest frame is a plasma characterized by (1.26), we need to transform the plasma frequency to the laboratory frame.

By using (1.17), the constitutive relations for a moving isotropic medium can be written in the form of **B** and **D** expressed in terms of **E** and **H**. The result is as follows:

$$\mathbf{B} = \mu' \overline{\mathbf{A}} \cdot \mathbf{H} - \mathbf{\Omega} \times \mathbf{E}, \tag{2.29a}$$

$$\mathbf{D} = \epsilon' \overline{\mathbf{A}} \cdot \mathbf{E} + \mathbf{\Omega} \times \mathbf{H}, \tag{2.29b}$$

where

$$\overline{\mathbf{A}} = \frac{1}{c\mu'} \overline{\mathbf{Q}}^{-1} = \frac{1-\beta^2}{1-n^2\beta^2} \left(\overline{\mathbf{I}} - \frac{n^2-1}{1-\beta^2} \boldsymbol{\beta\beta} \right) \tag{2.30a}$$

and

$$\mathbf{\Omega} = \frac{n^2-1}{1-n^2\beta^2} \boldsymbol{\beta}/c. \tag{2.30b}$$

When **v** is in the $\hat{z}$ direction, **A** becomes a diagonal matrix.

2.3b Moving Bianisotropic Media

We now consider the general case in which a bianisotropic medium with the constitutive relations 1.15 is in motion. We assume that the velocity is in the $\hat{z}$ direction. This assumption does not represent a restriction on the validity of the results that are obtained because we can always rotate our coordinate system so that the z axis points in the direction of motion. This coordinate rotation certainly affects the elements of the constitutive matrix. But after the transformation $\overline{\mathbf{C}}$ still takes the general form (1.15) except that the element values are different. By application of (2.18b), the constitutive matrices are determined to be

$$\overline{\mathbf{P}} = \begin{bmatrix} \gamma^2[p_{xx} - \beta(l_{xy} - m_{yx}) - \beta^2 q_{yy}] & \gamma^2[p_{xy} + \beta(l_{xx} + m_{yy}) + \beta^2 q_{yx}] & \gamma(p_{xz} + \beta m_{yz}) \\ \gamma^2[p_{yx} - \beta(l_{yy} + m_{xx}) + \beta^2 q_{xy}] & \gamma^2[p_{yy} + \beta(l_{yx} - m_{xy}) - \beta^2 q_{xx}] & \gamma(p_{yz} - \beta m_{xz}) \\ \gamma(p_{zx} - \beta l_{zy}) & \gamma(p_{zy} + \beta l_{zx}) & p_{zz} \end{bmatrix}, \tag{2.31a}$$

$$\bar{\mathbf{Q}}=\begin{bmatrix} \gamma^2[q_{xx}+\beta(m_{xy}-l_{yx})-\beta^2 p_{yy}] & \gamma^2[q_{xy}-\beta(m_{xx}+l_{yy})+\beta^2 p_{yx}] & \gamma(q_{xz}-\beta l_{yz}) \\ \gamma^2[q_{yx}+\beta(m_{yy}+l_{xx})+\beta^2 p_{xy}] & \gamma^2[q_{yy}-\beta(m_{yx}-l_{xy})-\beta^2 p_{xx}] & \gamma(q_{yz}+\beta l_{xz}) \\ \gamma(q_{zx}+\beta m_{zy}) & \gamma(q_{zy}-\beta m_{zx}) & q_{zz} \end{bmatrix}, \tag{2.31b}$$

$$\bar{\mathbf{L}}=\begin{bmatrix} \gamma^2[l_{xx}+\beta(p_{xy}+q_{yx})+\beta^2 m_{yy}] & \gamma^2[l_{xy}-\beta(p_{xx}-q_{yy})-\beta^2 m_{yx}] & \gamma(l_{xz}+\beta q_{yz}) \\ \gamma^2[l_{yx}+\beta(p_{yy}-q_{xx})-\beta^2 m_{xy}] & \gamma^2[l_{yy}-\beta(p_{yx}+q_{xy})+\beta^2 m_{xx}] & \gamma(l_{yz}-\beta q_{xz}) \\ \gamma(l_{zx}+\beta p_{zy}) & \gamma(l_{zy}-\beta p_{zx}) & l_{zz} \end{bmatrix} \tag{2.31c}$$

$$\bar{\mathbf{M}}=\begin{bmatrix} \gamma^2[m_{xx}-\beta(q_{xy}+p_{yx})+\beta^2 l_{yy}] & \gamma^2[m_{xy}+\beta(q_{xx}-p_{yy})-\beta^2 l_{yx}] & \gamma(m_{xz}-\beta p_{yz}) \\ \gamma^2[m_{yx}-\beta(q_{yy}-p_{xx})-\beta^2 l_{xy}] & \gamma^2[m_{yy}+\beta(q_{yx}+p_{xy})+\beta^2 l_{xx}] & \gamma(m_{yz}+\beta p_{xz}) \\ \gamma(m_{zx}-\beta q_{zy}) & \gamma(m_{zy}+\beta q_{zx}) & m_{zz} \end{bmatrix}. \tag{2.31d}$$

In (2.31), we omit primes on the constitutive parameters for the moving medium in the rest frame. The constitutive matrices of the moving bianisotropic medium are represented by $\bar{\mathbf{P}}$, $\bar{\mathbf{Q}}$, $\bar{\mathbf{L}}$, and $\bar{\mathbf{M}}$, which reduce to their rest frame values as $\beta=0$. We note that, if a bianisotropic medium satisfies the lossless conditions in its rest frame, the moving bianisotropic medium also satisfies the lossless conditions. We conclude that for moving lossless bianisotropic media the 6×6 constitutive matrix $\bar{\mathbf{C}}$ can contain 21 independent complex elements.

Formulas 2.31a–2.31d enable an observer to characterize a bianisotropic medium moving along the $\hat{z}$ direction. We observe that when a biisotropic medium is in motion it becomes bianisotropic. For a moving electrically gyrotropic medium with scalar permeability μ' and permittivity tensor

$$\bar{\epsilon}=\begin{bmatrix} \epsilon' & -i\epsilon'_g & 0 \\ i\epsilon'_g & \epsilon' & 0 \\ 0 & 0 & \epsilon'_z \end{bmatrix} \tag{2.32}$$

we obtain

$$\overline{\mathbf{C}}=\frac{\gamma^2}{c\mu'}\begin{bmatrix} n^2-\beta^2 & -in_g^2 & 0 & -i\beta n_g^2 & -\beta(n^2-1) & 0 \\ in_g^2 & n^2-\beta^2 & 0 & \beta(n^2-1) & -i\beta n_g^2 & 0 \\ 0 & 0 & n_z^2/\gamma^2 & 0 & 0 & 0 \\ -i\beta n_g^2 & -\beta(n^2-1) & 0 & 1-n^2\beta^2 & in_g^2\beta^2 & 0 \\ \beta(n^2-1) & -i\beta n_g^2 & 0 & -in_g^2\beta^2 & 1-n^2\beta^2 & 0 \\ 0 & 0 & 0 & 0 & 0 & 1/\gamma^2 \end{bmatrix}, \tag{2.33}$$

where

$$n_g^2=c^2\mu'\epsilon_g', \tag{2.34a}$$

$$n_z^2=c^2\mu'\epsilon_z'. \tag{2.34b}$$

The parameters that govern the gyrotropic nature of the medium must be carefully transformed. For example (Chawla and Unz[20]), the plasma frequency is a Lorentz-invariant $\omega_p'=\omega_p$, while the cyclotron frequency transforms as $\omega_c'=\gamma\omega_c$. Remember that the static magnetic field is in the direction of motion. If the magnetic field were perpendicular to the direction of motion, we would have $\omega_c'=\gamma^2\omega_c$ instead. The applied frequency ω must also be properly transformed.

2.3c Moving Uniaxial Media

Constitutive relations for a moving uniaxial medium are also derived from general formula 2.31. We assume that in the rest frame of the moving medium the constitutive relations are as follows:

$$\bar{\boldsymbol{\epsilon}}=\begin{bmatrix} \epsilon' & 0 & 0 \\ 0 & \epsilon' & 0 \\ 0 & 0 & \epsilon_z' \end{bmatrix}, \tag{2.35a}$$

$$\bar{\boldsymbol{\mu}}=\begin{bmatrix} \mu' & 0 & 0 \\ 0 & \mu' & 0 \\ 0 & 0 & \mu_z' \end{bmatrix}, \tag{2.35b}$$

where the z axis coincides with the optic axis. Note that isotropic media and electric or magnetic uniaxial media are all special cases of (2.35). The primes manifest that these quantities are associated with the rest frame of the medium. In the laboratory frame, where the media appear to be moving uniformly with velocity $\mathbf{v}$ along the $\hat{z}$ direction, the constitutive matrix of the medium is determined to be

$$\overline{\mathbf{C}} = \frac{1}{c\mu'}\left[\begin{array}{ccc|ccc} p & 0 & 0 & 0 & -l & 0 \\ 0 & p & 0 & l & 0 & 0 \\ 0 & 0 & p_z & 0 & 0 & 0 \\ \hline 0 & -l & 0 & q & 0 & 0 \\ l & 0 & 0 & 0 & q & 0 \\ 0 & 0 & 0 & 0 & 0 & q_z \end{array}\right]. \tag{2.36}$$

Note that p, q, l, p_z, and q_z are all dimensionless quantities. The constitutive matrix can be transformed into the **EH** and the **DB** representations by using (1.17) and (1.18); we obtain

$$\overline{\mathbf{C}}_{EH} = \left[\begin{array}{ccc|ccc} \epsilon & 0 & 0 & 0 & \xi & 0 \\ 0 & \epsilon & 0 & -\xi & 0 & 0 \\ 0 & 0 & \epsilon_z & 0 & 0 & 0 \\ \hline 0 & -\xi & 0 & \mu & 0 & 0 \\ \xi & 0 & 0 & 0 & \mu & 0 \\ 0 & 0 & 0 & 0 & 0 & \mu_z \end{array}\right], \tag{2.37}$$

$$\overline{\mathbf{C}}_{DB} = \left[\begin{array}{ccc|ccc} \kappa & 0 & 0 & 0 & \chi & 0 \\ 0 & \kappa & 0 & -\chi & 0 & 0 \\ 0 & 0 & \kappa_z & 0 & 0 & 0 \\ \hline 0 & -\chi & 0 & \nu & 0 & 0 \\ \chi & 0 & 0 & 0 & \nu & 0 \\ 0 & 0 & 0 & 0 & 0 & \nu_z \end{array}\right]. \tag{2.38}$$

The element values of the constitutive matrices in different representations are summarized in Table 2.1. Remember that the velocity $\mathbf{v}$ is in the $\hat{z}$ direction. The matrix elements are velocity-dependent. The primed quantities are measured in the rest frame of the media; if they are coordinate-dependent, we must also transform them to the laboratory frame. For instance, in the case of a moving isotropic plasma, both the plasma frequency and the applied frequency must be transformed.

Table 2.1
Constitutive Parameters for Moving Media

EB representation	$p=\gamma^2(n^2-\beta^2)$, $p_z=an^2$, $a=\epsilon_z'/\epsilon'$ $q=\gamma^2(1-n^2\beta^2)$, $q_z=1/b$, $b=\mu_z'/\mu'$ $l=\gamma^2\beta(n^2-1)$, $qp+l^2=n^2=c^2\mu'\epsilon'$
EH representation	$\epsilon=(qp+l^2)/c^2\mu' q=\epsilon'(1-\beta^2)/(1-n^2\beta^2)$, $\epsilon_z=\epsilon_z'$ $\mu=\mu'/q=\mu'(1-\beta^2)/(1-n^2\beta^2)$, $\mu_z=\mu_z'$ $\xi=-l/cq=-\beta(n^2-1)/c(1-n^2\beta^2)$
DB representation	$\kappa=c^2\mu'/p=c^2\mu'(1-\beta^2)/(n^2-\beta^2)$, $\kappa_z=1/\epsilon_z'$ $\nu=(qp+l^2)/p\mu'=c^2\epsilon'(1-\beta^2)/(n^2-\beta^2)$, $\nu_z=1/\mu_z'$ $\chi=cl/p=c\beta(n^2-1)/(n^2-\beta^2)$

From Table 2.1, we see that the constitutive matrix $\overline{\mathbf{C}}$ becomes diagonal if $l=0$ or $\mu'\epsilon'=1/c^2$ $(n=1)$, as can sometimes be achieved in a dispersive medium. For an anisotropic plasma subject to a strong magnetic field in the $\hat{z}$ direction, $\epsilon'=\epsilon_0$. In such a medium we also have $\mu'=\mu_0$. Thus $l=0$, $n=p=q=1$, and $\overline{\mathbf{C}}$ will be diagonal. This is an example of a moving medium that is not bianisotropic. All of the constitutive relations take forms 2.36–2.38 whether the moving medium in its rest frame is an isotropic medium, a uniaxial crystal, a uniaxial gyrotropic plasma, or a magnetic and electric uniaxial medium.

2.3d Accelerated Media

The constitutive relations of an accelerated medium are also bianisotropic in form. They are space-dependent and can be viewed as inhomogeneous. General formulations in arbitrary accelerating frames have been considered by some authors, and explicit forms for the constitutive relations have been proposed for rotating and linearly accelerated media.[3] Not only relative motion but also absolute motion of both observer and medium play a crucial role in determining the constitutive relations.

PROBLEMS

2.1. To consider transformation of time intervals, let a clock be in S', which moves along the $\hat{z}$ direction of S. The time interval of the clock as read by S' is $\Delta t' = t_2' - t_1'$, called the *proper time interval*. The corresponding time interval of the clock as read by S is $\Delta t = t_2 - t_1$, called the *coordinate time interval*. We write the LT for the space and time intervals by using (2.5):

$$c\,\Delta t' = \gamma(c\,\Delta t - \beta\,\Delta z),$$

$$\Delta z' = \gamma(\Delta z - \beta c\,\Delta t).$$

The clock is stationary in S', and hence $\Delta z' = 0$. Show that $\Delta t = \gamma\,\Delta t'$. Observe that the coordinate time interval is always larger than the proper time interval. This phenomenon is known as *time dilation*.

2.2. In the early stages of special relativity, the sudden disappearance of an absolute time scale led to the well-known "twin paradox." The paradox as stated was that one of a pair of twins left home, traveled at a uniform (high) speed in some direction for a certain period of time, and then returned home and found himself younger than his brother. By the symmetry argument that motion is relative, it was argued that neither twin should have grown older than the other, and the validity of special relativity was challenged. In the following discussion we show that both of the twins agree that one is older than the other and the problem is not a symmetric one.

Let both A and B be at the origin in frame S; B starts to move at $t=0$ with speed v in the positive $\hat{z}$ direction of S. As A reads time t, B moves back with speed v. Consider the following events:

Event 1. Twin B is at $z = vt$ when A reads time t. In frame S, this event is described by (ct, vt), where ct is the time coordinate of the event with dimension of length, and vt is the space coordinate of the event.

Event 2. As A reads time $2t$, both A and B are at $z=0$. In frame S, this event is described by $(2ct, 0)$.

Consider two other frames of references, S' and S''. Frame S' moves with velocity v in the positive direction of the z axis; S'' moves with velocity v in the negative direction of the z axis. Show that the space-time coordinates for the two events in frames S' and S'' are as listed in Table 2.2.

Table 2.2

Time-Space Coordinates of the Two Events in the Three Frames of Reference[a]

Event	Observer S	Observer S'	Observer S''
1	$ct[1,\beta]$	$ct[1/\gamma,0]$	$ct[\gamma(1+\beta^2),2\gamma\beta]$
2	$ct[2,0]$	$ct[2\gamma,-2\gamma\beta]$	$ct[2\gamma,2\gamma\beta]$

[a]The first part in brackets denotes time coordinates multiplied by c, and the second part denotes space coordinates.

(a) Using the space-time coordinates of the two events, show that, by time dilation, twin A agrees that the total proper time interval for twin B is $2t/\gamma$, while his own coordinate time interval is $2t$.

(b) During the initial period before turning around, B is in S', and the elapsed time according to B is t/γ. Show that, according to B, the elapsed time during the final period after turning around is also t/γ. Thus twin B agrees with twin A that his time span is $2t/\gamma$, while that of A is $2t$.

(c) Show that an observer moving uniformly relative to A, especially in frames S' and S'', concludes that the proper elapsed time of B is less than that of A by a factor $1/\gamma$.

(d) Suppose that twin B started his journey right after his birth and traveled with a speed $v=0.8c$. If he comes back at 30 years of age, how old is his twin A?

(e) The problem is inherently asymmetrical; one twin has to turn, and it is this twin that experiences less proper time, $2t/\gamma$. If B does not turn, then, as A reads proper time $2t$, the coordinate time reading for B at two different locations, $z=0$ and $z=2vt$, is $2\gamma t$ because of the dilation of time. After turning around and meeting A again at $z=0$, the proper time reading of B has been shown to be $2t/\gamma$. The effect of turning around causes a time difference of $2t(\gamma-1/\gamma)=2\gamma\beta^2 t$. Show that this "lost time" is equal to the time coordinate difference between S' and S'' for Event 1 in Table 2.2.

(f) It is interesting to imagine how B experiences the period of losing time during his turning around. Consider a third event, occurring when A reads time t at $z=0$ in S. Find the time-space coordinates for Event 3 in S' and S''. Show that, according to S, Events 1 and 3 are simultaneous; according to S', Event 1 is earlier than Event 3; and according to S'', Event 3 is earlier than Event 1. At the turning time, twin B changes his frame from S' to S''.

Show that B loses track of anything that happened at $z=0$ during a time period $2\gamma\beta^2 ct$.

(g) One may interpret the loss of time as being due to the change of velocity from β to $-\beta$. Yet this lost time is also directly proportional to t, namely, to how long B has traveled. Why?

2.3. The star alpha Centauri is distant 4.3 light years from the Earth. Observer B leaves Earth in a rocket ship that travels toward this star at acceleration g. Halfway from α Cen (2.15 light years from Earth), B turns off the forward acceleration and accelerates backward toward Earth at g, so that the rocket arrives at α Cen with zero speed and turns back. On the return trip, halfway between these places, B again changes the direction of acceleration. Observer B arrives at Earth with zero speed. Show that B takes 7 years to complete the round trip, while the elapsed time on Earth is 12 years.

2.4. Assume that S' moves with velocity v_1 relative to S and S'' moves with velocity v_2 relative to S, both along the z axis of S. Write the LT between S' and S, and the LT between S'' and S. Eliminate the space-time coordinates z and t of S from the two LTs and show that the resulting LT implies a relative velocity u between S' and S'', where

$$u=\frac{v_1-v_2}{1-v_1v_2/c^2}.$$

This is an additive law for two velocities along the same direction. Generalize this procedure and deduce an additive law for two velocities in different directions.

2.5. Show that a rigid rod moving along its longitudinal direction with velocity v appears to be shortened by a factor $1/\gamma$. This phenomenon is known as the *Lorentz contraction*. Let the rod be at rest in S'. Its two end points are at $z'=0$ and $z'=z_1'$. In the laboratory frame S, the rod is moving along $\hat{z}$ with v. Its length is measured by recording the positions of its two end points simultaneously, $t_2=t_1$. Show that, by LT 2.5, $\Delta z=z_1'/\gamma$, where $\Delta z=z_2-z_1$ is the rod length as measured in S. Note that from the point of view of observer S', who moves with the rod, the laboratory observer is not measuring the two end points simultaneously.

Consider that the end points form two events. Show that, according to S', if z_1 is measured at $t'=0$ in S then z_2 is measured at $ct'=-\gamma\beta z_2$ in S. For a rod moving from left to right, if we measure the position of the right end point first and the left end point next, we naturally obtain a result smaller than the actual length of the rod. Thus S' expects that S will claim a shorter length.

2.6. A rigid square board with two sides parallel to the z axis and two sides perpendicular to the z axis is moving uniformly along the z axis with respect to an observer S at the origin. Do you expect S to perceive a rectangular board because of the effect of Lorentz contraction? Show that, in fact, as viewed by observer S, the board is still square but is turned by an angle resulting from the relative motion. Determine the angle of rotation. Note that if S performs length measurements of the four sides, he will find the board in rectangular shape.

2.7. Consider the transformation from observer S' to observer S of the length of a rod which may or may not be rigid. From the point of view of S, the length of the rod at time t_0 is defined as the difference between space positions of the two end points as measured simultaneously at t_0 in his frame of reference:

$$\mathbf{d}(t_0) = \mathbf{Q}(t_0) - \mathbf{P}(t_0),$$

where $\mathbf{Q}(t_0)$ and $\mathbf{P}(t_0)$ denote position readings of the two end points. But a simultaneous measurement at two space points in S is not simultaneous in S'. Consider the general case when the length of the rod is time-variant, and let the measurement performed be represented by two events. Their space-time coordinates are listed in Table 2.3.

Table 2.3
Time-Space Coordinates of Two End Points

Event	Observer S	Observer S'
1	$[ct_0, \mathbf{P}(t_0)]$	$[ct_0', \mathbf{P}'(t_0')]$
2	$[ct_0, \mathbf{Q}(t_0)]$	$[ct', \mathbf{Q}'(t')]$

By using the space-time Lorentz transformation formulas, show that

$$c(t' - t_0') = -\boldsymbol{\beta} \cdot \mathbf{X},$$

$$\mathbf{d} = \bar{\boldsymbol{\alpha}}^{-1} \cdot \mathbf{X},$$

$$\mathbf{X} = \mathbf{Q}'(t') - \mathbf{P}'(t_0').$$

Note that, when the rod is not rigid and stationary in S', $\mathbf{X}$ is not the result of a length measurement performed in S' because the two end points are not measured simultaneously and hence do not conform to the definition of length.

(a) Show that, when the rod is rigid and stationary in S', the rod as viewed from S possesses length

$$\mathbf{d}=\bar{\alpha}^{-1}\cdot\mathbf{d}'=\frac{1}{\gamma}\mathbf{d}'_{\|}+\mathbf{d}'_{\perp},$$

where $\mathbf{d}'$ represents the length of the rod in S' and the subscripts $\|$ and $\perp$ denote components of $\mathbf{d}'$ parallel and perpendicular, to the velocity. Thus the parallel component of the rod is contracted, whereas the perpendicular component is unchanged. The rod appears to be shortened and rotated by an angle.

(b) When the rod is rigid and is moving uniformly with velocity $\mathbf{v}'$ in S', the length of the rod in S' at t'_0 is

$$\mathbf{d}'(t'_0)=\mathbf{Q}'(t'_0)-\mathbf{P}'(t'_0).$$

We can write $\mathbf{X}$ in terms of $\mathbf{d}'$ and $\mathbf{v}'$:

$$\mathbf{X}=\mathbf{v}'(t'-t'_0)+\mathbf{d}'.$$

Show that, in terms of $\mathbf{d}'$,

$$\mathbf{d}=\bar{\alpha}^{-1}\cdot\left(\mathbf{d}'-\boldsymbol{\beta}'\frac{\boldsymbol{\beta}\cdot\mathbf{d}'}{1+\boldsymbol{\beta}\cdot\boldsymbol{\beta}'}\right),$$

where $\boldsymbol{\beta}'=\mathbf{v}'/c$. Thus, in addition to Lorentz contraction, there is another change of length in the rigid rod because of its motion in S'.

(c) The rod is short and is time-variant. Since it is short, we can expand $t'-t'_0$ to

$$c(t'-t'_0)=\sum_{l=1}^{n}A_l\delta^l,$$

where $\delta=\boldsymbol{\beta}\cdot\mathbf{d}'$ designates a small quantity. Expand $\mathbf{X}$ to the nth order in δ, and show that

$$\mathbf{X}=\mathbf{d}'+\sum_{k=1}^{\infty}\frac{\mathbf{Q}'^{(k)}(t'_0)}{k!}(ct'-ct'_0)^k$$

$$=\mathbf{d}'+\sum_{k=1}^{n}\frac{\mathbf{Q}'^{(k)}}{k!}\left(\sum_{l=1}^{n}A_l\delta^l\right)^k,$$

where $\mathbf{Q}'^{(k)}(t_0)$ is the kth derivative of $\mathbf{Q}'(t')$ with respect to ct' at $t'=t_0'$. Once we determine A_l to all orders in δ, we can also determine $\mathbf{d}=\bar{\boldsymbol{\alpha}}^{-1}\cdot\mathbf{X}$ to all orders in δ. Using $c(t'-t_0')=-\boldsymbol{\beta}\cdot\mathbf{X}$, show that

$$\delta+(1+\boldsymbol{\beta}\cdot\dot{\mathbf{Q}}')\sum_{l=1}^{n}A_l\delta^l=-\sum_{k=2}^{n}\frac{\boldsymbol{\beta}\cdot\mathbf{Q}'^{(k)}}{k!}\left(\sum_{l=1}^{n}A_l\delta^l\right)^k,$$

where $\dot{\mathbf{Q}}'$ is the first derivative of $\mathbf{Q}'(t')$ relative to ct' at $t'=t_0'$. Comparing coefficients for δ^k, $k=1,2,\ldots,n$, determine the coefficients A_k in the following way:

$$k=1:\quad (1+\boldsymbol{\beta}\cdot\dot{\mathbf{Q}}')A_1=-1,$$

$$k=2:\quad (1+\boldsymbol{\beta}\cdot\dot{\mathbf{Q}}')A_2=-\tfrac{1}{2}(\boldsymbol{\beta}\cdot\mathbf{Q}'^{(2)})A_1^2,$$

$$k=3:\quad (1+\boldsymbol{\beta}\cdot\dot{\mathbf{Q}}')A_3=-(\boldsymbol{\beta}\cdot\mathbf{Q}'^{(2)})A_1A_2-\frac{1}{3!}(\boldsymbol{\beta}\cdot\mathbf{Q}'^{(3)})A_1^3.$$

Prove that, to second order in δ, $\mathbf{d}$ reads as

$$\mathbf{d}=\bar{\boldsymbol{\alpha}}^{-1}\cdot(\bar{\mathrm{I}}+A_1\dot{\mathbf{Q}}'\boldsymbol{\beta})\cdot[\mathrm{I}+\tfrac{1}{2}A_1^2\delta\mathbf{Q}'^{(2)}\boldsymbol{\beta}]\cdot\mathbf{d}'$$

and, to third order in δ,

$$\mathbf{d}=\bar{\boldsymbol{\alpha}}^{-1}\cdot(\bar{\mathrm{I}}+A_1\dot{\mathbf{Q}}'\boldsymbol{\beta})\cdot\left[\bar{\mathrm{I}}+\tfrac{1}{2}A_1^2\delta((1-A_1^2\delta\boldsymbol{\beta}\cdot\mathbf{Q}'^{(2)})\mathbf{Q}'^{(2)}+\tfrac{1}{3}A_1\delta\mathbf{Q}'^{(3)})\boldsymbol{\beta}\right]\cdot\mathbf{d}'.$$

2.8. Derive the transformation formula for $\mathbf{E}$ and $c\mathbf{B}$ as shown in (2.16b). Find the corresponding FOLT and GT formulas. What are the inverse transformations of these formulas?

2.9. A positive charge moves along the z axis in frame S. Using transformation law 2.16b, find the $\mathbf{B}$ field attributable to the moving charge. Does your result agree with Ampère's law?

2.10. Show that a moving current loop generates an electric dipole moment by using transformation law 2.13. Indicate the polarization of the dipole. For simplicity, consider a square loop.

2.11. In a certain reference frame, a static uniform electric field E_0 is parallel to the z axis and a static uniform magnetic field $cB_0=2E_0$ forms a $30°$ angle with respect to $\hat{z}$. Determine the relative velocity of a reference frame in which the electric and magnetic fields are parallel.

2.12. An observer S observes a uniform electric field in the $\hat{x}$ direction, $\mathbf{E}=\hat{x}E_0$, and a uniform magnetic field in the $\hat{y}$ direction, $\mathbf{B}=\hat{y}B_0$. Let $E_0>cB_0$. Find an observer S' moving relative to S with velocity v in the z direction, so that he observes only an electric field. Determine the electric field strength and the velocity v. Can you find an observer moving with a velocity less than c who observes only a magnetic field?

2.13. What are the constitutive relations for a moving biisotropic medium and for a biaxial medium moving along one of its principal axes?

2.14. The space-time transformation from a frame rotating with angular velocity Ω around the z axis to an inertial frame is given nonrelativistically by

$$x=x'\cos\Omega t'+y'\sin\Omega t',$$

$$y=-x'\sin\Omega t'+y'\cos\Omega t',$$

$$z=z',$$

$$t=t'.$$

Show that, in order to preserve the form invariance of Maxwell's equations, the current charge densities and the field vectors must transform as follows:

$$\mathbf{J}=\mathbf{J}'-\rho'\boldsymbol{\Omega}\times\mathbf{r}',$$

$$\rho=\rho',$$

$$\mathbf{E}=\mathbf{E}'+(\boldsymbol{\Omega}\times\mathbf{r}')\times\mathbf{B}',$$

$$\mathbf{H}=\mathbf{H}'-(\boldsymbol{\Omega}\times\mathbf{r}')\times\mathbf{D}',$$

$$\mathbf{D}=\mathbf{D}',$$

$$\mathbf{B}=\mathbf{B}',$$

where $\boldsymbol{\Omega}=\hat{z}\Omega$, and $\mathbf{r}$ is the position vector. Find the constitutive relations for an isotropic medium as viewed from the rotating frame, under the assumption that the velocity $\boldsymbol{\Omega}\times\mathbf{r}=\boldsymbol{\Omega}\times\mathbf{r}'$ is small.

3

Plane Waves in Homogeneous Media

The simplest solution to Maxwell's equations in source-free regions is the plane wave solution. The behavior of a plane wave in a homogeneous medium is characterized by a dispersion relation that relates the components of the wave vector **k** to the properties of the medium and to the angular frequency of the wave. The wave vector **k** occupies a unique and fundamental position in the study of plane waves. We shall explore the meaning of the **k** vector and the definitions of other quantities derived from the **k** vector. In deriving the dispersion relations, we find that it is more convenient to choose a coordinate system for each **k** vector. According to Maxwell's equations, both field vectors **D** and **B** are orthogonal to the **k** vector, although they may not necessarily be orthogonal to each other. A coordinate system formed by the **k** vector and the plane containing **D** and **B** is called the *kDB system* and is useful for obtaining dispersion relations and for discussing results. With this approach, we examine plane waves in various media.

3.1 DISPERSION RELATIONS

In source-free regions, for time-harmonic fields with an $e^{-i\omega t}$ dependence, Maxwell's equations are as follows:

$$\nabla \times \mathbf{E} = i\omega \mathbf{B}, \tag{3.1a}$$

$$\nabla \cdot \mathbf{B} = 0, \tag{3.1b}$$

$$\nabla \times \mathbf{H} = -i\omega \mathbf{D}, \tag{3.1c}$$

$$\nabla \cdot \mathbf{D} = 0. \tag{3.1d}$$

We consider in general a bianisotropic medium with constitutive relations

$$c\mathbf{D} = \overline{\mathbf{P}} \cdot \mathbf{E} + \overline{\mathbf{L}} \cdot c\mathbf{B}, \tag{3.2a}$$

$$\mathbf{H} = \overline{\mathbf{M}} \cdot \mathbf{E} + \overline{\mathbf{Q}} \cdot c\mathbf{B}. \tag{3.2b}$$

Equations 3.1 and 3.2 form a complete set of equations that are solvable for the electromagnetic field quantities.

When the medium is homogeneous, plane wave solutions of the form $e^{i(\mathbf{k}\cdot\mathbf{r}-\omega t)}$ are admissible. The vector **k** is called the *wave vector*, the *propagation vector*, or simply the **k** *vector*. For the plane wave solution, Maxwell's equations become

$$\mathbf{k}\times\mathbf{E}=\omega\mathbf{B}, \tag{3.3a}$$

$$\mathbf{k}\cdot\mathbf{B}=0, \tag{3.3b}$$

$$\mathbf{k}\times\mathbf{H}=-\omega\mathbf{D}, \tag{3.3c}$$

$$\mathbf{k}\cdot\mathbf{D}=0. \tag{3.3d}$$

Geometrically, **k** is perpendicular to both **B** and **D**; **B** is perpendicular to **k** and **E**; and **D** is perpendicular to **k** and **H**.

3.1a Dispersion Relations for Isotropic Media

Consider an isotropic medium, in which **D** and **E** are parallel, **B** and **H** are parallel, $\bar{P}=c\epsilon\bar{\mathbf{I}}$, $\bar{\mathbf{Q}}=(1/c\mu)\mathbf{I}$, and $\bar{\mathbf{L}}=\bar{\mathbf{M}}=0$. Then

$$\mathbf{D}=\epsilon\mathbf{E}, \tag{3.4a}$$

$$\mathbf{H}=\frac{1}{\mu}\mathbf{B}. \tag{3.4b}$$

Cross-multiplying (3.3a) by **k** and using (3.4b) and (3.3c) to eliminate **B** and **H**, we obtain

$$\mathbf{k}\times(\mathbf{k}\times\mathbf{E})=-\omega^2\mu\mathbf{D}. \tag{3.5}$$

From (3.4a) and (3.3d), we have $\mathbf{k}\cdot\mathbf{E}=0$, and (3.5) becomes

$$(k^2-\omega^2\mu\epsilon)\mathbf{E}=0.$$

Requiring nontrivial solutions for **E**, we obtain the dispersion relation

$$k^2=\omega^2\mu\epsilon, \tag{3.6}$$

which provides a relation between the magnitude of **k**, denoted by k, and the angular frequency ω. The angular frequency ω has the dimension of inverse time, the wave vector **k** has the dimension of inverse distance, and the square root of the product $\mu\epsilon$ has the dimension of inverse velocity. The dimensionless quantity

$$n\equiv c\sqrt{\mu\epsilon}=\frac{ck}{\omega} \tag{3.7}$$

is the *refractive index* of the medium. From (3.6), we compute

$$u \equiv \frac{\omega}{k} = (\mu\epsilon)^{-1/2} \tag{3.8}$$

as the velocity of an electromagnetic wave in a nondispersive isotropic medium. Thus the larger the refractive index, the slower is the wave velocity inside the medium. When the medium is dispersive, the dispersion relation may be a complicated function between k and ω.

The dispersion relation, similar to conservation theorems in mechanics, expresses a relationship between the energy and the momentum of an electromagnetic wave. Equation 3.6 states that $n^2\omega^2/c^2 - k^2 = 0$. From a quantum-mechanical point of view, which we shall pursue in Chapter 7, the energy of a photon is $n\hbar\omega$ and the momentum is $\hbar k$, where $\hbar = 1.05 \times 10^{-34}$ J-s is Planck's constant divided by 2π. If we compare this with a similar relation in relativistic mechanics, where $E^2 - p^2c^2 = m_0^2c^4$ for a particle with rest mass m_0 moving with velocity u, we may conclude that the rest mass of a photon is zero.

In (3.6), if neither μ nor ϵ is a function of ω or k, the relation is linear. In general, however, especially for dispersive media, this is not true. In anisotropic and bianisotropic media, directional dependence is also important. We can regard the dispersion relation as a function involving the three components of $\mathbf{k}$ (k_x, k_y, k_z) and the angular frequency ω.

3.1b Polarization

The polarization of a wave is conventionally defined by the time variation of the electric field vector **E** at a fixed point in space. By (3.3d), the **D** vector is always perpendicular to the **k** vector in a homogeneous medium. In free space or in an isotropic medium, **E** is parallel to **D** and perpendicular to **k**. In anisotropic or bianisotropic media, however, **E** is not necessarily perpendicular to **k** because **E** and **D** are no longer parallel to each other. We shall discuss polarization based on the time variation of **D**. Whenever necessary, we use the term **D** polarization in order to distinguish this type from the conventional polarization based on **E**.

From a fixed observation point in space, the tip of the **D** vector moves with time. If the tip moves along a straight line, the wave is linearly polarized. When the locus of the tip is a circle, the wave is circularly polarized. For an elliptically polarized wave, the tip of **D** describes an ellipse. The tip can move either clockwise or counterclockwise for circularly or elliptically polarized waves. If the right-hand thumb points in the direction of propagation while the fingers point in the direction of the tip motion, the wave is right-hand polarized. The wave is left-hand polarized when it is described by the left-hand thumb and fingers.

To facilitate a mathematical discussion of polarization, we decompose the **D** vector of a wave into two components perpendicular to each other. Assume that the direction of **k** is along the z axis. For a specific point in space, we write

$$\mathbf{D}=\left(\hat{x}|D_x|e^{i\psi_x}+\hat{y}|D_y|e^{i\psi_y}\right)e^{-i\omega t},$$

where $|D_x|$ and $|D_y|$ are magnitudes of **D** in the $\hat{x}$ and $\hat{y}$ directions, while ψ_x and ψ_y denote their phases. The instantaneous space-time **D** field is equal to the real part of the complex convention

$$\text{Re}(\mathbf{D})=\hat{x}d_x+\hat{y}d_y=\hat{x}|D_x|\cos(\omega t-\psi_x)+\hat{y}|D_y|\cos(\omega t-\psi_y),$$

where d_x and d_y denote instantaneous values of **D** along the x and y axes. The locus of the tip of Re (**D**) is determined by eliminating time t between the two components of Re (**D**).

Linear Polarization. When ψ_x and ψ_y differ by an integer multiple of 2π, the two components are in phase. The wave is linearly polarized, and the straight-line locus traverses the first and the third quadrants (Fig. 3.1*a*). When ψ_x and ψ_y differ by an odd integer multiple of π, the two components are 180° out of phase, and the straight-line locus traverses the second and fourth quadrants (Fig. 3.1*b*).

Circular Polarization. The magnitudes of the two components are equal, $|D_x|=|D_y|=d_0$, and the phases differ by 90°. Consider the case $\psi_y-\psi_x=\pi/2$. We have

$$\text{Re}(\mathbf{D})=d_0[\hat{x}\cos(\omega t-\psi_x)+\hat{y}\sin(\omega t-\psi_x)].$$

It can be seen that, while the x component is at the maximum, the y component is zero. As time progresses, the y component increases and the x component decreases. The tip of Re (**D**) rotates counterclockwise from the positive $\hat{x}$ direction to the positive $\hat{y}$ direction (Fig. 3.1*c*). Elimination of t yields a circle of radius d_0; $d_x^2+d_y^2=d_0^2$. Thus the wave is right-hand circularly polarized. Similar reasoning shows that with $\psi_y-\psi_x=-\pi/2$ the wave is left-hand circularly polarized (Fig. 3.1*d*).

Elliptical Polarization. In general, a polarized wave has elliptical polarization; that is, when time is eliminated from the two components of Re (**D**), the resultant equation describes an ellipse.

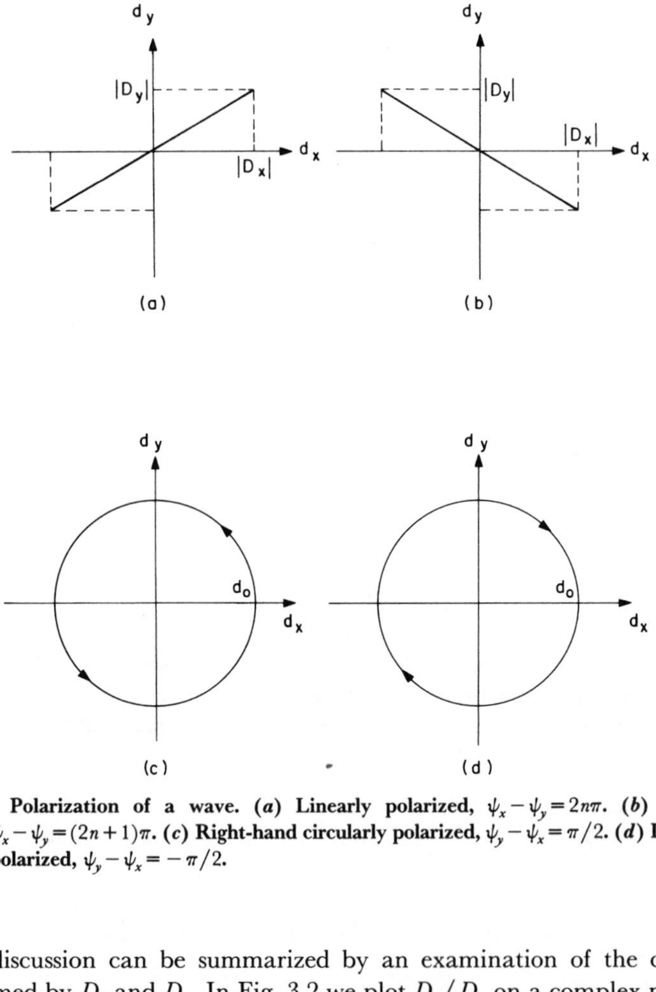

Figure 3.1 Polarization of a wave. (*a*) Linearly polarized, $\psi_x - \psi_y = 2n\pi$. (*b*) Linearly polarized, $\psi_x - \psi_y = (2n+1)\pi$. (*c*) Right-hand circularly polarized, $\psi_y - \psi_x = \pi/2$. (*d*) Left-hand circularly polarized, $\psi_y - \psi_x = -\pi/2$.

This discussion can be summarized by an examination of the complex ratio formed by D_y and D_x. In Fig. 3.2 we plot D_y/D_x on a complex plane. If the phase of this ratio is $n\pi$, the wave is linearly polarized. If the magnitude is unity and the phase is $\pi/2$, the wave is right-hand circularly polarized. If the magnitude is unity and the phase is $-\pi/2$, the wave is left-hand circularly polarized. Otherwise the wave is elliptically polarized. The polarization is right-hand if the phase is between 0 and π, and left-hand if the phase is between π and 2π.

Although the polarization that has been discussed is for **D**, it applies also to **E**. It is inside anisotropic and bianisotropic media where **D** and **E** are not parallel that the definitions of **D** and **E** polarization begin to differ. Note also that waves may be partially polarized or not polarized at all.[12]

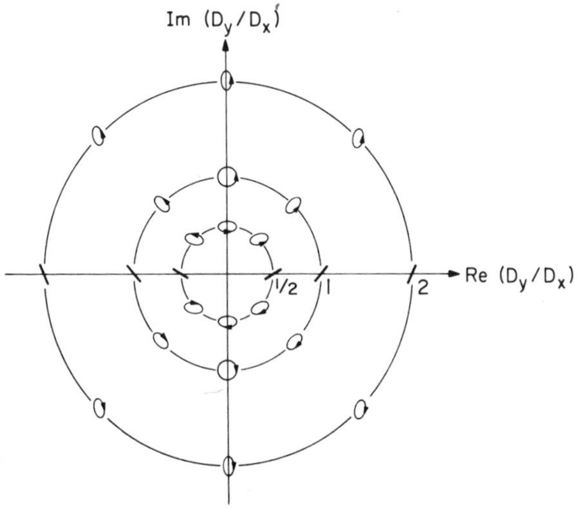

Figure 3.2 Complex D_y/D_x plane. The polarization of a wave is illustrated for various magnitudes and phases of D_y/D_x.

3.1c Waves in Uniaxial Media

Consider the case of a uniaxial medium, which is the simplest anisotropic medium. In its principal coordinate system, $\overline{\mathbf{L}}=\overline{\mathbf{M}}=\overline{0}$, $\overline{\mathbf{Q}}=(1/c\mu)\overline{\mathbf{I}}$, and $\overline{\mathbf{P}}=c\bar{\epsilon}$. Thus we have

$$\mathbf{D}=\bar{\epsilon}\cdot\mathbf{E}, \tag{3.9a}$$

$$\bar{\epsilon}=\begin{bmatrix}\epsilon & 0 & 0\\ 0 & \epsilon & 0\\ 0 & 0 & \epsilon_z\end{bmatrix}. \tag{3.9b}$$

The medium is positive uniaxial if $\epsilon_z>\epsilon$, and negative uniaxial if $\epsilon_z<\epsilon$. We substitute (3.9) in (3.5) and note that $\mathbf{k}\cdot\mathbf{E}=(1-\epsilon_z/\epsilon)k_zE_z$ to obtain

$$-k^2\mathbf{E}+\mathbf{k}\left(1-\frac{\epsilon_z}{\epsilon}\right)k_zE_z=-\omega^2\mu\bar{\epsilon}\cdot\mathbf{E}.$$

In matrix form, we write

$$\begin{bmatrix}-k^2+\omega^2\mu\epsilon & 0 & (1-\epsilon_z/\epsilon)k_xk_z\\ 0 & -k^2+\omega^2\mu\epsilon & (1-\epsilon_z/\epsilon)k_yk_z\\ 0 & 0 & -k_x^2-k_y^2-\epsilon_zk_z^2/\epsilon+\omega^2\mu\epsilon_z\end{bmatrix}\begin{bmatrix}E_x\\ E_y\\ E_z\end{bmatrix}=0. \tag{3.10}$$

Setting the determinant of the matrix equal to zero yields the dispersion relations for the uniaxial medium:

$$k^2 = \omega^2 \mu \epsilon, \tag{3.11a}$$

$$k_x^2 + k_y^2 + \frac{\epsilon_z}{\epsilon} k_z^2 = \omega^2 \mu \epsilon_z. \tag{3.11b}$$

Equation 3.11a is the dispersion relation for *ordinary waves*; (3.11b), for *extraordinary waves*. For an extraordinary wave, the magnitude of the **k** vector is a function of its direction of propagation. For an ordinary wave the magnitude of **k** is independent of its direction of propagation. Ordinary and extraordinary waves are the two permissible normal modes or characteristic waves in a uniaxial medium.

Assume that the direction of **k** makes an angle θ with respect to the optic axis $\hat{z}$. Equation 3.11b then becomes

$$k^2\left(\sin^2\theta + \frac{\epsilon_z}{\epsilon}\cos^2\theta\right) = \omega^2 \mu \epsilon_z. \tag{3.12}$$

This can be solved for k, the magnitude of the **k** vector for extraordinary waves. Thus the ordinary wave propagates with phase velocity

$$u_0 = \pm(\mu\epsilon)^{-1/2}, \tag{3.13a}$$

the extraordinary wave with phase velocity

$$u_e = \pm\left(\frac{\sin^2\theta}{\mu\epsilon_z} + \frac{\cos^2\theta}{\mu\epsilon}\right)^{1/2} \tag{3.13b}$$

The $\pm$ sign in (3.13) implies that the velocities of propagation in opposite directions are equal in magnitudes. This result of two characteristic waves propagating at different velocities in a medium is called *birefringence*. Along the optic axis both waves have the same velocity. Note that, when the medium is isotropic, $\epsilon_z = \epsilon$ and (3.11a) and (3.11b) are identical. The two waves degenerate into one and propagate at the same velocity.

The polarizations of the two types of wave are determined from (3.10). Inserting (3.11a) into (3.10), we see that $\mathbf{E}_z$ must be equal to zero. Thus the ordinary wave does not possess a z component; **D** and **E** are both perpendicular to the optic axis (Fig. 3.3*a*). Since **D** and **E** are also perpendicular to **k**, the ordinary wave is linearly polarized with electric vectors perpendicular to the plane determined by the optic axis and the **k** vector.

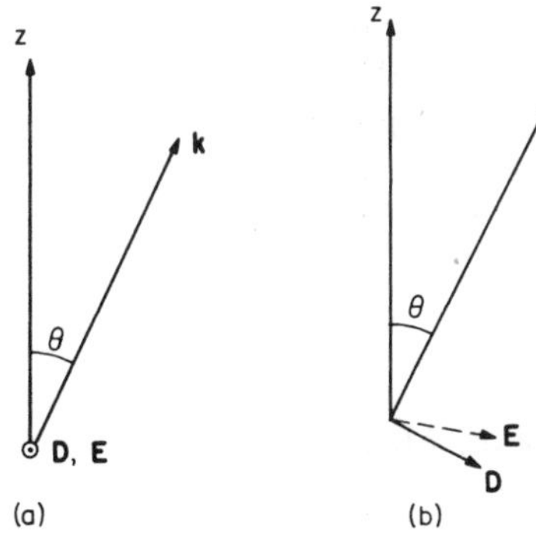

Figure 3.3 Polarization of ordinary and extraordinary waves in positive uniaxial media. (*a*) Ordinary wave: D and E are perpendicular to the plane determined by the optic axis *z* and the wave vector k. (*b*) Extraordinary wave: D and E are parallel to the plane determined by the optic axis *z* and the wave vector k.

Inserting (3.11b) into (3.10), we determine the polarization for extraordinary waves. Without loss of generality, we can restrict the **k** vector to be in the $y-z$ plane, namely, $k_x=0$. This can always be achieved by rotating the coordinate around the z axis. We observe that $E_x=0$, and E_y and E_z are related, in view of (3.3d) and constitutive relation 3.9, by

$$\epsilon \sin\theta E_y + \epsilon_z \cos\theta E_z = 0. \tag{3.14}$$

This equation can also be obtained from (3.10) and (3.12). We see that, for fixed ϵ and E_y, the larger the ϵ_z, the smaller is the E_z. The reason is that **D** is always perpendicular to **k**; the more easily the medium is polarizable in the $\hat{z}$ direction, the smaller E_z is needed in order to obtain the same D_z. Thus the extraordinary wave is linearly polarized with both **E** and **D** lying in the plane determined by the optic axis and the **k** vector (Fig. 3.3*b*).

The direction of **k** is the same as $\mathbf{D}\times\mathbf{B}$. For ordinary waves, **D** and **E** are parallel and so are **B** and **H**. Poynting's vector $\mathbf{E}\times\mathbf{H}$ is also in the direction of **k**. For extraordinary waves, however, **D** and **E** are not parallel. Thus Poynting's vector is not in the direction of **k**, implying that the phase and group velocities are in different directions. Using Maxwell's equations 3.3, we calculate all field vectors for both ordinary and extraordinary waves. The results are listed in Table 3.1.

When a wave enters a uniaxial medium, it breaks up into two linearly polarized characteristic waves. All **E**-field components perpendicular to the $\mathbf{k}-\hat{z}$ plane constitute the ordinary wave; the rest, the extraordinary wave. The two waves propagate at different velocities, a phenomenon known as *double refraction*. Consider the simple case of a wave incident upon a uniaxial crystal in the $\hat{y}$ direction. Suppose that it is linearly polarized in the $x-z$ plane but not necessarily parallel to either the x or the z axis. The part of the incident wave that is transmitted is composed of ordinary and extraordinary waves propagating at different speeds. Suppose that the medium has a finite thickness. Then the wave transmitted on the other side will no longer be linearly polarized, but will, in general, be elliptically polarized. By properly adjusting the thickness and orientation of the medium, we can obtain any polarization that we desire from an incident wave of known polarization.

Table 3.1
Characteristic Waves in Uniaxial Media[a]

Wave Characteristics	Ordinary Waves	Extraordinary Waves
D	$\begin{bmatrix} 1 \\ 0 \\ 0 \end{bmatrix}$	$\begin{bmatrix} 0 \\ \cos\theta \\ -\sin\theta \end{bmatrix}$
E	$\begin{bmatrix} 1/\epsilon \\ 0 \\ 0 \end{bmatrix}$	$\begin{bmatrix} 0 \\ \cos\theta/\epsilon \\ -\sin\theta/\epsilon_z \end{bmatrix}$
B	$\begin{bmatrix} 0 \\ k\cos\theta/\omega\epsilon \\ -k\sin\theta/\omega\epsilon \end{bmatrix}$	$\begin{bmatrix} -k(\cos^2\theta/\epsilon+\sin^2\theta/\epsilon_z)/\omega \\ 0 \\ 0 \end{bmatrix}$
H	$\begin{bmatrix} 0 \\ k\cos\theta/\omega\mu\epsilon \\ -k\sin\theta/\omega\mu\epsilon \end{bmatrix}$	$\begin{bmatrix} -k(\cos^2\theta/\epsilon+\sin^2\theta/\epsilon_z)\omega\mu \\ 0 \\ 0 \end{bmatrix}$
k	$\omega\sqrt{\mu\epsilon}$	$\omega\sqrt{\mu\epsilon_z}\left(\frac{\epsilon_z}{\epsilon}\cos^2\theta+\sin^2\theta\right)^{-1/2}$
Dispersion Relation	$k^2=\omega^2\mu\epsilon$	$k_z^2+\frac{\epsilon}{\epsilon_z}(k_x^2+k_y^2)=\omega^2\mu\epsilon$

[a] The optic axis is the z axis; the **k** vector is in the $y-z$ plane, $\mathbf{k}=\hat{z}k\cos\theta+\hat{y}k\sin\theta$.

3.1d Dispersion Relations for Bianisotropic Media

In general, we can determine the dispersion relations by eliminating any three field vectors from (3.2) and (3.3) to obtain an equation for one field vector. The equation is homogeneous and is a vector equation composed of three scalar equations. The condition of a nontrivial solution gives rise to the dispersion relation.

Let us eliminate **D**, **H**, and **B** from (3.2) and (3.3). We define an operator $\bar{\mathbf{k}}$ such that $\bar{\mathbf{k}} \cdot \mathbf{A} \equiv \mathbf{k} \times \mathbf{A}$ for any vector **A** and obtain

$$\left\{ \frac{\omega^2}{c^2} \bar{\mathbf{P}} + \frac{\omega}{c} (\bar{\mathbf{L}} \cdot \bar{\mathbf{k}} + \bar{\mathbf{k}} \cdot \bar{\mathbf{M}}) + \bar{\mathbf{k}} \cdot \bar{\mathbf{Q}} \cdot \bar{\mathbf{k}} \right\} \cdot \mathbf{E} = 0. \tag{3.15}$$

For nontrivial solutions of **E**, the determinant of the matrix in (3.15) operating on **E** must be equal to zero. Hence

$$\left| \frac{\omega^2}{c^2} \bar{\mathbf{P}} + \frac{\omega}{c} (\bar{\mathbf{L}} \cdot \bar{\mathbf{k}} + \bar{\mathbf{k}} \cdot \bar{\mathbf{M}}) + \bar{\mathbf{k}} \cdot \bar{\mathbf{Q}} \cdot \bar{\mathbf{k}} \right| = 0. \tag{3.16}$$

This is the dispersion relation for plane waves in a homogeneous bianisotropic medium, from which the relations between components of **k** and ω can be determined. The solution of (3.16) is simple and straightforward in theory but very complicated in practice. This becomes apparent if we try to derive the dispersion relations for isotropic media from (3.16). In Section 3.3, we develop a *kDB* system and use it to derive the dispersion relations for the various media.

3.2 WAVE VECTOR k

3.2a The k Vector

The **k** vector occupies a unique and fundamental position in the study of electromagnetic waves. Consider a plane wave of the type $e^{i\mathbf{k}\cdot\mathbf{r}-i\omega t}$. The exponential dependence $\mathbf{k}\cdot\mathbf{r}-\omega t$ at a particular space-time point gives the phase of the wave at that point. At a fixed time t_0, the change of phase is governed by $\mathbf{k}\cdot\mathbf{r}$. Along the direction of **k**, the total phase of the wave changes by 2π as $kr=2\pi$, which is equivalent to the spatial distance of a wavelength. The wavelength, defined by $\lambda=2\pi/k$, measures the spatial periodicity of an electromagnetic wave. The magnitude of **k** is called the *wave number*, which is equal to the number of wavelengths in a distance 2π.

As time progresses, a phase front of constant $\mathbf{k}\cdot\mathbf{r}-\omega t$ along the direction of $\mathbf{k}$ propagates with a velocity $u=\omega/k$, which is the phase velocity of the wave. The time for a constant phase front to move a unit distance is given by the reciprocal of u, that is, $\tau_p=k/\omega$, where τ_p is the phase delay per unit length of the wave.

In dispersive media, k is a nonlinear function of ω, $k=k(\omega)$. Waves of different frequencies propagate at different phase velocities. Consider a group of plane waves with different angular frequencies propagating in a dispersive medium, and assume that all angular frequencies are in the neighborhood of a center frequency ω_0. We can expand the wave number k around ω_0:

$$k(\omega)=k(\omega_0)+(\omega-\omega_0)\left(\frac{dk(\omega)}{d\omega}\right)_{\omega=\omega_0}$$

$$+\frac{1}{2!}(\omega-\omega_0)^2\left(\frac{d^2k(\omega)}{d\omega^2}\right)_{\omega=\omega_0}+\cdots.$$

Consider a narrow-band signal, where only the first two terms are important. The space-time dependence of each frequency component becomes $\exp[-i(\omega-\omega_0)(t-r[dk/d\omega]_{\omega=\omega_0})]\exp(ikr-i\omega_0 t)$. The signal can be viewed as propagating with phase delay k/ω_0 and group delay τ_g, where $\tau_g=dk(\omega)/d\omega$. The group velocity v_g is defined as $v_g=1/[dk(\omega)/d\omega]_{\omega=\omega_0}$. When k is linear in ω, the phase delay and the group delay are equal, and so are the phase velocity and the group velocity.[98]

For a dispersive medium, we can construct a k-ω diagram, plotting k as a function of ω. Figure 3.4*a* shows the k-ω diagram for a plasma with ϵ as given in (1.26). The slope of the straight line from the origin to a point on the curve at $\omega=\omega_0$ represents the phase delay at that frequency. The slope of the tangent to the curve at $\omega=\omega_0$ represents the group delay at that frequency. We can also plot an ω-k diagram with k as the abscissa and ω as the ordinate (Fig. 3.4*b*). The slope of the straight line from the origin to a point on the curve at $\omega=\omega_0$ then represents the phase velocity at $\omega=\omega_0$, and the slope of the tangent to the curve at $\omega=\omega_0$ represents the group velocity. For a nondispersive medium, the ω-k and the k-ω curves are straight lines starting from the origin, and the phase delay and the group delay are the same, as are the phase velocity and the group velocity.

When we do not restrict the direction of $\mathbf{k}$, we can generalize these definitions. Consider the exponential factor $(\mathbf{k}\cdot\mathbf{r}-\omega t)$. In a Cartesian coordinate system, we can write the $\mathbf{k}$ vector as

$$\mathbf{k}=\hat{x}k_x+\hat{y}k_y+\hat{z}k_z, \tag{3.17}$$

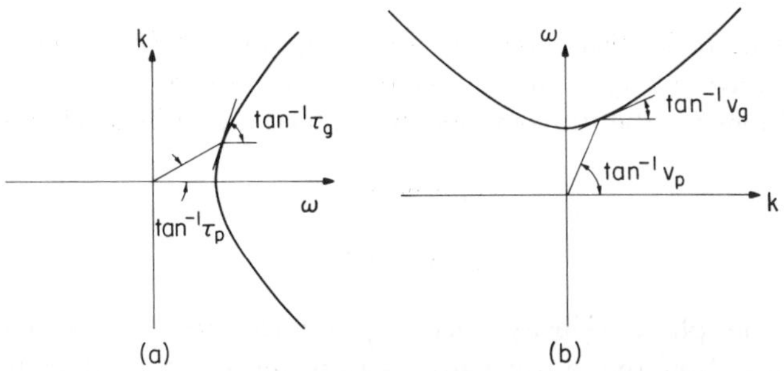

Figure 3.4 Dispersion curves for an isotropic plasma. (*a*) $\omega - k$ diagram. (*b*) $k - \omega$ diagram.

and a position vector

$$\mathbf{r} = \hat{x}x + \hat{y}y + \hat{z}z, \tag{3.18}$$

so that

$$\mathbf{k} \cdot \mathbf{r} = k_x x + k_y y + k_z z, \tag{3.19}$$

where $\hat{x}$, $\hat{y}$, and $\hat{z}$ denote unit vectors along the x, y, and z directions of the Cartesian coordinate system, and k_x, k_y, and k_z are the Cartesian components of the **k** vector. For a given **k** vector, a constant phase front is determined by $\mathbf{k} \cdot \mathbf{r} =$ constant, which indicates that the front is perpendicular to the **k** vector (Fig. 3.5). Since at all times the constant phase front must be perpendicular to **k**, we conclude that this phase front propagates in the direction of **k**.

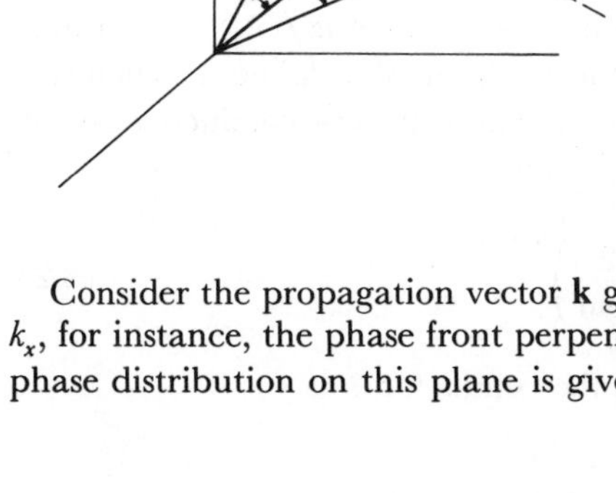

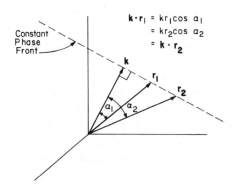

Figure 3.5 The k vector and a constant phase front that is perpendicular to k.

Consider the propagation vector **k** given by (3.17). In the direction of $\hat{x}$, or k_x, for instance, the phase front perpendicular to $\hat{x}$ is not a constant one. The phase distribution on this plane is given by $k_y y + k_z z$, and the phase velocity and wavelength in this direction are ω/k_x and $2\pi/k_x$. Thus the wave appears to have a larger phase velocity and wavelength in the $\hat{x}$ direction. The definitions for the ith components of phase and group velocities are

$$v_{pi}=\omega/k_i, \tag{3.20}$$

$$v_{gi}=\partial\omega/\partial k_i. \tag{3.21}$$

Evidently the phase velocity vector v_{pi} reduces to $u=\omega/k$ only in the direction of **k**. Note that a wave group can be formed from plane waves with different frequencies or from plane waves with different **k** vectors. Thus the difference in the phase and group velocities may be caused by media that are dispersive or non-isotropic. When a medium is non-isotropic, the magnitude of **k** may vary as a function of its direction.

The components of a **k** vector may not all be real, even in a lossless medium where the magnitude of **k** is real. Consider an isotropic medium with dispersion relation 3.6:

$$k^2=k_x^2+k_y^2+k_z^2=\omega^2\mu\epsilon. \tag{3.22}$$

We see that the components of the **k** vector, k_x, k_y, and k_z, can have complex values even for real ϵ and μ. The imaginary part of a complex component in a particular direction indicates that the wave either grows or decays exponentially in that direction. Let the **k** vector of a plane wave possess two components, k_y and k_x, such that k_y is imaginary, $k_y=i\alpha_y$, and $k_x^2-\alpha_y^2=\omega^2\mu\epsilon$. The wave then behaves as $e^{ik_x x}\,e^{-\alpha_y y}$. It propagates in the $\hat{x}$ direction with phase delay k_x/ω and decays in the $\hat{y}$ direction exponentially with rate α_y. We call the wave *evanescent* in the $\hat{y}$ direction. It can also be called a *nonuniform plane wave* propagating in the $\hat{x}$ direction because the constant phase front, which is a plane perpendicular to the $\hat{x}$ direction, no longer has constant wave amplitudes.

When the medium is lossy, $\omega^2\mu\epsilon$ is complex. The wave number k is complex too. A wave propagating in a certain direction also decays exponentially in that direction. Consider a conducting medium with $\epsilon=\epsilon'+i\sigma/\omega$. The imaginary part of the wave number $k=\omega[\mu(\epsilon'+i\sigma/\omega)]^{1/2}$ determines how fast the wave attenuates inside the medium. We define a penetration depth d_p to be $d_p=1/\mathrm{Im}(k)$. When the medium is very conductive so that $\epsilon\approx i\sigma/\omega$, the wave number $k\approx(1+i)/\delta$, and

$$\delta=\left(\frac{2}{\omega\mu\sigma}\right)^{1/2}, \tag{3.23}$$

where δ is also called the *skin depth* of the conductor.

3.2b Phase Invariance and Doppler Effect

Consider a receiver and a transmitter in relative motion, and assume that the receiver receives a plane wave from the transmitter. According to the transmitter, the wave is described by

$$\begin{bmatrix} \mathbf{E}(r,t) \\ \mathbf{B}(r,t) \end{bmatrix} = \begin{bmatrix} \mathbf{E}_0 \\ \mathbf{B}_0 \end{bmatrix} e^{i(\mathbf{k}\cdot\mathbf{r}-\omega t)}. \tag{3.24}$$

Let the receiver move with velocity $\mathbf{v}$ with respect to the transmitter, and let primes denote quantities associated with the receiver. According to the Lorentz transformation formulas (see Chapter 2), we have

$$\mathbf{r} = \bar{\boldsymbol{\alpha}}\cdot\mathbf{r}' + \gamma\boldsymbol{\beta}ct',$$

$$ct = \gamma(ct' + \boldsymbol{\beta}\cdot\mathbf{r}'),$$

for space-time coordinates, and

$$\begin{bmatrix} \mathbf{E} \\ c\mathbf{B} \end{bmatrix} = \gamma\begin{bmatrix} \bar{\boldsymbol{\alpha}}^{-1} & -\bar{\boldsymbol{\beta}} \\ \bar{\boldsymbol{\beta}} & \bar{\boldsymbol{\alpha}}^{-1} \end{bmatrix}\cdot\begin{bmatrix} \mathbf{E}' \\ c\mathbf{B}' \end{bmatrix}.$$

for field vectors. Making use of these transformation formulas, we can write the expression for the plane wave in the frame of the receiver as

$$\begin{bmatrix} \mathbf{E}'(\mathbf{r}',t') \\ \mathbf{B}'(\mathbf{r}',t') \end{bmatrix} = \begin{bmatrix} \mathbf{E}_0' \\ \mathbf{B}_0' \end{bmatrix} e^{i(\mathbf{k}'\cdot\mathbf{r}'-\omega' t')}. \tag{3.25}$$

The primed quantities can all be expressed in terms of the unprimed quantities. For the angular frequency and the wave vector, we have

$$\mathbf{k}' = \bar{\boldsymbol{\alpha}}\cdot\mathbf{k} - \frac{\gamma\omega\boldsymbol{\beta}}{c}, \tag{3.26a}$$

$$\frac{\omega'}{c} = \gamma\left(\frac{\omega}{c} - \boldsymbol{\beta}\cdot\mathbf{k}\right). \tag{3.26b}$$

Transformation formula 3.26 is identical to that for space and time coordinates with $\mathbf{r}$ replaced by $\mathbf{k}$ and ct replaced by ω/c. With transformation formula 3.26, the phase of the plane wave in both frames is an invariant quantity. This invariance of phase, which enables us to deduce transformation formula 3.26, is referred to as the principle of phase invariance.

The aberration effect is a consequence of (3.26a). The perpendicular component of $\mathbf{k}'$ is equal to that of $\mathbf{k}$, while the parallel component is changed by the motion. Consider an observer on Earth looking at a star at the zenith. Since the Earth is moving with respect to the star, a $\mathbf{k}'$ component antiparallel to $\boldsymbol{\beta}$ is generated. Thus the observer must tilt his telescope in the direction of the Earth's motion, just as, on a windless rainy day, a bicycle rider always tilts his umbrella in the forward direction. It is straightforward to determine from (3.26a) a relation for the angles between $\boldsymbol{\beta}$ and $\mathbf{k}'$ and between $\boldsymbol{\beta}$ and $\mathbf{k}$. Let θ denote the angle between $\mathbf{k}$ and $\boldsymbol{\beta}$, and θ' the angle between $\mathbf{k}'$ and $\boldsymbol{\beta}$. Recall that $\bar{\boldsymbol{\alpha}}=\bar{\mathbf{I}}+(\gamma-1)\boldsymbol{\beta\beta}/\beta^2$. Cross-multiplying and dot-multiplying (3.26a) by $\boldsymbol{\beta}$, we obtain

$$k'\sin\theta'=k\sin\theta,$$

$$k'\cos\theta'=\gamma k\cos\theta-\gamma\beta\omega/c.$$

In an isotropic medium, $k=n\omega/c$, elimination of k' and k from these two equations gives

$$\tan\theta'=\frac{\tan\theta}{\gamma(1-\beta\sec\theta/n)}. \tag{3.27}$$

This is the relativistic formula for aberration.

The Doppler effect is a consequence of (3.26b). Using the dispersion relation for isotropic media and letting the angle between $\mathbf{k}$ and $\boldsymbol{\beta}$ be θ, we find from (3.26b) that

$$\omega'=\gamma\omega(1-n\beta\cos\theta). \tag{3.28}$$

When the receiver is receding from the transmitter, $\boldsymbol{\beta}$ and $\mathbf{k}$ are in the same direction. The frequency is shifted downward or red-shifted. When the receiver is approaching the transmitter, the frequency is shifted upward or blue-shifted. When the receiver is moving perpendicular to $\mathbf{k}$, we have the transverse Doppler shift $\omega'=\gamma\omega$, which is a purely relativistic effect.

3.2c k Surfaces

The dispersion relations provide a functional relationship among components of the $\mathbf{k}$ vector and the angular frequency ω. The equations are usually quadratic in k for each characteristic wave when there is more than one. We write, in general,

$$f(k_x,k_y,k_z,\omega)=0. \tag{3.29}$$

In three-dimensional space with base vectors k_x, k_y, and k_z, called **k** *space*, a quadratic equation describes a two-dimensional hypersurface. The surface is called the *wave surface* or simply the **k** *surface*.

In the isotropic case with dispersion relation 3.6, the **k** surface is a sphere with radius $\omega\sqrt{\mu\epsilon}$. In the uniaxial case with dispersion relations 3.11, the **k** surface for ordinary waves is still a sphere. For extraordinary waves, as seen from (3.11b) or (3.12), the **k** surface is an ellipsoid (Fig. 3.6). The ellipsoid in three-dimensional **k** space is a surface of revolution around the k_z axis. The length of the principal axis along k_z is equal to $\omega\sqrt{\mu\epsilon}$, while that transverse to k_z is equal to $\omega\sqrt{\mu\epsilon_z}$.

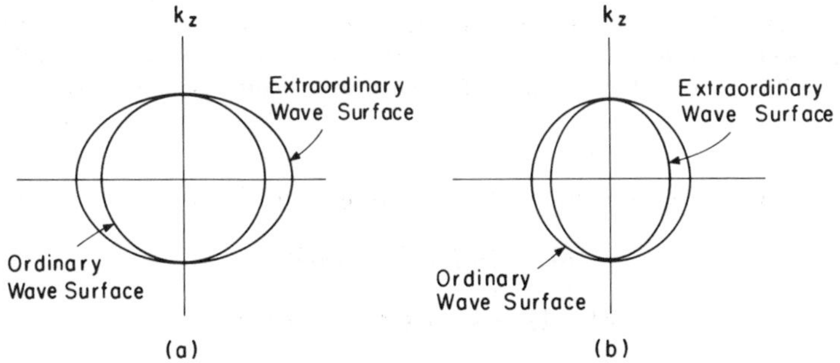

Figure 3.6 Ordinary and extraordinary wave surfaces for uniaxial media. (*a*) For positive uniaxial media. (*b*) For negative uniaxial media.

We observe that the magnitude of **k**, as described by the **k** surface, may vary as a function of its direction. In a particular direction, the **k** vector intersects the **k** surface at a point. The magnitude of **k** in this direction is proportional to the length from the origin to this point. The phase velocity of a wave in the medium is equal to ω/k. The group velocity pertaining to the wave surface is defined by (3.21):

$$v_{gi}=\frac{\partial\omega}{\partial k_i}=-\frac{\partial f/\partial k_i}{\partial f/\partial\omega}. \tag{3.30}$$

Since a vector with components $\partial f/\partial k_i$ is normal to the surface $f(k_x,k_y,k_z,\omega)=0$, we conclude that, as a wave propagates with phase velocity u along the direction of **k** that intersects the wave surface at a point, the group velocity is in a direction normal to the wave surface at that point. As an example, consider a positive uniaxial medium (Fig. 3.7); whereas the phase velocity and the group velocity are in the same direction for the ordinary wave, they are in different directions for the extraordinary waves.

We now show that the direction of group velocity coincides with the direction of flow of the time-average Poynting's vector. First, we take a

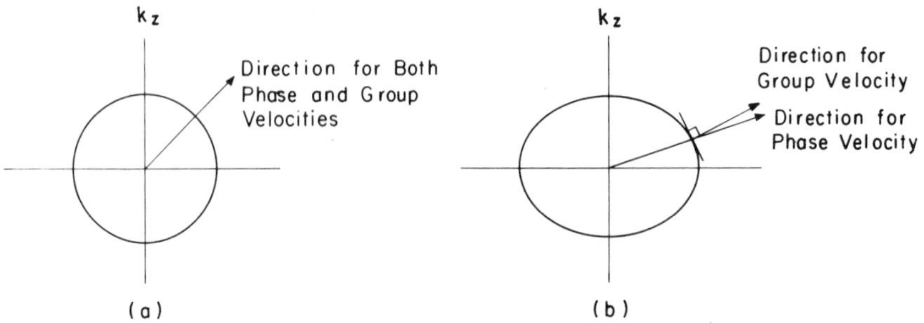

Figure 3.7 Directions for phase velocity and group velocity for a given k vector in a positive uniaxial medium. (*a*) Ordinary wave: phase velocity and group velocity in the same direction. (*b*) Extraordinary wave: phase velocity and group velocity in different directions.

differential of (3.3a) and the complex conjugate of (3.3c):

$$\delta\mathbf{k}\times\mathbf{E}+\mathbf{k}\times\delta\mathbf{E}=\omega\delta\mathbf{B}, \tag{3.31}$$

$$\delta\mathbf{k}\times\mathbf{H}^*+\mathbf{k}\times\delta\mathbf{H}^*=-\omega\delta\mathbf{D}^*. \tag{3.32}$$

We dot-multiply (3.31) by $\mathbf{H}^*$ and (3.32) by $\mathbf{E}$, and then subtract. Using the identity $\mathbf{A}\cdot(\mathbf{B}\times\mathbf{C})=\mathbf{B}\cdot(\mathbf{C}\times\mathbf{A})=\mathbf{C}\cdot(\mathbf{A}\times\mathbf{B})$, we obtain

$$2\delta\mathbf{k}\cdot(\mathbf{E}\times\mathbf{H}^*)=\omega(\mathbf{H}^*\cdot\delta\mathbf{B}+\mathbf{E}\cdot\delta\mathbf{D}^*)+\delta\mathbf{E}\cdot(\mathbf{k}\times\mathbf{H}^*)-\delta\mathbf{H}^*\cdot(\mathbf{k}\times\mathbf{E}).$$

Again, we apply (3.3a) and (3.3c) to the last two terms to obtain

$$2\delta\mathbf{k}\cdot(\mathbf{E}\times\mathbf{H}^*)=\omega(\mathbf{H}^*\cdot\delta\mathbf{B}+\mathbf{E}\cdot\delta\mathbf{D}^*-\mathbf{D}^*\cdot\delta\mathbf{E}-\mathbf{B}\cdot\delta\mathbf{H}^*). \tag{3.33}$$

For a bianisotropic medium, we introduce constitutive relations 3.2. Using the lossless conditions of bianisotropic media, we observe that the right-hand side of (3.33) is a pure imaginary number. Since the time-average Poynting's vector is equal to the real part of $\mathbf{E}\times\mathbf{H}^*$, we conclude that

$$\delta\mathbf{k}\cdot\langle\mathbf{S}\rangle=0. \tag{3.34}$$

The change of $\mathbf{k}$ is tangential to the wave surface. Thus, at a point on the $\mathbf{k}$ surface, the direction of the time-average Poynting's vector is normal to the wave surface at that point. The group velocity is also normal to the $\mathbf{k}$ surface. We conclude that the group velocity coincides with the direction of Poynting's vector flow.

Instead of dealing directly with the $\mathbf{k}$ vector, we can define a dimensionless vector by dividing $\mathbf{k}$ by $\omega\sqrt{\mu_0\epsilon_0}$:

$$\mathbf{n} \equiv \frac{\mathbf{k}}{\omega\sqrt{\mu_0\epsilon_0}}. \tag{3.35}$$

We call $\mathbf{n}$ the index vector or simply $\mathbf{n}$ vector, since the magnitude of $\mathbf{n}$ characterizes the index of refraction of a medium. In terms of the components of the vector $\mathbf{n}$, dispersion relations like (3.11a) multiplied by (3.11b) are called *Fresnel equations*.

3.2d Ray Vectors and Ray Surfaces

The direction of propagation of a wave becomes ambiguous in a complicated medium. From Poynting's theorem, we learned that the power flow of an electromagnetic field is governed by Poynting's vector, $\mathbf{S} = \mathbf{E} \times \mathbf{H}$. The direction of power flow is perpendicular to both $\mathbf{E}$ and $\mathbf{H}$. This is the direction that we observe when we pass a ray of light through the medium. We have also learned that the direction of propagation of the constant phase is along $\mathbf{k}$. By (3.3b) and (3.3d), $\mathbf{k}$ is perpendicular to both $\mathbf{D}$ and $\mathbf{B}$. In a bianisotropic medium, the directions of the wave vector $\mathbf{k}$ and the Poynting's vector $\mathbf{S}$ do not, in general, coincide.

The Poynting's power flow direction is characterized by the ray vector $\mathbf{s}$. We define the magnitude of $\mathbf{s}$ by

$$\mathbf{s} \cdot \mathbf{k} = 1, \tag{3.36}$$

where $\mathbf{s}$ is perpendicular to both $\mathbf{E}$ and $\mathbf{H}$:

$$\mathbf{s} \cdot \mathbf{E} = 0, \tag{3.37a}$$

$$\mathbf{s} \cdot \mathbf{H} = 0. \tag{3.37b}$$

The ray vector $\mathbf{s}$ has the dimension of length. We have used the wave vector $\mathbf{k}$ to describe a plane wave; the governing equations are expressed in (3.3). We can also use $\mathbf{s}$ to arrive at a similar set of equations. We cross-multiply (3.3a) by $\mathbf{s}$ and use the vector identity $\mathbf{s} \times (\mathbf{k} \times \mathbf{E}) = \mathbf{k}(\mathbf{s} \cdot \mathbf{E}) - (\mathbf{k} \cdot \mathbf{s})\mathbf{E}$. In view of (3.36) and (3.37a), we obtain from (3.3a)

$$\mathbf{s} \times \mathbf{B} = -\frac{\mathbf{E}}{\omega}. \tag{3.38a}$$

Similarly, from (3.3c), by using (3.36) with (3.37b), we obtain

$$\mathbf{s} \times \mathbf{D} = \frac{\mathbf{H}}{\omega}. \tag{3.38b}$$

We compare (3.37) and (3.38) with (3.3) and observe the duality of the two sets of equations.

Similarly, we can define a ray surface. Equations for the ray surface in a bianisotropic medium can be determined in exactly the same manner as those for the wave surface. Also, by observing the duality between (3.3) and (3.38) and the constitutive relations in **EH** base and in **DB** base, we can obtain equations of the ray surfaces by changing $\bar{\epsilon}\to\bar{\kappa}$, $\bar{\mu}\to\bar{\nu}$, $\bar{\xi}\to\bar{\chi}$, $\bar{\zeta}\to\bar{\gamma}$, $\omega\to 1/\omega$, and $\mathbf{k}\to\mathbf{s}$.

For a uniaxial medium, we have, for the ordinary wave,

$$s_x^2+s_y^2+s_z^2=\frac{1}{\omega^2\mu\epsilon} \tag{3.39}$$

and, for the extraordinary wave,

$$s_x^2+s_y^2+\frac{\epsilon}{\epsilon_z}s_z^2=\frac{1}{\omega^2\mu\epsilon_z}. \tag{3.40}$$

The ray surface is plotted in Fig. 3.8. The oblate spheroidal wave surface for the positive uniaxial medium becomes a prolate spheroidal ray surface, and the prolate spheroid for the negative uniaxial medium becomes an oblate spheroid.

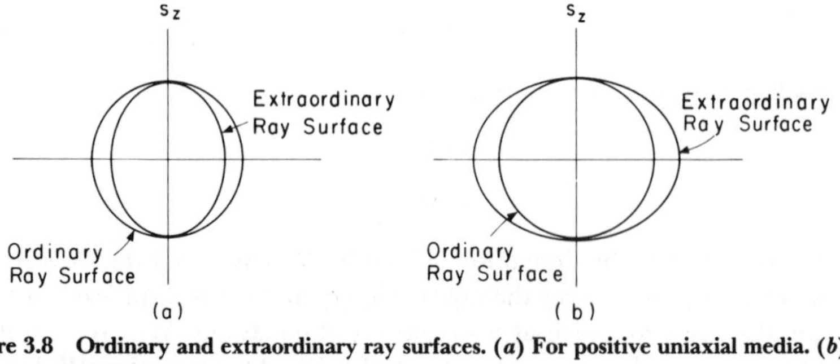

Figure 3.8 Ordinary and extraordinary ray surfaces. (*a*) For positive uniaxial media. (*b*) For negative uniaxial media.

Let us now discuss some general properties of the ray surface. We showed in (3.34) that $\mathbf{s}\cdot\delta\mathbf{k}=0$; that is, the normal to the wave surface gives the direction of the corresponding ray vector. Similarly, we can show that the normal to the ray surface gives the direction of the corresponding $\mathbf{k}$ vector; this follows from (3.36), $\mathbf{k}\cdot\mathbf{s}=1$. Differentiation yields $\mathbf{k}\cdot\delta\mathbf{s}+\mathbf{s}\cdot\delta\mathbf{k}=0$. Therefore $\mathbf{k}\cdot\delta\mathbf{s}=0$.

We next show that the ray surface describes a constant phase. This is true because the phase of a wave along a ray can be written as

$$\psi = \int \mathbf{k}\cdot d\mathbf{l} = \int \mathbf{k}\cdot \frac{\mathbf{s}}{s}\, dl = \frac{l}{s},$$

where l denotes the length of the segment along the ray path. In geometrical optics, the quantity $\psi/(\omega/c)$ is the *eikonal* of the wave. When l is equal to s multiplied by a constant, the eikonal is equal to c/ω times the constant. The ray surface gives the magnitudes of s in all directions. It follows that the ray surface is also a constant-phase surface.

3.3 THE *kDB* SYSTEM AND CHARACTERISTIC WAVES

3.3a the *kDB* System

From (3.3d) and (3.3b), we observe that **k** is perpendicular to both **D** and **B**. The plane perpendicular to **k** and containing **D** and **B** is called the *DB* plane. We can mount two perpendicular unit vectors $\hat{e}_1$ and $\hat{e}_2$ on the *DB* plane in addition to the **k** vector to form a coordinate system. The unit vector along **k** is $\hat{e}_3$. This coordinate system is called the *kDB system*. Many problems involving general anisotropic and bianisotropic media are solved easily in this coordinate system.

In the *kDB* system, it is convenient to write the constitutive relations of a bianisotropic medium in the **DB** representation:

$$E_i = \kappa_{ij} D_j + \chi_{ij} B_j, \tag{3.41a}$$

$$H_i = \gamma_{ij} D_j + \nu_{ij} B_j. \tag{3.41b}$$

We use subscripts 1, 2, and 3 to denote components in the *kDB* system. The 3×3 constitutive matrices can be displayed in any coordinate system. From Maxwell's equations 3.3b and 3.3d we see that $D_3 = B_3 = 0$. After **E** and **H** have been eliminated by using constitutive relations 3.41, Maxwell's equations 3.3a and 3.3c become

$$\begin{bmatrix} u+\chi_{21} & \chi_{22} \\ -\chi_{11} & u-\chi_{12} \end{bmatrix} \begin{bmatrix} B_1 \\ B_2 \end{bmatrix} = \begin{bmatrix} -\kappa_{21} & -\kappa_{22} \\ \kappa_{11} & \kappa_{12} \end{bmatrix} \begin{bmatrix} D_1 \\ D_2 \end{bmatrix}, \tag{3.42a}$$

$$\begin{bmatrix} u-\gamma_{21} & -\gamma_{22} \\ \gamma_{11} & u+\gamma_{12} \end{bmatrix} \begin{bmatrix} D_1 \\ D_2 \end{bmatrix} = \begin{bmatrix} \nu_{21} & \nu_{22} \\ -\nu_{11} & -\nu_{12} \end{bmatrix} \begin{bmatrix} B_1 \\ B_2 \end{bmatrix}, \tag{3.42b}$$

where $u=\omega/k$ is the phase velocity along the **k** direction. All components are expressed in the *kDB* system. The transformation of constitutive relations to the *kDB* system will be discussed in subsequent developments.

Dispersion relations are obtained by eliminating either **D** or **B** from (3.42a) and (3.42b) and setting the determinant of the homogeneous equations equal to zero. In the case of an isotropic medium, $\chi_{ij}=\gamma_{ij}=0$, $\kappa_{ij}=\delta_{ij}/\epsilon$, and $\nu_{ij}=\delta_{ij}/\mu$. Equations 3.42 become

$$u\begin{bmatrix} B_1 \\ B_2 \end{bmatrix} = \frac{1}{\epsilon}\begin{bmatrix} 0 & -1 \\ 1 & 0 \end{bmatrix}\begin{bmatrix} D_1 \\ D_2 \end{bmatrix}, \tag{3.43a}$$

$$u\begin{bmatrix} D_1 \\ D_2 \end{bmatrix} = \frac{1}{\mu}\begin{bmatrix} 0 & 1 \\ -1 & 0 \end{bmatrix}\begin{bmatrix} B_1 \\ B_2 \end{bmatrix}. \tag{3.43b}$$

Elimination of **B** gives

$$\left(u^2-\frac{1}{\mu\epsilon}\right)\begin{bmatrix} D_1 \\ D_2 \end{bmatrix}=0. \tag{3.44}$$

Thus the dispersion relation for an isotropic medium is $u^2=1/\mu\epsilon$, which is identical to (3.6). We observe from (3.44) that either D_1 or D_2 or both D_1 and D_2 can be nonzero. The wave can have any polarization: linear, circular, or elliptical. In anisotropic and bianisotropic media, however, this is not the case. Upon entering the media, waves of arbitrary polarization will be broken into superpositions of the characteristic waves that the media can support.

In isotropic media, the three vectors **k**, **D**, and **B** are mutually perpendicular and form a right-hand coordinate system. We can choose $\hat{e}_1$ of the *kDB* coordinate system to be in the direction of **D**, and $\hat{e}_2$ to be in the direction of **B**. For anisotropic and bianisotropic media, the directions of $\hat{e}_1$ and $\hat{e}_2$ are determined by the internal symmetrical properties of the media. In the principal coordinate system that characterizes the media, the constitutive relations take particularly simple forms. We use x, y, and z to denote the coordinate axes of the principal coordinate system.

To establish the *kDB* system for a general medium, we make use of the principal system. The *kDB* system is defined by unit vectors $\hat{e}_1$, $\hat{e}_2$, and $\hat{e}_3$. The unit vector $\hat{e}_3$ points in the direction of **k**. The unit vector $\hat{e}_2$ lies in the plane determined by $\hat{z}$ and **k** and is perpendicular to $\hat{e}_3$. The unit vector $\hat{e}_1$ is

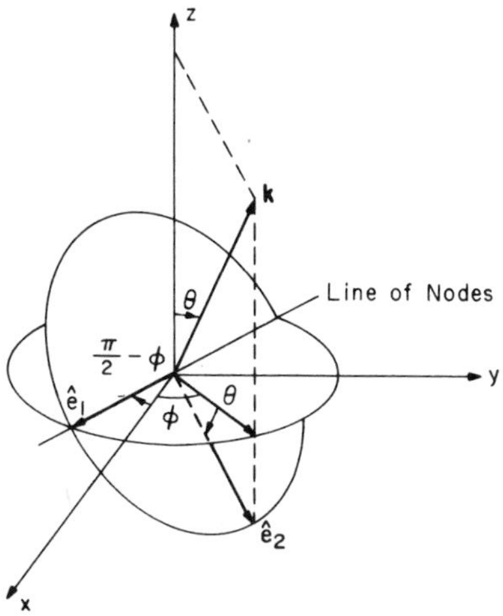

Figure 3.9 Establishing the *kDB* system.

perpendicular to the plane determined by $\hat{z}$ and $\mathbf{k}$ and is therefore perpendicular to both $\hat{e}_3$ and $\hat{e}_2$. We can generate the *kDB* coordinate axes from the principal coordinate axes by two successive rotations (Fig. 3.9).

To determine a transformation formula between the *kDB* system and the principle coordinate system, we first consider a vector $\mathbf{A}$. Let the components of $\mathbf{A}$ in the principal coordinate system be denoted as A_x, A_y, and A_z, and the components in the *kDB* system be denoted as A_1, A_2, and A_3. We then write

$$\mathbf{A}=\hat{x}A_x+\hat{y}A_y+\hat{z}A_z, \tag{3.45a}$$

$$\mathbf{A}_k=\hat{e}_1A_1+\hat{e}_2A_2+\hat{e}_3A_3. \tag{3.45b}$$

The subscript k denotes that vector $\mathbf{A}$ is represented in *kDB*. Assuming that the $\mathbf{k}$ vector makes angles θ and ϕ with the z and x axes, as shown in Fig. 3.9, we can write the $\mathbf{k}$ vector in both the *kDB* and the principal systems as

$$\begin{aligned}\mathbf{k}&=\hat{e}_3k\\ &=\hat{x}k\sin\theta\cos\phi+\hat{y}k\sin\theta\sin\phi+\hat{z}k\cos\theta.\end{aligned} \tag{3.46}$$

The line along $\hat{e}_1$ is called the line of nodes. The $x-y$ plane is rotated initially clockwise around z by $\pi/2-\phi$, and then around the line of nodes by θ. The resulting plane perpendicular to $\mathbf{k}$ is the *DB* plane. The unit vector $\hat{e}_2$ is determined from $\hat{e}_2=\mathbf{k}\times\hat{e}_1$. The transformation of the vector $\mathbf{A}$

from xyz to kDB is given by

$$\begin{bmatrix} A_1 \\ A_2 \\ A_3 \end{bmatrix} = \begin{bmatrix} \sin\phi & -\cos\phi & 0 \\ \cos\theta\cos\phi & \cos\theta\sin\phi & -\sin\theta \\ \sin\theta\cos\phi & \sin\theta\sin\phi & \cos\theta \end{bmatrix} \begin{bmatrix} A_x \\ A_y \\ A_z \end{bmatrix}. \tag{3.47}$$

Equation 3.47 is a direct consequence of (3.45). Dot-multiplying (3.45a) by $\hat{e}_1$, we have

$$\begin{aligned} A_1 &= \hat{e}_1\cdot\hat{x}A_x + \hat{e}_1\cdot\hat{y}A_y + \hat{e}_1\cdot\hat{z}A_z \\ &= \sin\phi A_x - \cos\phi A_y. \end{aligned}$$

Similarly,

$$\begin{aligned} A_2 &= \hat{e}_2\cdot\hat{x}A_x + \hat{e}_2\cdot\hat{y}A_y + \hat{e}_2\cdot\hat{z}A_z \\ &= \cos\theta\cos\phi A_x + \cos\theta\sin\phi A_y - \sin\theta A_z, \\ A_3 &= \hat{e}_3\cdot\hat{x}A_x + \hat{e}_3\cdot\hat{y}A_y + \hat{e}_3\cdot\hat{z}A_z \\ &= \sin\theta\cos\phi A_x + \sin\theta\sin\phi A_y + \cos\theta A_z. \end{aligned}$$

Writing in matrix form, we obtain (3.47). It is a simple matter to prove that

$$\begin{bmatrix} A_x \\ A_y \\ A_z \end{bmatrix} = \begin{bmatrix} \sin\phi & \cos\theta\cos\phi & \sin\theta\cos\phi \\ -\cos\phi & \cos\theta\sin\phi & \sin\theta\sin\phi \\ 0 & -\sin\theta & \cos\theta \end{bmatrix} \begin{bmatrix} A_1 \\ A_2 \\ A_3 \end{bmatrix}. \tag{3.48}$$

Equation 3.48 can be obtained either by carrying out a procedure similar to that leading to (3.47) or by finding the inverse of (3.47).

To display constitutive relations 3.41 for a bianisotropic medium in the kDB system, all field vectors $\mathbf{E}, \mathbf{D}, \mathbf{B}, \mathbf{H}$ are transformed from the principal system to the kDB system just as the vector $\mathbf{A}$ in (3.47) was transformed. We denote the transformation matrix $\overline{\mathbf{T}}$:

$$\mathbf{A}_k = \overline{\mathbf{T}}\cdot\mathbf{A}, \tag{3.49a}$$

$$\overline{\mathbf{T}} = \begin{bmatrix} \sin\phi & -\cos\phi & 0 \\ \cos\theta\cos\phi & \cos\theta\sin\phi & -\sin\theta \\ \sin\theta\cos\phi & \sin\theta\sin\phi & \cos\theta \end{bmatrix}. \tag{3.49b}$$

It can be seen that $\overline{\mathbf{T}}^{-1}=\overline{\mathbf{T}}^t$, and by definition the transformation is an orthogonal transformation. The matrix elements of the constitutive matrices in the *kDB* system are related to those in the principal system as follows:

$$\overline{\kappa}_k=\overline{\mathbf{T}}\cdot\overline{\kappa}\cdot\overline{\mathbf{T}}^t; \qquad \overline{\kappa}=\overline{\mathbf{T}}^t\cdot\overline{\kappa}_k\cdot\overline{\mathbf{T}}, \tag{3.50a}$$

$$\overline{\nu}_k=\overline{\mathbf{T}}\cdot\overline{\nu}\cdot\overline{\mathbf{T}}^t; \qquad \overline{\nu}=\overline{\mathbf{T}}^t\cdot\overline{\nu}_k\cdot\overline{\mathbf{T}}, \tag{3.50b}$$

$$\overline{\chi}_k=\overline{\mathbf{T}}\cdot\overline{\chi}\cdot\overline{\mathbf{T}}^t; \qquad \overline{\chi}=\overline{\mathbf{T}}^t\cdot\overline{\chi}_k\cdot\overline{\mathbf{T}}, \tag{3.50c}$$

$$\overline{\gamma}_k=\overline{\mathbf{T}}\cdot\overline{\gamma}\cdot\overline{\mathbf{T}}^t; \qquad \overline{\gamma}=\overline{\mathbf{T}}^t\cdot\overline{\gamma}_k\cdot\overline{\mathbf{T}}. \tag{3.50d}$$

Note that this derivation of the relations between the principal system and the *kDB* system applies to any Cartesian system and the *kDB* system.

3.3b Waves in Biaxial Media

In biaxial media, the three principal dielectric constants are different. In the principal coordinate system,

$$\overline{\kappa}=\begin{bmatrix}\kappa_x & 0 & 0\\ 0 & \kappa_y & 0\\ 0 & 0 & \kappa_z\end{bmatrix}, \tag{3.51a}$$

$$\overline{\nu}=\nu\overline{\mathbf{I}}, \tag{3.51b}$$

$$\overline{\chi}=\overline{\gamma}=0. \tag{3.51c}$$

The $\overline{\kappa}$ matrix is also called the *impermittivity tensor*. To relate to the permittivities, we note that $\kappa_x=1/\epsilon_x$, $\kappa_y=1/\epsilon_y$, and $\kappa_z=1/\epsilon_z$. The *impermeability* ν is the reciprocal of μ. By using (3.50), the constitutive matrices can be transformed to the *kDB* system, and we obtain

$$\overline{\kappa}_k=\begin{bmatrix}\kappa_{11} & \kappa_{12} & \kappa_{13}\\ \kappa_{21} & \kappa_{22} & \kappa_{23}\\ \kappa_{31} & \kappa_{32} & \kappa_{33}\end{bmatrix} \tag{3.52a}$$

$$\overline{\nu}_k=\nu\overline{\mathbf{I}}, \tag{3.52b}$$

$$\overline{\chi}_k=\overline{\gamma}_k=\overline{0}. \tag{3.52c}$$

where in (3.52a)

$$\kappa_{11}=\kappa_x\sin^2\phi+\kappa_y\cos^2\phi$$

$$\kappa_{12}=\kappa_{21}=(\kappa_x-\kappa_y)\cos\theta\sin\phi\cos\phi$$

$$\kappa_{22}=(\kappa_x\cos^2\phi+\kappa_y\sin^2\phi)\cos^2\theta+\kappa_z\sin^2\theta$$

$$\kappa_{13}=\kappa_{31}=(\kappa_x-\kappa_y)\sin\theta\sin\phi\cos\phi$$

$$\kappa_{23}=\kappa_{32}=(\kappa_x\cos^2\phi+\kappa_y\sin^2\phi-\kappa_z)\sin\theta\cos\theta$$

$$\kappa_{33}=(\kappa_x\cos^2\phi+\kappa_y\sin^2\phi)\sin^2\theta+\kappa_z\cos^2\theta$$

In the *kDB* system, by the elimination of $\mathbf{B}_k$ from (3.42a) and (3.42b), we obtain a homogeneous equation for $\mathbf{D}_k$:

$$\begin{bmatrix} u^2-\nu\kappa_{11} & -\nu\kappa_{12} \\ -\nu\kappa_{12} & u^2-\nu\kappa_{22} \end{bmatrix}\begin{bmatrix} D_1 \\ D_2 \end{bmatrix}=0. \tag{3.53}$$

The magnetic field $\mathbf{B}$ is related to $\mathbf{D}$ by (3.42a):

$$\begin{bmatrix} B_1 \\ B_2 \end{bmatrix}=\frac{1}{u}\begin{bmatrix} -\kappa_{12} & -\kappa_{22} \\ \kappa_{11} & \kappa_{12} \end{bmatrix}\begin{bmatrix} D_1 \\ D_2 \end{bmatrix}. \tag{3.54}$$

Note that $\kappa_{21}=\kappa_{12}$. In (3.53), the matrix that operates on $\mathbf{D}_k$ is no longer diagonal. It reduces to the uniaxial case with a diagonal matrix when $\kappa_{12}=0$, namely, $\kappa_x=\kappa_y$.

The phase velocities of the characteristic waves are immediately obtained by setting the determinant operating on $\mathbf{D}$ equal to zero:

$$u^2=\frac{\nu}{2}\left[(\kappa_{11}+\kappa_{22})\pm\sqrt{(\kappa_{11}-\kappa_{22})^2+4\kappa_{12}^2}\right]. \tag{3.55}$$

We obtain two $\mathbf{D}_k$ vectors corresponding to the two values for u^2. On the *DB* plane, expressed in terms of the *kDB* base vectors $\hat{e}_1$ and $\hat{e}_2$, we have

$$\frac{D_2}{D_1}=\frac{\nu\kappa_{12}}{u^2-\nu\kappa_{22}}=\frac{2\kappa_{12}}{\kappa_{11}-\kappa_{22}\pm\sqrt{(\kappa_{11}-\kappa_{22})^2+4\kappa_{12}^2}}. \tag{3.56}$$

Thus the two characteristic waves are linearly polarized with the $\mathbf{D}$ vectors orthogonal to each other.

Let

$$\tan 2\psi = \frac{2\kappa_{12}}{\kappa_{11} - \kappa_{22}}. \tag{3.57}$$

We observe from (3.56) that

$$\frac{D_2}{D_1} = \tan\psi \quad \text{or} \quad -\cot\psi. \tag{3.58}$$

In Fig. 3.10 the two vectors are depicted on the *DB* plane. The velocities of both waves are functions of θ and ϕ. In view of (3.52a), we see that none of the **E** vectors for the two waves lies on the *DB* plane. Each **E** vector has a component in the **k** direction. Since all other field vectors **D**, **B**, and **H** lie on the *DB* plane, the energy propagation directions are different from the **k** direction. We call the two characteristic waves for a biaxial medium Type I and Type II waves; both are extraordinary waves.

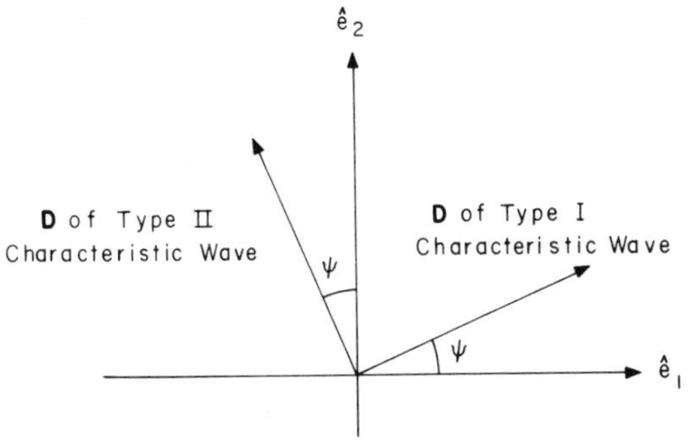

Figure 3.10 Both types of characteristic waves in a biaxial medium are linearly polarized.

In the case of uniaxial media, $\kappa_{12}=0$. Equation 3.53 becomes

$$\begin{bmatrix} u^2 - \nu\kappa & 0 \\ 0 & u^2 - \nu(\kappa\cos^2\theta + \kappa_z \sin^2\theta) \end{bmatrix} \begin{bmatrix} D_1 \\ D_2 \end{bmatrix} = 0.$$

The **D** field vectors for the two characteristic waves coincide with the base vectors $\hat{e}_1$ and $\hat{e}_2$, and the Type I wave becomes an ordinary wave. The dispersion relations in terms of the **k** vector in the principal coordinate system can be obtained by noting that $u=\omega/k$ and that $k_x = k\sin\theta\cos\phi$,

$k_y = k\sin\theta\sin\phi$, and $k_z = k\cos\theta$. This equation gives, for the ordinary wave,

$$k^2 = \frac{\omega^2}{\nu\kappa}$$

and, for the extraordinary wave,

$$k_x^2 + k_y^2 + \frac{\kappa k_z^2}{\kappa_z} = \frac{\omega^2}{\nu\kappa_z}.$$

These last two equations are identical to (3.11) because the impermeability is the reciprocal of the permeability, $\nu = 1/\mu$, and the impermittivity tensor $\bar{\kappa}$ is the inverse of the permittivity tensor $\bar{\epsilon}, \kappa = 1/\epsilon, \kappa_z = 1/\epsilon_z$.

We note that the procedure for deriving dispersion relations with the *kDB* system is applicable also to lossy media, where $\mathbf{k}$ is a complex vector. When the primary objective is to derive dispersion relations in terms of $\mathbf{k}$, angles θ and ϕ may be regarded as intermediate objects that are disposable after use is made of the *kDB* system.

3.3c Waves in Gyrotropic Media

Consider a gyrotropic medium possessing the following constitutive matrices:

$$\bar{\kappa} = \begin{bmatrix} \kappa & i\kappa_g & 0 \\ -i\kappa_g & \kappa & 0 \\ 0 & 0 & \kappa_z \end{bmatrix}, \tag{3.59a}$$

$$\bar{\nu} = \nu\bar{\mathbf{I}}, \tag{3.59b}$$

$$\bar{\chi} = \bar{\gamma} = 0. \tag{3.59c}$$

An example of such a medium is an anisotropic plasma with the external magnetic field in the $\hat{z}$ direction.

To determine the phase velocities and polarizations of the characteristic waves, we transform to the *kDB* system for a plane wave with wave vector $\mathbf{k}$. In the *kDB* system

$$\bar{\kappa}_k = \begin{bmatrix} \kappa & i\kappa_g\cos\theta & 2i\kappa_g\sin\theta \\ -i\kappa_g\cos\theta & \kappa\cos^2\theta + \kappa_z\sin^2\theta & (\kappa-\kappa_z)\sin\theta\cos\theta \\ -2i\kappa_g\sin\theta & (\kappa-\kappa_z)\sin\theta\cos\theta & \kappa\sin^2\theta + \kappa_z\cos^2\theta \end{bmatrix}. \tag{3.60}$$

Using (3.42), we obtain a homogeneous equation for $\mathbf{D}$:

$$\begin{bmatrix} u^2-\nu\kappa & -i\nu\kappa_g\cos\theta \\ i\nu\kappa_g\cos\theta & u^2-\nu(\kappa\cos^2\theta+\kappa_z\sin^2\theta) \end{bmatrix}\begin{bmatrix} D_1 \\ D_2 \end{bmatrix}=0. \tag{3.61}$$

The phase velocities of the two characteristic waves are found to be

$$u^2=\frac{\nu}{2}\left[\kappa+\kappa\cos^2\theta+\kappa_z\sin^2\theta\pm\sqrt{(\kappa-\kappa_z)^2\sin^4\theta+4\kappa_g^2\cos^2\theta}\,\right]. \tag{3.62}$$

The components of the two field vectors are related by

$$\frac{D_2}{D_1}=\frac{-2i\kappa_g\cos\theta}{(\kappa-\kappa_z)\sin^2\theta\pm\sqrt{(\kappa-\kappa_z)^2\sin^4\theta+4\kappa_g^2\cos^2\theta}}. \tag{3.63}$$

We can define an angle ψ such that

$$\tan 2\psi=\frac{2\kappa_g\cos\theta}{(\kappa-\kappa_z)\sin^2\theta}. \tag{3.64}$$

Equation 3.63 then states that

$$\frac{D_2}{D_1}=-i\tan\psi \qquad \text{(Type I wave)}, \tag{3.65a}$$

$$=i\cot\psi \qquad \text{(Type II wave)}. \tag{3.65b}$$

Both ratios are imaginary. Thus both characteristic waves are elliptically polarized. Components D_1 and D_2 for each wave are unequal in magnitude and $\pi/2$ out of phase. Assume that $\kappa>\kappa_z$ and κ_g is positive. The Type I wave is left-hand polarized; the Type II wave, right-hand polarized. When propagating perpendicular to the $\hat{z}$ direction, both waves are linearly polarized and have different velocities. This birefringence is the Cotton-Mouton effect. When κ_g is zero, the medium becomes uniaxial.

Consider the case of propagation along the $\hat{z}$ direction. Since $\theta=0$, (3.61) becomes

$$\begin{bmatrix} u^2-\nu\kappa & -i\nu\kappa_g \\ i\nu\kappa_g & u^2-\nu\kappa \end{bmatrix}\begin{bmatrix} D_1 \\ D_2 \end{bmatrix}=0. \tag{3.66}$$

The velocity of propagation is determined to be $u^2 = \nu(\kappa \pm \kappa_g)$. The corresponding polarization is $D_2/D_1 = \mp i$, and hence both characteristic waves are circularly polarized. The left-hand circularly polarized wave has a velocity $(\nu\kappa + \nu\kappa_g)^{1/2}$, and the right-hand wave has a velocity $(\nu\kappa - \nu\kappa_g)^{1/2}$.

Upon entering the gyrotropic medium along the $\hat{z}$ direction, a linearly polarized wave is broken up into two circularly polarized waves propagating at different velocities. To simplify the discussion, let the wave be polarized along $\hat{e}_1$. In terms of circularly polarized waves, we then have

$$\begin{aligned}\mathbf{D} &= \hat{e}_1 d \\ &= \frac{d}{2}(\hat{e}_1 + \hat{e}_2 i) + \frac{d}{2}(\hat{e}_1 - \hat{e}_2 i),\end{aligned}$$

where d is a real number. After traveling a distance z_0 inside the medium, the two waves are phase-shifted by different amounts:

$$\begin{aligned}\mathbf{D} &= \frac{d}{2}(\hat{e}_1 + \hat{e}_2 i)e^{i\phi_r} + \frac{d}{2}(\hat{e}_1 - \hat{e}_2 i)e^{i\phi_l} \\ &= \hat{e}_1 \frac{d}{2}(e^{i\phi_r} + e^{i\phi_l}) + i\hat{e}_2 \frac{d}{2}(e^{i\phi_r} - e^{i\phi_l}),\end{aligned}$$

where

$$\phi_r = \frac{\omega z_0}{\sqrt{\nu(\kappa - \kappa_g)}}$$

and

$$\phi_l = \frac{\omega z_0}{\sqrt{\nu(\kappa + \kappa_g)}}.$$

From the ratio for the two components of $\mathbf{D}$, we find

$$\frac{D_2}{D_1} = -\tan\frac{\phi_r - \phi_l}{2}.$$

The two components are in phase, and the wave is linearly polarized. According to Fig. 3.2, the field vector $\mathbf{D}$ has been rotated clockwise by an angle $(\phi_r - \phi_l)/2$. It is important to note that, if the same linearly polarized wave is propagated along the $-\hat{z}$ direction for the same distance, the field vector is rotated in exactly the same manner by the same amount.

The phenomenon of rotation of a linearly polarized field vector when passing through a gyrotropic medium is known as *Faraday rotation*. The electrons circulating along the magnetic field lines are responsible for this effect. Faraday rotation also occurs in ferrites in the presence of external magnetic fields; there the effect is caused by precession of spin axes around the magnetic field. A parallel analysis can be carried out for ferrites by using a magnetically anisotropic model with an impermeability tensor.

3.3d Waves in Bianisotropic Media

Consider bianisotropic media with the following constitutive relations:

$$\mathbf{E}=\begin{bmatrix} \kappa & 0 & 0 \\ 0 & \kappa & 0 \\ 0 & 0 & \kappa_z \end{bmatrix}\cdot\mathbf{D}+\begin{bmatrix} \chi & 0 & 0 \\ 0 & \chi & 0 \\ 0 & 0 & \chi_z \end{bmatrix}\cdot\mathbf{B}, \tag{3.67a}$$

$$\mathbf{H}=\begin{bmatrix} \gamma & 0 & 0 \\ 0 & \gamma & 0 \\ 0 & 0 & \gamma_z \end{bmatrix}\cdot\mathbf{D}+\begin{bmatrix} \nu & 0 & 0 \\ 0 & \nu & 0 \\ 0 & 0 & \nu_z \end{bmatrix}\cdot\mathbf{B}. \tag{3.67b}$$

When $\bar{\chi}=\bar{\gamma}$, this relation reduces to that used by Dzyaloshinskii[32] in his description of magnetoelectric media.

In the *kDB* system, the constitutive matrix $\bar{\boldsymbol{\kappa}}_k$ becomes

$$\bar{\boldsymbol{\kappa}}_k=\begin{bmatrix} \kappa & 0 & 0 \\ 0 & \kappa\cos^2\theta+\kappa_z\sin^2\theta & (\kappa-\kappa_z)\sin\theta\cos\theta \\ 0 & (\kappa-\kappa_z)\sin\theta\cos\theta & \kappa\sin^2\theta+\kappa_z\cos^2\theta \end{bmatrix}. \tag{3.68}$$

A similar form holds for the three other matrices, $\bar{\chi}_k$, $\bar{\gamma}_k$, and $\bar{\nu}_k$. Inserting the corresponding constitutive parameters in (3.42) and eliminating $\mathbf{B}_k$, we obtain

$$\begin{bmatrix} \kappa_\theta(u^2\nu+\nu_\theta\chi\gamma-\kappa\nu\nu_\theta) & u\kappa_\theta(\nu_\theta\chi-\nu\gamma_\theta) \\ u\kappa(\nu_\theta\gamma-\nu\chi_\theta) & \kappa(u^2\nu_\theta+\nu\chi_\theta\gamma_\theta-\kappa_\theta\nu\nu_\theta) \end{bmatrix}\begin{bmatrix} D_1 \\ D_2 \end{bmatrix}=0, \tag{3.69}$$

where we use the short notation:

$$\kappa_\theta = \kappa\cos^2\theta + \kappa_z \sin^2\theta,$$

$$\nu_\theta = \nu\cos^2\theta + \nu_z \sin^2\theta,$$

$$\chi_\theta = \chi\cos^2\theta + \chi_z \sin^2\theta,$$

$$\gamma_\theta = \gamma\cos^2\theta + \gamma_z \sin^2\theta.$$

Solution of u and D from (3.69), although lengthy, is straightforward. We shall now discuss several special cases.

Consider a lossless magnetoelectric medium in which $\bar{\gamma} = \bar{\chi}$ are both real. We see from (3.69) that the characteristic waves are linearly polarized. Along the direction of $\hat{z}$ (3.69) becomes

$$\begin{bmatrix} u^2 - \kappa\nu + \chi^2 & 0 \\ 0 & u^2 - \kappa\nu + \chi^2 \end{bmatrix} \begin{bmatrix} D_1 \\ D_2 \end{bmatrix} = 0. \tag{3.70}$$

This is a degenerate case. The characteristic waves can have any polarization, and the phase velocity is $u = (\kappa\nu - \chi^2)^{1/2}$. Note that we must have $\kappa\nu > \chi^2$; otherwise the velocity becomes imaginary.

Consider the case with both $\boldsymbol{\chi}$ and $\boldsymbol{\gamma}$ imaginary and the bianisotropic medium lossless. Let $\bar{\boldsymbol{\chi}} \to i\bar{\boldsymbol{\chi}}$; then the lossless condition requires $\bar{\boldsymbol{\gamma}} = -i\bar{\boldsymbol{\chi}}$. We see from (3.69) that the characteristic waves are elliptically polarized. Along the direction of $\hat{z}$ (3.69) becomes

$$\begin{bmatrix} u^2 - \kappa\nu + \chi^2 & i2\chi u \\ -i2\chi u & u^2 - \kappa\nu + \chi^2 \end{bmatrix} \begin{bmatrix} D_1 \\ D_2 \end{bmatrix} = 0. \tag{3.71}$$

The velocity of propagation is determined from $u^2 - \kappa\nu + \chi^2 = \pm 2\chi u$, and the corresponding polarization is $D_2/D_1 = \pm i$. Thus both characteristic waves are circularly polarized. As in the case of gyrotropic media, a linearly polarized wave entering this medium along the $\hat{z}$ direction is broken up into two characteristic waves that propagate at different velocities. The net result is a rotation of the field vector **D**. A profound difference exists, however, between this rotation and Faraday rotation. Comparison of (3.69) with (3.61) reveals that the off-diagonal elements in (3.61) change sign when we change θ from 0 to π, whereas those in (3.69) remain unaltered.

The significance of this difference can be demonstrated as follows. Consider a linearly polarized wave that passes through a slab of a gyrotropic medium along the $\hat{z}$ direction. Assume that, upon exiting, its field vector is rotated 45°. If the wave is reflected by a mirror and reenters the slab, after the whole journey the field vector is rotated a total 90°. Consider the same experiment with the gyrotropic medium replaced by a bianisotropic medium as discussed above. On its return path after being reflected by the mirror, the field vector is rotated back to its original position and the net result is no rotation at all. Because of this difference, we call this rotatory power *optical activity* to distinguish it from the Faraday effect. As we shall see in Chapter 7, the optical activity is reciprocal, whereas the Faraday effect is nonreciprocal. The nonreciprocal nature of the Faraday effect is due to the fact that it is induced by an external magnetic field and that magnetic field vectors are pseudo vectors.[63]

3.3e Waves in Moving Uniaxial Media

Consider a bianisotropic medium characterized by the constitutive relations

$$\bar{\kappa} = \begin{bmatrix} \kappa & 0 & 0 \\ 0 & \kappa & 0 \\ 0 & 0 & \kappa_z \end{bmatrix}, \tag{3.72a}$$

$$\bar{\nu} = \begin{bmatrix} \nu & 0 & 0 \\ 0 & \nu & 0 \\ 0 & 0 & \nu_z \end{bmatrix}, \tag{3.72b}$$

$$\bar{\chi} = \bar{\gamma}^{+} = \begin{bmatrix} 0 & \chi & 0 \\ -\chi & 0 & 0 \\ 0 & 0 & 0 \end{bmatrix}. \tag{3.72c}$$

As discussed in Chapter 2, moving isotropic and uniaxial media possess constitutive relations of this form.

In the *kDB* system, the constitutive matrices become

$$\boldsymbol{\kappa}_k = \begin{bmatrix} \kappa & 0 & 0 \\ 0 & \kappa\cos^2\theta + \kappa_z\sin^2\theta & (\kappa-\kappa_z)\sin\theta\cos\theta \\ 0 & (\kappa-\kappa_z)\sin\theta\cos\theta & \kappa\sin^2\theta + \kappa_z\cos^2\theta \end{bmatrix}, \quad (3.73a)$$

$$\bar{\nu}_k = \begin{bmatrix} \nu & 0 & 0 \\ 0 & \nu\cos^2\theta + \nu_z\sin^2\theta & (\nu-\nu_z)\sin\theta\cos\theta \\ 0 & (\nu-\nu_z)\sin\theta\cos\theta & \nu\sin^2\theta + \nu_z\cos^2\theta \end{bmatrix}, \quad (3.73b)$$

$$\bar{\chi}_k = \bar{\gamma}_k^{+} = \begin{bmatrix} 0 & \chi\cos\theta & \chi\sin\theta \\ -\chi\cos\theta & 0 & 0 \\ -\chi\sin\theta & 0 & 0 \end{bmatrix}. \quad (3.73c)$$

Substituting in (3.42) and eliminating **B**, we obtain the following equation for **D**:

$$\begin{bmatrix} 1 - \dfrac{(u-\chi\cos\theta)^2}{\kappa(\nu\cos^2\theta+\nu_z\sin^2\theta)} & 0 \\ 0 & 1 - \dfrac{(u-\chi\cos\theta)^2}{\nu(\kappa\cos^2\theta+\kappa_z\sin^2\theta)} \end{bmatrix} \begin{bmatrix} D_1 \\ D_2 \end{bmatrix} = 0. \quad (3.74)$$

The phase velocities of the two characteristic waves are easily obtained from (3.74). Other field components are found from (3.42b) and the constitutive relations. The results are listed in Table 3.2.

In the case of moving uniaxial media, we find the values for the constitutive parameters from Table 2.1. Note that $k_z^2 = k^2\cos^2\theta$ and $k_x^2 + k_y^2 = k^2\sin^2\theta$. Written explicitly in terms of components of the **k** vector, the dispersion

Table 3.2
Characteristic Waves in Moving Uniaxial Media

Wave Characteristics	Type I Wave	Type II Wave
$\mathbf{D}_k$	$\begin{bmatrix} 1 \\ 0 \\ 0 \end{bmatrix}$	$\begin{bmatrix} 0 \\ 1 \\ 0 \end{bmatrix}$
$\mathbf{B}_k$	$\begin{bmatrix} 0 \\ \dfrac{\kappa}{u-\chi\cos\theta} \\ 0 \end{bmatrix}$	$\begin{bmatrix} \dfrac{-(u-\chi\cos\theta)}{\nu} \\ 0 \\ 0 \end{bmatrix}$
$\mathbf{E}_k$	$\begin{bmatrix} \dfrac{\kappa u}{\nu-\chi\cos\theta} \\ 0 \\ 0 \end{bmatrix}$	$\begin{bmatrix} 0 \\ \dfrac{u(u-\chi\cos\theta)}{\nu} \\ \left[\dfrac{\chi\kappa_z+u(\kappa-\kappa_z)\cos\theta}{u}\right]\sin\theta \end{bmatrix}$
$\mathbf{H}_k$	$\begin{bmatrix} 0 \\ u \\ \left[-\chi+\dfrac{\kappa(\nu_z-\nu)\cos\theta}{u-\chi\cos\theta}\right]\sin\theta \end{bmatrix}$	$\begin{bmatrix} -u \\ 0 \\ 0 \end{bmatrix}$
u	$\chi\cos\theta\pm\sqrt{\kappa(\nu\cos^2\theta+\nu_z\sin^2\theta)}$	$\chi\cos\theta\pm\sqrt{\nu(\kappa\cos^2\theta+\kappa_z\sin^2\theta)}$
Dispersion Relation	$k_x^2+k_y^2+\dfrac{\nu}{\nu_z}k_z^2-\dfrac{1}{\kappa\nu_z}(\omega-\chi k_z)^2=0$	$k_x^2+k_y^2+\dfrac{\kappa}{\kappa_z}k_z^2-\dfrac{1}{\nu\kappa_z}(\omega-\chi k_z)^2=0$

relation for the Type I wave is

$$k_x^2+k_y^2+\frac{\nu k_z^2}{\nu_z}-\frac{(\omega-\chi k_z)^2}{\kappa\nu_z}=0, \tag{3.75a}$$

which propagates with velocity

$$u=\chi\cos\theta\pm\sqrt{\kappa(\nu\cos^2\theta+\nu_z\sin^2\theta)}\ . \tag{3.75b}$$

The dispersion relation for the Type II wave is

$$k_x^2+k_y^2+\frac{\kappa k_z^2}{\kappa_z}-\frac{(\omega-\chi k_z)^2}{\nu\kappa_z}=0, \tag{3.76a}$$

which propagates with velocity

$$u=\chi\cos\theta\pm\sqrt{\nu(\kappa\cos^2\theta+\kappa_z\sin^2\theta)}\ . \tag{3.76b}$$

It is interesting to note the $\pm$ signs in (3.75b) and (3.76b). The plus sign corresponds to waves propagating in the direction of medium motion; the negative sign, to waves propagating in the opposite direction of medium motion. The values of the parameters κ, κ_z, ν, ν_z, and χ are given in Table 2.1. In Fig. 3.11, we plot the **k** surfaces for a moving isotropic medium with $n=2$ in its rest frame and for a moving uniaxial medium with $n=2$ and $a=b=2$. For $1-n^2\beta^2>0$, the surface is an ellipse rotating around the k_z axis; this is the nonrelativistic case. As $1-n^2\beta^2<0$, the **k** surface becomes a hyperbola rotating around the k_z axis; this is the relativistic case. We call this high-velocity region the *Čerenkov zone*. The velocity that separates the nonrelativistic zone and the Čerenkov zone is $\beta=\pm 1/n$, which is equal to the velocity of light in the rest frame of the moving medium.

To facilitate further discussion, we make use of Table 2.1 and write (3.75a) and (3.76a) explicitly in terms of β dependence. After some manipulations, we obtain

$$k_x^2+k_y^2+b\frac{1-n^2\beta^2}{1-\beta^2}\left(k_z-\frac{n+\beta}{n\beta+1}\frac{\omega}{c}\right)\left(k_z-\frac{n-\beta}{n\beta-1}\frac{\omega}{c}\right)=0, \tag{3.77a}$$

$$k_x^2+k_y^2+a\frac{1-n^2\beta^2}{1-\beta^2}\left(k_z-\frac{n+\beta}{n\beta+1}\frac{\omega}{c}\right)\left(k_z-\frac{n-\beta}{n\beta-1}\frac{\omega}{c}\right)=0. \tag{3.77b}$$

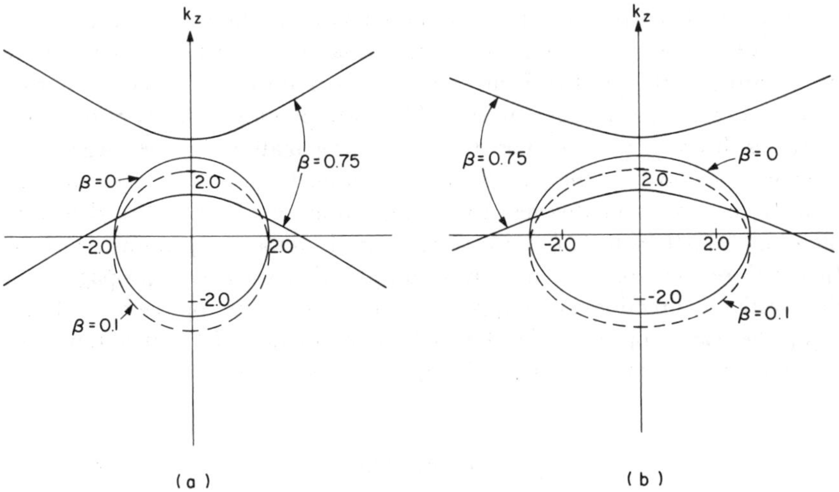

Figure 3.11 **(*a*) The k surface for a moving isotropic medium with $a=b=1$ and $n=2$ in its rest frame. (*b*) The k surface for a moving uniaxial medium with $a=b=2$ and $n=2$ in its rest frame.**

We examine two cases. First, consider a wave propagating in the $\hat{x}$ direction perpendicular to the medium velocity, $k_y=k_z=0$. The **k** vectors become

$$\mathbf{k}=\hat{x}k_x=\pm\hat{x}\frac{\omega}{c}\left(b\frac{n^2-\beta^2}{1-\beta^2}\right)^{1/2}, \tag{3.78a}$$

$$\mathbf{k}=\hat{x}k_x=\pm\hat{x}\frac{\omega}{c}\left(a\frac{n^2-\beta^2}{1-\beta^2}\right)^{1/2}. \tag{3.78b}$$

The $\pm$ sign distinguishes waves propagating in the positive and the negative $\hat{x}$ directions. As β approaches 1, k becomes infinity. Thus the velocity along the $\hat{x}$ direction is zero when the medium velocity approaches the velocity of light in vacuum.

Second, consider a wave propagating in the direction of medium motion, $k_x=k_y=0$. The two types of wave degenerate into one, and the **k** vectors become

$$\mathbf{k}=\hat{z}\frac{n+\beta}{n\beta+1}\frac{\omega}{c}, \tag{3.79a}$$

$$\mathbf{k}=\hat{z}\frac{n-\beta}{n\beta-1}\frac{\omega}{c}. \tag{3.79b}$$

Equation 3.79a corresponds to waves propagating in the positive $\hat{z}$ direction, and (3.79b) to waves propagating in the negative $\hat{z}$ direction. For the wave propagating in the positive $\hat{z}$ direction, we observe that, as β increases from 0 to 1, k decreases from $n\omega/c$ to ω/c. The corresponding velocity of the wave increases from c/n to c. For the wave propagating in the negative $\hat{z}$ direction, we observe that, as β increases from 0 to $1/n$, k changes from $-n\omega/c$ to $-\infty$, and the velocity changes from $-c/n$ to 0. As β further increases from $1/n$ to 1, k reverses sign and decreases from infinity to ω/c. In the Čerenkov zone the negatively propagating wave now propagates in the positive $\hat{z}$ direction. As β approaches 1, the velocity approaches c. In all cases, the wave appears to be dragged by the motion of the medium. This phenomenon is referred to as the *Fizeau-Fresnel drag*.

PROBLEMS

3.1. From source-free Maxwell's equations in free space, derive the wave equation

$$(\nabla^2 + k^2)\mathbf{E} = 0.$$

Show that $\mathbf{E} = \hat{x}e^{ikx}$ satisfies the wave equation. Do you regard it as an electromagnetic wave? By definition, an electromagnetic wave satisfies all of Maxwell's equations.

3.2. When $|D_x| = |D_y| = d_0$, but $\psi_x - \psi_y = \phi$ is not an integer multiplier of $\pi/2$, what is the polarization of the wave propagating the $\hat{z}$ direction? Show that the locus for the tip of the $\mathbf{D}$ vector, after eliminating the time dependence from its two components, is

$$d_x^2 + d_y^2 - 2d_x d_y \cos\phi = d_0^2 \sin^2\phi.$$

Plot this equation on the $d_x - d_y$ plane, and show that it is an ellipse inclined 45° with respect to the d_x and d_y axes. Discuss the direction of rotation of the tip of $\mathbf{D}$ when $\phi = \pi/3$, $4\pi/3$, and $7\pi/3$.

3.3. Show that any elliptically polarized wave can be decomposed into a right-hand circularly polarized wave and a left-hand circularly polarized wave.

3.4. Show that in a uniaxial crystal the ray directions for ordinary and extraordinary rays make an angle α such that

$$\tan\alpha = \frac{(\kappa - \kappa_z)\sin\theta\cos\theta}{\kappa\cos^2\theta + \kappa_z\sin^2\theta}.$$

By differentiating with respect to θ, show that maximum α occurs at

$$\tan\alpha_{\max} = \frac{n_e^2 - n_0^2}{2n_0 n_e},$$

where $n_e^2 = c^2\mu/\kappa_z$ and $n_0^2 = c^2\mu/\kappa$.

3.5. Consider a conductive uniaxial medium with

$$\bar{\boldsymbol{\epsilon}} = \begin{bmatrix} \epsilon & 0 & 0 \\ 0 & \epsilon & 0 \\ 0 & 0 & \epsilon_z \end{bmatrix} \quad \text{and} \quad \bar{\boldsymbol{\sigma}} = \begin{bmatrix} \sigma & 0 & 0 \\ 0 & \sigma & 0 \\ 0 & 0 & \sigma_z \end{bmatrix}.$$

Find dispersion relations for this medium. Explain the operation of a polaroid with this model by assuming $\sigma_z/\sigma \ll 1$. Show that a piece of polaroid turns any wave into a linearly polarized wave.

3.6. *Fresnel ellipsoid* is defined for an anisotropic medium by $\epsilon_{ij}x_ix_j = 1$, where ϵ_{ij} is expressed in the principal coordinates. The inverse of the permittivity tensor $\bar{\boldsymbol{\epsilon}}$ is $\bar{\boldsymbol{\kappa}}$, which is called the impermittivity tensor. If we define an ellipsoid in terms of $\bar{\boldsymbol{\kappa}}$ instead of $\bar{\boldsymbol{\epsilon}}$ in the principal coordinate system of the medium and write $\kappa_{ij}x_ix_j = 1$, we have a tensor ellipsoid. Construct the Fresnel ellipsoid and tensor ellipsoid for a uniaxial medium and a biaxial medium. Expressed in the principal coordinate system, the principal refractive indices are usually used in these definitions by replacing ϵ_{ij} with $n_i^2\delta_{ij}$ and κ_{ij} by δ_{ij}/n_i^2, in which case the tensor ellipsoid is also called an index ellipsoid or a reciprocal ellipsoid.

3.7. *Fermat's principle* in geometrical optics is a variational principle, which states that the eikonal, defined by $\int \mathbf{k}\cdot d\mathbf{l}$, is a minimum for an actual optical ray between any two points. Use this principle and the dispersion relation $\mathbf{k} = \mathbf{n}(\omega/c)$ to show that the rays in an inhomogeneous medium are bent in the direction of increasing refractive index.

3.8. Find the Doppler shift of a plane wave in an isotropic plasma medium. Plot ω' as a function of ω, and discuss your results.

3.9. The dispersion relations for a medium in motion can be derived from the dispersion relations for the medium at rest, by using the Lorentz transformation formulas for $\mathbf{k}$ and ω. Use (3.26) to find the dispersion relations for a uniaxial medium moving along its optic axis. Check your results with (3.75) and (3.76).

3.10. Determine the dispersion relations for a uniaxial medium with

$$\bar{\bar{\kappa}}=\begin{bmatrix}\kappa & 0 & 0\\ 0 & \kappa & 0\\ 0 & 0 & \kappa_z\end{bmatrix}, \qquad \bar{\bar{\nu}}=\begin{bmatrix}\nu & 0 & 0\\ 0 & \nu & 0\\ 0 & 0 & \nu_z\end{bmatrix},$$

and discuss your results.

3.11. Derive the dispersion relations for a biaxial medium and plot the **k** surfaces.

3.12. Use the *kDB* system to determine the dispersion relations for a biisotropic medium. Discuss your results.

3.13. In a ferrite, the magnetic moment **M** roughly obeys the relationship $d\mathbf{M}/dt = g\mu_0\mathbf{M}\times\mathbf{H}$, where g is the gyromagnetic ratio. When a $\hat{z}$-directed dc magnetic field H_0 (zero order) and an rf magnetic field $\mathbf{H}_1$ (first order) are present, the total fields take the form $\mathbf{H}=\hat{z}H_0+\mathbf{H}_1$, $\mathbf{M}=\hat{z}M_0+\mathbf{M}_1$, and $\mathbf{B}=\mu_0(\mathbf{H}+\mathbf{M})$. Find dispersion relations for the first-order fields. Show that Faraday rotation exists in the ferrite.

3.14. Determine the dispersion relations for an optically active quartz crystal with the constitutive relations given by Problem 1.11b, and discuss your results.

4

Reflection and Transmission

In studying the reflection and transmission of electromagnetic waves, it is instructive to distinguish between dispersion analysis and amplitude analysis. In dispersion analysis, we inquire into what types of waves are excited by the incident wave and what they look like. Phase matching offers a simple and powerful approach to answering these questions. The laws of reflection and transmission, the phenomena of total reflection and evanescence, and Doppler effects at moving boundaries follow directly from the results of phase matching. In amplitude analysis, we calculate the amplitudes of various waves in terms of the amplitude of the incident wave. First, we study the reflection and transmission of an incident wave at a plane boundary, stationary and in motion. We then generalize to plane stratified media, introducing propagation matrices to facilitate the calculations. The propagation matrices are used to calculate the transmission and reflection coefficients and to relate wave amplitudes in one region of a stratified medium to those in other regions. The reflection coefficients are also cast in closed form in terms of continuous fractions ready for analytical studies and for machine computation.

4.1 PHASE MATCHING

4.1a Laws of Reflection and Transmission

Consider two homogeneous media separated by a plane boundary surface situated at $x=0$ (Fig. 4.1). A plane wave is incident upon the boundary from the medium in region 0. A reflected wave is generated in this medium, and a transmitted wave is generated in the transmitted region t. The space-time dependences of these three waves are as follows:

$$\text{Incident} \qquad e^{i\mathbf{k}\cdot\mathbf{r}-i\omega t}, \tag{4.1a}$$

$$\text{Reflected} \qquad e^{i\mathbf{k}_r\cdot\mathbf{r}-i\omega t}, \tag{4.1b}$$

$$\text{Transmitted} \qquad e^{i\mathbf{k}_t\cdot\mathbf{r}-i\omega t}. \tag{4.1c}$$

The boundary condition requires the tangential **E** and **H** fields to be continuous across the surface for all y, z, and t. Thus we must have

$$k_y = k_{ry} = k_{ty}, \tag{4.2a}$$

$$k_z = k_{rz} = k_{tz}, \tag{4.2b}$$

$$\omega = \omega_r = \omega_t. \tag{4.2c}$$

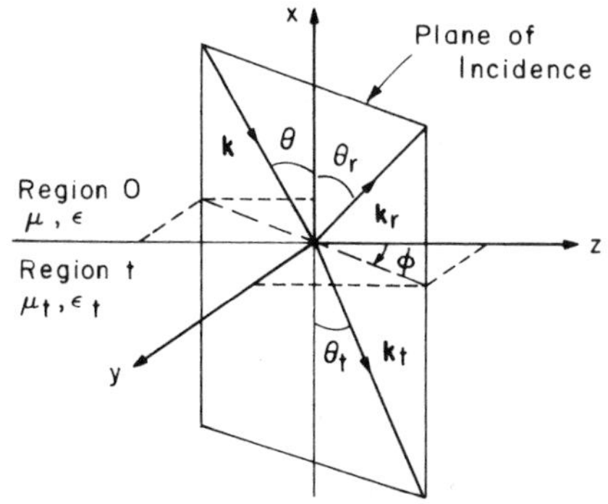

Figure 4.1 **A plane wave incident upon a plane boundary at $x=0$.**

According to (4.2c), there is no change in frequency for the reflected and transmitted waves. According to (4.2a) and (4.2b), tangential components of all **k** vectors are continuous across the boundary. We deduce the laws of reflection and transmission from (4.2a) and (4.2b):

1. Incident, reflected, and transmitted wave vectors lie in one plane, called the *plane of incidence*.
2. The angle of reflection θ_r is related to the angle of incidence θ by

$$k(\theta)\sin\theta = k_r(\theta_r)\sin\theta_r. \tag{4.3}$$

3. The angle of transmission (or refraction) is related to the angle of incidence by Snell's law:

$$k(\theta)\sin\theta = k_t(\theta_t)\sin\theta_t. \tag{4.4}$$

The sine of the angle is defined by the ratio of the tangential components of the wave vector to its magnitude for all three waves. These results hold for general bianisotropic media.

When both media are isotropic, we have

$$k = k_r = \omega\sqrt{\mu\epsilon} = \frac{\omega}{c}n, \tag{4.5a}$$

$$k_t = \omega\sqrt{\mu_t\epsilon_t} = \frac{\omega}{c}n_t. \tag{4.5b}$$

The magnitudes of all three **k** vectors are angle-independent. Equation 4.3 yields $\theta = \theta_r$. Thus the angle of reflection is equal to the angle of incidence.

In view of (4.5), (4.4) becomes

$$\frac{\sin\theta}{\sin\theta_t} = \frac{n_t}{n}, \tag{4.6}$$

which is Snell's law. Note that the laws of reflection and transmission have been cast in terms of angles of reflection and transmission. They are direct consequences of a simple statement: *tangential components of* **k** *vectors are continuous across the boundary*.

4.1b Phase Matching with k Surfaces

We illustrate the continuity of the tangential **k** vector by a simple graph (Fig. 4.2). Assuming that the plane of incidence lies in the $x-z$ plane, we plot the **k** surfaces for the two isotropic media. Let the **k** surface for the medium in the incident region have a shorter radius than that in medium t. Note that the plot is constructed on the $k_x - k_z$ plane in **k** space. Although the k_z axis has nothing to do with the physical z axis, it provides the feeling of a physical boundary. Therefore we use dashed lines for the **k** surface of medium zero below the k_z line, and similarly for the **k** surface of medium t above the k_z line. We plot the incident **k** vector from the surface toward the origin of the **k** surface. Requiring that the tangential k_z components be equal, we construct the **k** vectors for the reflected and transmitted waves.

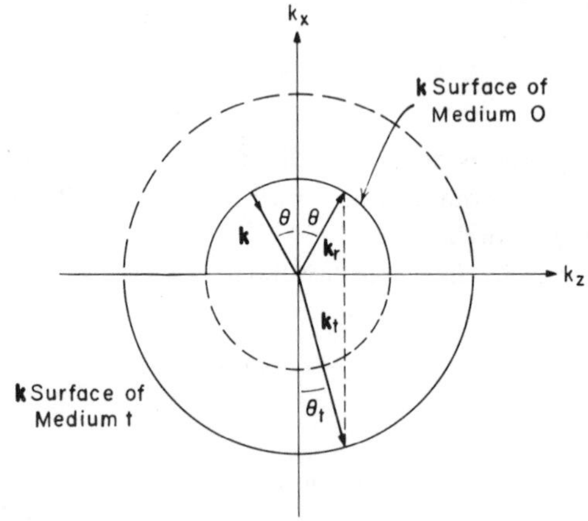

Figure 4.2 A k-space representation of incident, reflected, and transmitted k vectors. By phase-matching conditions, the k_z components of the three k vectors are equal. The incident k vector can also be drawn from the center toward the k surface.

It is not always possible, however, to construct the reflected **k** vector or the transmitted **k** vector. Suppose that the medium in region zero is denser than medium t (we mean that $n > n_t$). Then the radius of the wave surface in medium t is shorter than that in medium zero (Fig. 4.3). By the phase-matching condition, we see that, as k_z of the incident wave becomes larger than k_t, there is no intersection with the small sphere because this amounts to requiring that one component of a vector be greater than its magnitude—an impossibility unless the vector is complex. The wave surface in medium t is described by

$$k_{tx}^2 + k_z^2 = k_t^2. \tag{4.7a}$$

Since $k_z > k_t$,

$$k_{tx} = \pm\left(k_t^2 - k_z^2\right)^{1/2} = \pm i\alpha_{tx} \tag{4.7b}$$

is purely imaginary. Remember that the wave in region t is characterized by $e^{ik_{tx}x + ik_z z - i\omega t}$. For $k_z > k_t$, it becomes $e^{\alpha_{tx}x + ik_z z - i\omega t}$. Thus the transmitted wave decays exponentially in the negative $\hat{x}$ direction and propagates along the $\hat{z}$ direction with phase velocity ω / k_z. This can be regarded as a plane wave with varying amplitude on the constant phase front, and it is a nonuniform plane wave. Since the nonuniform plane wave has its maximum amplitude at the boundary surface and decays exponentially away from the surface, it can be called a *surface wave*. A wave having an exponential decay factor in a particular direction is called *evanescent* in that direction; thus the surface

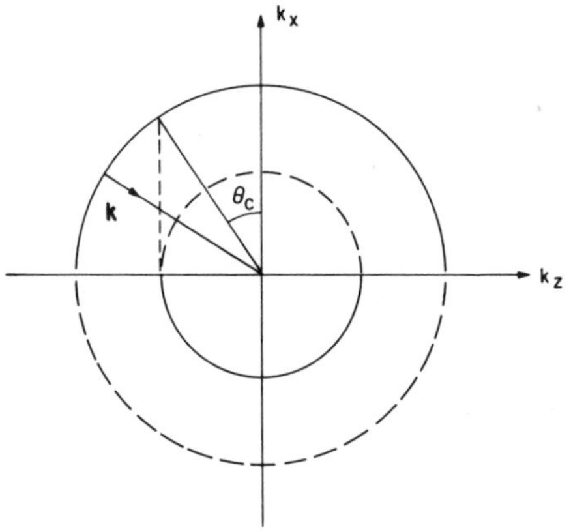

Figure 4.3 The medium in the incident region has the larger k sphere, and the medium in the transmitted region has a smaller k sphere. There is no real $\mathbf{k}_t$ corresponding to the incident $\mathbf{k}_i$ if the incident angle is greater than the critical angle θ_c.

wave is evanescent in the $-\hat{x}$ direction. Since evanescence in the transmitted wave begins when $k_t = k_z = k\sin\theta_c$, the angle θ_c is the *critical angle*. When the wave amplitudes are solved and the energy flow of the waves is determined, we shall find that there is no time-average power penetrating the medium t when the incident angle is greater than θ_c. For this reason, this phenomenon is considered to be *total reflection*.

Construction of the **k** surfaces enables us to visualize the **k** vectors in the reflected and transmitted regions (Figs. 4.2 and 4.3). This technique can also be applied to media that are not isotropic. Consider the phenomenon of double refraction by a positive uniaxial medium (Fig. 4.4). First, let the optic axis of the medium be perpendicular to the plane of incidence. The two transmitted wave vectors are shown in Fig. 4.4*a*. The power-flow directions for the ordinary and the extraordinary waves are the same as the direction of the wave vectors. Next, let the optic axis be parallel to the plane of incidence. The two transmitted wave vectors are shown in Fig. 4.4*b*. By the nature of the wave surface, the power-flow direction **s** for the extraordinary wave is no longer the same as the direction of **k**. Note that by proper source excitation we can excite either the ordinary or the extraordinary wave. For instance, if the incident wave is linearly polarized perpendicularly to the plane of incidence, only ordinary waves are excited.

Consider another case with an extraordinary wave excited in a uniaxial medium and incident upon the interface of an isotropic medium. Let the optic axis be in the plane of incidence and make an angle with the boundary. The wave surface for the extraordinary wave, as well as that for

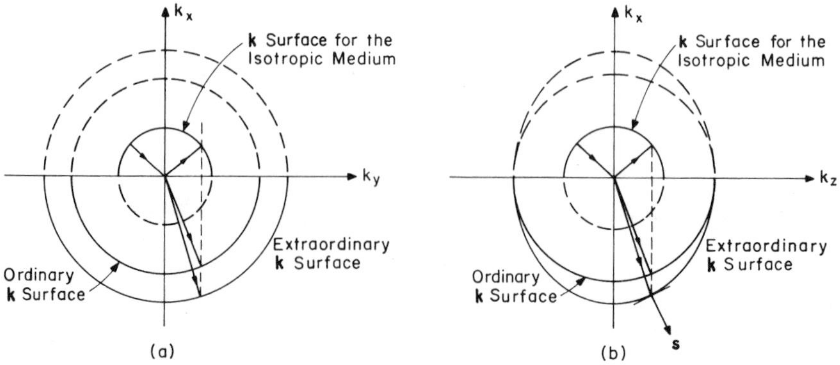

Figure 4.4 **A plane wave incident upon a uniaxial medium. (*a*) Phase matching for plane of incidence perpendicular to the optic axis of a positive uniaxial medium. (*b*) Phase matching for plane of incidence parallel to the optic axis of a positive uniaxial medium. The direction of s is different from that of k for the extraordinary wave.**

the isotropic medium, is shown in Fig. 4.5. In Figs. 4.1–4.4 we have drawn the incident **k** vector with arrows pointing toward the origin. Although we use dotted lines for half of the spheres to convey the feeling of a physical boundary, we must realize that the plot is for **k** space and not physical space. In Fig. 4.5 we draw the **k** vector from center to surface. We see that the reflected wave vector, instead of pointing in the positive k_x and k_z directions, is now pointing in the positive k_z and negative k_x directions. The power-flow direction for the reflected wave is, however, pointing in the positive k_x and k_z directions. Thus the reflected wave, while carrying energy away from the interface, has its phase front propagating toward the interface. This is called a *backward wave* with respect to the normal at the interface.

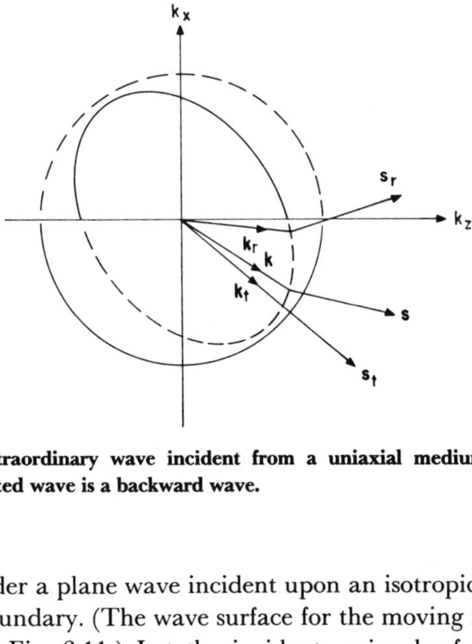

Figure 4.5 An extraordinary wave incident from a uniaxial medium upon an isotropic medium. The reflected wave is a backward wave.

Finally, consider a plane wave incident upon an isotropic medium moving parallel to its boundary. (The wave surface for the moving medium has been determined; see Fig. 3.11.) Let the incident region be free space and the medium have refractive index $n=2$ in its rest frame. By using **k** surfaces in phase matching, we see that with $\beta=0.1$ the transmitted wave is dragged in the direction of motion for incident waves having either positive or negative k_z components. For $\beta=0.75$, which is in the Čerenkov zone, an incident **k** vector with a negative $\hat{z}$ component will give rise to a transmitted **s** vector with a positive $\hat{z}$ component, and the power is carried by the medium in its direction of motion.

4.1c Moving Boundaries

In the last example, we dealt with a moving medium. Since the boundary surface is stationary, there is no Doppler shift in frequency. When the boundary surface is in motion, the frequencies of the waves change. Note that a moving boundary does not necessarily imply a moving medium. A shock wave front in a stationary medium furnishes an example: although the medium properties behind and in front of the shock are different, the medium is stationary.

Let the plane boundary shown in Fig. 4.1 be moving with velocity

$$\mathbf{v} = \hat{x}v_x + \hat{y}v_y + \hat{z}v_z.$$

At $t=0$, the boundary is at $x=0$. At any time t, the boundary is at $x=v_x t$. We require that the boundary conditions [now (1.7)] be satisfied for the field components; hence all exponents in (4.1) must be equal:

$$k_x(v_x t) + k_y(y + v_y t) + k_z(z + v_z t) - \omega t$$

$$= k_{rx}(v_x t) + k_{ry}(y + v_y t) + k_{rz}(z + v_z t) - \omega_r t$$

$$= k_{tx}(v_x t) + k_{ty}(y + v_y t) + k_{tz}(z + v_z t) - \omega_t t.$$

Since these equalities hold for all y, z, and t, we conclude that

$$k_y = k_{ry} = k_{ty}, \tag{4.8a}$$

$$k_z = k_{rz} = k_{tz}, \tag{4.8b}$$

$$k_x v_x - \omega = k_{rx} v_x - \omega_r = k_{tx} v_x - \omega_t. \tag{4.8c}$$

In view of (4.8a) and (4.8b), the laws of reflection and transmission are not changed. The change in frequency for all three waves is determined only by the normal component of the velocity.

Consider the simple case of a wave normally incident from an isotropic medium upon another isotropic medium moving in the $+\hat{x}$ direction toward the wave. The incident wave vector is

$$\mathbf{k} = -\hat{x}n\frac{\omega}{c}, \tag{4.9a}$$

and the reflected wave vector is

$$\mathbf{k}_r = \hat{x}n\frac{\omega_r}{c}. \tag{4.9b}$$

The second medium is moving; therefore it is bianisotropic. The **k** vector for the transmitted wave propagating in a direction opposite to that of medium motion is given by (3.79b):

$$\mathbf{k}_t = -\hat{x}\,\frac{\omega_t}{c}\,\frac{n_t-\beta}{1-n_t\beta}, \tag{4.9c}$$

where $\beta = v_x/c$. Introducing (4.9) into (4.8), we obtain

$$\omega_r = \omega\frac{1+n\beta}{1-n\beta}, \tag{4.10a}$$

$$\omega_t = \omega\frac{(1-n_t\beta)(1+n\beta)}{1-\beta^2}, \tag{4.10b}$$

$$k_r = -k\frac{1+n\beta}{1-n\beta}, \tag{4.10c}$$

$$k_t = k\frac{(n_t-\beta)(1+n\beta)}{n(1-\beta^2)}, \tag{4.10d}$$

where both ω_r and k_r are independent of n_t, the refractive index for the transmitted medium. The reflected wave is Doppler-shifted toward the high-frequency side, and its wave number is also changed by the same amount. Its phase velocity is unchanged. For the transmitted wave, the frequency is shifted downward when $n_t > n$. Its velocity is

$$\frac{\omega_t}{k_t} = n\frac{\omega}{k}\,\frac{1-n_t\beta}{n_t-\beta}, \tag{4.11}$$

which again demonstrates the drag effect of the medium.

Thus far, the discussions of wave behavior at a boundary surface can be considered as a dispersion analysis, in which we are concerned with how a wave behaves if it exists. We shall now turn to calculations of wave amplitudes.

4.2 REFLECTION AND TRANSMISSION COEFFICIENTS

4.2a TE Wave

An incident wave with any polarization can be decomposed into TE and TM wave components. The TE wave is linearly polarized perpendicular to the plane of incidence, and is called *perpendicularly polarized, horizontally polarized*, or simply an *s-wave*. The TM wave is linearly polarized with the electric vector parallel to the plane of incidence, and is called *parallel polarized, vertically polarized*, or simply a *p-wave*. Because of the linearity of Maxwell's equations, we can treat the two polarizations separately and superimpose them to obtain the final answer.

Consider a TE wave incident on a plane boundary separating two isotropic media at $x=0$. Let the plane of incidence be the $x-z$ plane (Fig. 4.6). The electric field vectors for the incident, reflected, and transmitted waves are as follows:

$$\mathbf{E}_i=\hat{y}E_0e^{-ik_xx+ik_zz-i\omega t}, \tag{4.12a}$$

$$\mathbf{E}_r=\hat{y}R^{\mathrm{TE}}E_0e^{ik_xx+ik_zz-i\omega t}, \tag{4.12b}$$

$$\mathbf{E}_t=\hat{y}T^{\mathrm{TE}}E_0e^{-ik_{tx}x+ik_zz-i\omega t}, \tag{4.12c}$$

where E_0 is the amplitude of the incident wave, R^{TE} the amplitude reflection coefficient, and T^{TE} the amplitude transmission coefficient. Because of phase-matching conditions, all three waves have the same k_z. The k_x values for the incident and reflected waves are the same, because of the dispersion relation of the medium in region zero. In (4.12), the amplitudes of the components of the **k** vectors, k_x, k_{tx}, *and* k_z, are all positive. The boundary is stationary, so there is no Doppler shift in frequency. The magnetic field vectors **H** corresponding to (4.12) are determined from Faraday's law:

$$\mathbf{H}_i=-\frac{E_0}{\omega\mu}(\hat{z}k_x+\hat{x}k_z)e^{-ik_xx+ik_zz-i\omega t}, \tag{4.13a}$$

$$\mathbf{H}_r=-R^{\mathrm{TE}}\frac{E_0}{\omega\mu}(-\hat{z}k_x+\hat{x}k_z)e^{ik_xx+ik_zz-i\omega t}, \tag{4.13b}$$

$$\mathbf{H}_t=-T^{\mathrm{TE}}\frac{E_0}{\omega\mu_t}(\hat{z}k_{tx}+\hat{x}k_z)e^{-ik_{tx}x+ik_zz-i\omega t}. \tag{4.13c}$$

The boundary conditions at $x=0$ require that E_y and H_z be continuous. We have

$$1+R^{\mathrm{TE}}=T^{\mathrm{TE}}, \tag{4.14a}$$

$$\frac{k_x}{\mu}(-1+R^{\mathrm{TE}})=-\frac{k_{tx}}{\mu_t}T^{\mathrm{TE}}. \tag{4.14b}$$

In matching the boundary conditions, we did not use the conditions that normal **B** and normal **D** are continuous across the boundary, because these

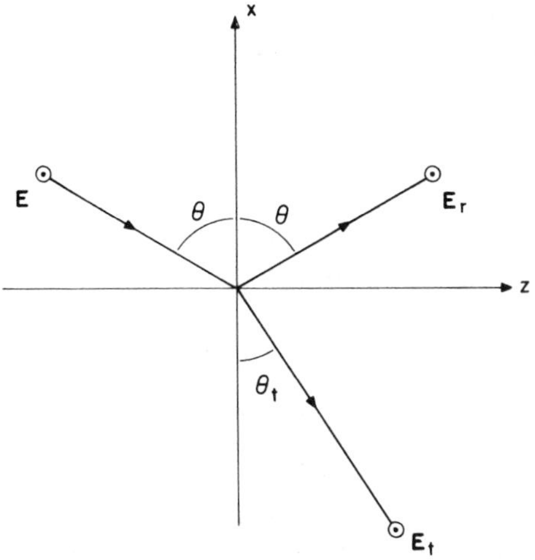

Figure 4.6 TE wave incident upon a plane boundary. The electric field vector is perpendicular to the plane of incidence.

two boundary conditions are not independent of the two tangential **E** and **H** continuous conditions, just as Gauss' two laws are not independent of Faraday's and Ampère's laws. In this case, we can see that the condition of normal **B** continuous yields the same equation (4.14a) and there is no normal **D** component.

The reflection and transmission coefficients R^{TE} and T^{TE} are easily determined from (4.14):

$$R^{\mathrm{TE}} = \frac{1-(\mu k_{tx}/\mu_t k_x)}{1+(\mu k_{tx}/\mu_t k_x)}, \tag{4.15}$$

$$T^{\mathrm{TE}} = \frac{2}{1+(\mu k_{tx}/\mu_t k_x)}. \tag{4.16}$$

When medium t is a perfect conductor, $R^{\mathrm{TE}} = -1$. At the boundary, the reflected E_y reverses its direction, and the total tangential electric field vanishes. A perfect conductor may be treated as the limiting case of a good conductor, and we can show that E_y undergoes a phase change of $-\pi$ upon reflection. When medium t is a good conductor instead of a perfect conductor, a transmitted wave is generated. Since $|k_{tx}| \gg |k_z|$, we see from (4.13c) that $\mathbf{H}_t$ has a very large z component and a very small x component. The **k** vector of the transmitted wave is almost perpendicular to the surface independent of the incident angle of the incident wave.

It is worthwhile to note that the sum of the reflection coefficient and the transmission coefficient does not equal one, $R^{\mathrm{TE}} + T^{\mathrm{TE}} \neq 1$, as is evident from (4.14a). This does not imply any violation of energy conservation, however, because we can show that the sum of the power reflection coefficient and the power transmission coefficient equals one. The power reflection coefficient, also called *reflectivity*, is defined as the ratio of the normal component of the reflected Poynting's vector power to the normal component of the incident Poynting's vector power:

$$r^{\mathrm{TE}} = \frac{\hat{x} \cdot \mathbf{S}_r}{-\hat{x} \cdot \mathbf{S}_i}.$$

By the same token, we define the power transmission coefficient, or *transmissivity*, as

$$t^{\mathrm{TE}} = \frac{-\hat{x} \cdot \mathbf{S}_t}{-\hat{x} \cdot \mathbf{S}_i},$$

where $\mathbf{S}_t$ is the transmitted Poynting's vector. The Poynting's vectors are immediately calculated from (4.12) and (4.13):

$$\mathbf{S}_i = \left(-\hat{x} \frac{k_x}{\omega\mu} + \hat{z} \frac{k_z}{\omega\mu} \right) |E_0|^2, \tag{4.17a}$$

$$\mathbf{S}_r = \left(\hat{x} \frac{k_x}{\omega\mu} + \hat{z} \frac{k_z}{\omega\mu} \right) |R^{\mathrm{TE}}|^2 |E_0|^2, \tag{4.17b}$$

$$\mathbf{S}_t = \left(-\hat{x} \frac{k_{tx}}{\omega\mu_t} + \hat{z} \frac{k_z}{\omega\mu_t} \right) |T^{\mathrm{TE}}|^2 |E_0|^2. \tag{4.17c}$$

The reflectivity and the transmissivity are found to be

$$r^{\mathrm{TE}} = |R^{\mathrm{TE}}|^2, \tag{4.18a}$$

$$t^{\mathrm{TE}} = \frac{\mu k_{tx}}{\mu_t k_x} |T^{\mathrm{TE}}|^2. \tag{4.18b}$$

By introducing (4.15) and (4.16), we can show that for real k_{tx}, $r^{\mathrm{TE}} + t^{\mathrm{TE}} = 1$. Note that at total reflection the incident angle is greater than the critical angle and $k_{tx} = i\alpha_{tx}$. The transmissivity is purely imaginary, and the time-average transmitted power is zero. The reflection coefficient R^{TE} at total reflection is phase-shifted by $2\phi^{\mathrm{TE}}$, where

$$\phi^{\mathrm{TE}} = -\tan^{-1} \frac{\mu\alpha_{tx}}{\mu_t k_x}, \tag{4.19}$$

and the magnitude of R^{TE} is unity. Thus all of the power is reflected.

It is interesting to compare the two cases: (i) at total reflection, the incident power is totally reflected; and (ii) when the transmitting medium is a perfect conductor, the incident power is also totally reflected. As far as the reflected TE wave is concerned, it makes no difference whether the wave is reflected by a perfect conductor or by a dielectric boundary if both boundaries also render the reflected wave with identical phases. Consider a perfect conducting boundary at $x=-d$. The wave is phase-shifted by $-\pi$ at the perfect conducting boundary. The path length from $x=0$ to $x=-d$ provides another phase shift, $2k_x d$, for the wave. If d is such that

$$2k_x d-\pi=-2\tan^{-1}\frac{\mu\alpha_{tx}}{\mu_t k_x}, \tag{4.20}$$

then the reflected TE wave at $x>0$ experiences the same amount of phase shift in both cases. Experiments by Goos and Hänschen[43] have demonstrated that a beam of light is laterally shifted when totally reflected at a dielectric boundary. We must note, however, that, although this argument can be used to display a lateral shift (Fig. 4.7) for the reflected **k** vector, the analysis is true only for plane waves, and there is no way to observe a lateral shift experimentally for plane waves.

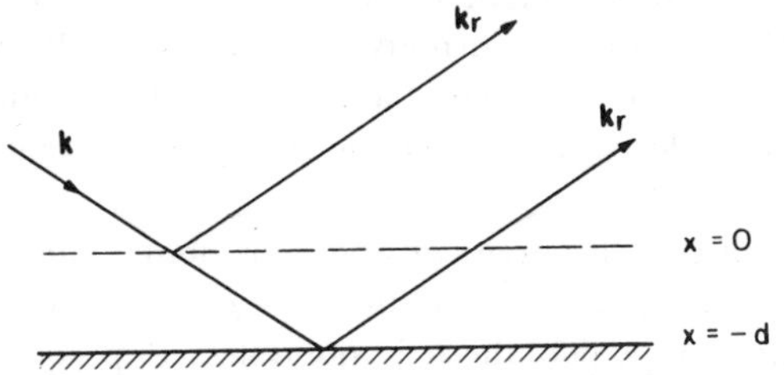

Figure 4.7 Total internal reflection at a dielectric boundary $x=0$. This case is equivalent to reflection by a perfect conducting boundary at $x=-d$.

4.2b TM Wave

The reflection and transmission of TM waves by a plane boundary can be carried out in a manner similar to the treatment of TE waves. We can also use the dual property of **E** and **H** in source-free Maxwell's equations to write the answers directly. The two curl equations are the dual of each other in the sense that, if we make the replacement $\mathbf{E}\rightarrow\mathbf{H}$, $\mathbf{H}\rightarrow-\mathbf{E}$, and $\mu\rightleftharpoons\epsilon$, we can obtain one equation from the other. This duality can be used to transform all results obtained for TE waves to TM waves. For the reflection and transmission coefficients, we have

$$R^{\mathrm{TM}}=\frac{1-(\epsilon k_{tx}/\epsilon_t k_x)}{1+(\epsilon k_{tx}/\epsilon_t k_x)}, \tag{4.21}$$

$$T^{\mathrm{TM}}=\frac{2}{1+(\epsilon k_{tx}/\epsilon_t k_x)}. \tag{4.22}$$

For the reflectivity and the transmissivity, we have

$$r^{\mathrm{TM}}=|R^{\mathrm{TM}}|^2, \tag{4.23a}$$

$$t^{\mathrm{TM}}=\frac{\epsilon k_{tx}}{\epsilon_t k_x}|T^{\mathrm{TM}}|^2. \tag{4.23b}$$

At total reflection, the phase shift for the reflected wave is $2\phi^{\mathrm{TM}}$, where

$$\phi^{\mathrm{TM}}=-\tan^{-1}\frac{\epsilon\alpha_{tx}}{\epsilon_t k_x}. \tag{4.24}$$

But when the reflector is a perfect conductor, the reflection coefficient becomes $+1$, as is seen from (4.21). In particular, note that R^{TM} and T^{TM} are for the magnetic field components H_y. At the surface of a perfect conductor, the tangential magnetic field doubles its strength in order to support the induced surface currents. The phase shift for the reflected magnetic field H_y is zero. The dual of a perfect electric conductor is a perfect magnetic conductor, for which the boundary condition requires a vanishing tangential **H**.

We shall now discuss the Brewster angle for nonmagnetic media. Assume that both media are dielectrics and that $\mu_t=\mu=\mu_0$. The Brewster angle is the incident angle at which there is no reflected power. In view of the reflectivity for TE and TM waves, this amounts to requiring that

$$k_{tx}=k_x \tag{4.25a}$$

for TE waves and

$$\epsilon k_{tx}=\epsilon_t k_x \tag{4.25b}$$

for TM waves. Since $k_{tx}=k_x$ only if both dielectrics have identical permittivities, we see that no Brewster angle can exist for TE waves. For TM waves, condition 4.25b amounts to

$$k_t\cos\theta_b=k\cos\theta_t, \tag{4.26}$$

where θ_b denotes the Brewster angle. Also, by Snell's law,

$$k \sin\theta_b = k_t \sin\theta_t. \tag{4.27}$$

It can be observed from (4.26) and (4.27) that

$$\theta_b + \theta_t = \frac{\pi}{2}. \tag{4.28}$$

Since the reflected angle is equal to the incident angle, the reflected wave vector $\mathbf{k}_r$ and the transmitted wave vector $\mathbf{k}_t$ are perpendicular to each other. Physically we can explain this by visualizing the dielectric media as consisting of dipoles that are excited by the incident wave and reradiate at the same frequency. Each individual dipole has a radiation pattern that is maximum in a direction perpendicular to the dipole axis and null along the dipole axis. For a TM wave excitation, all dipoles oscillate parallel to the plane of incidence along the **E**-field lines. At the Brewster angle of incidence, the reflected $\mathbf{k}_r$ vector is perpendicular to the transmitted k_t vector and in the same direction of dipole oscillation in the transmitted medium (Fig. 4.8). Thus no TM wave is reflected.

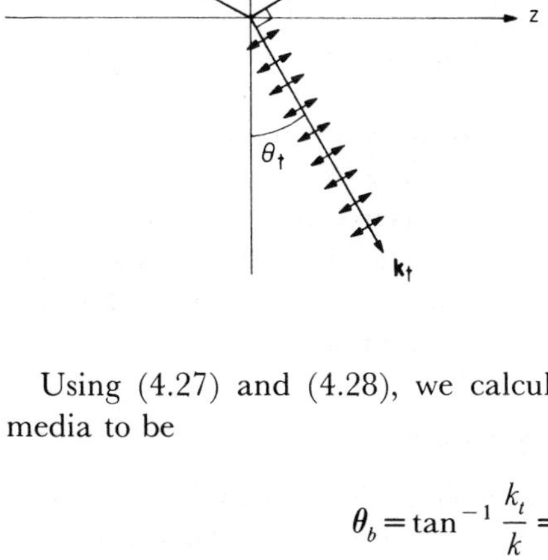

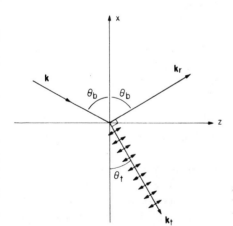

Figure 4.8 Interpretation of the Brewster angle with dipole models for the transmitted wave. When $\theta_b + \theta_t = \pi/2$, all dipoles in the transmitting medium oscillate parallel to $\mathbf{k}_r$.

Using (4.27) and (4.28), we calculate the Brewster angle for dielectric media to be

$$\theta_b = \tan^{-1}\frac{k_t}{k} = \tan^{-1}\sqrt{\frac{\epsilon_t}{\epsilon}}. \tag{4.29}$$

At this angle, all TM waves are transmitted. In Fig. 4.9, r^{TE} and r^{TM} are plotted as functions of incident angles. In general, on a solid dielectric surface, the TE waves reflect more than the TM waves.

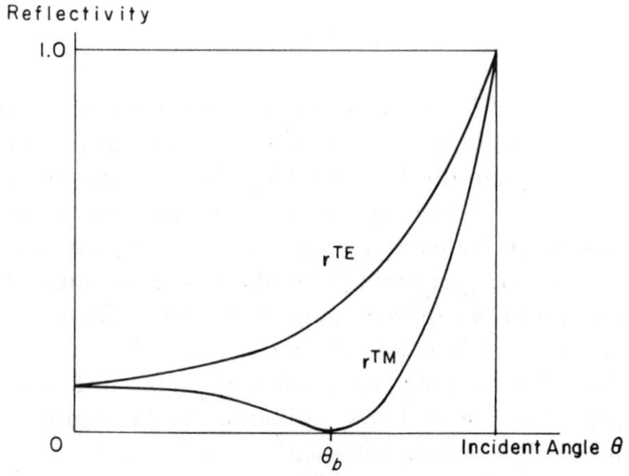

Figure 4.9 Reflection of TE and TM wave incident from free space on a dielectric boundary. The TM wave is always less reflected than the TE wave.

At the Brewster angle, TM waves are totally transmitted. When the incident angle is larger than the critical angle, all TE and TM waves are totally reflected. We can compare these phenomena at an isotropic dielectric interface as follows:

1. Total reflection occurs at a range of incident angles for which $\theta > \theta_c = \sin^{-1}(\epsilon_t/\epsilon)^{1/2}$. Total transmission of TM waves occurs only at the Brewster angle $\theta_b = \tan^{-1}(\epsilon_t/\epsilon)^{1/2}$.

2. Total reflection occurs only when the incident medium is denser than the transmitted medium. The Brewster angle occurs for any two media.

3. When an unpolarized wave is totally reflected, the reflected wave is still unpolarized. When the TM wave components of an unpolarized wave are totally transmitted, the reflected wave contains only TE waves. Thus the Brewster angle is also referred to as the polarization angle.

4.2c Moving Media

Consider a plane wave normally incident upon a dielectric medium moving toward the wave. Let the boundary be moving in the $\hat{z}$ direction with velocity v, and the electric field be linearly polarized in the $\hat{x}$ direction. We write

$$\mathbf{E}_i = \hat{x}E_i = \hat{x}E_0 e^{ikz - i\omega t}, \qquad k = -\frac{\omega}{c}, \tag{4.30a}$$

$$\mathbf{E}_r = \hat{x}E_r = \hat{x}RE_0 e^{ik_r z - i\omega_r t}, \qquad k_r = \frac{\omega_r}{c}, \tag{4.30b}$$

$$\mathbf{E}_t = \hat{x}E_t = \hat{x}TE_0 e^{ik_t z - i\omega_t t}, \qquad k_t = \frac{n-\beta}{n\beta - 1}\frac{\omega_t}{c}, \tag{4.30c}$$

where $\beta = v/c$. The dispersion relation presented in (4.30c) for the moving medium was derived in (3.79b). The magnetic field $\mathbf{B}$ is obtained from Faraday's law:

$$c\mathbf{B}_i = \hat{y}\frac{ck}{\omega}E_i = -\hat{y}E_i, \tag{4.31a}$$

$$c\mathbf{B}_r = \hat{y}\frac{ck_r}{\omega_r}E_r = \hat{y}E_r, \tag{4.31b}$$

$$c\mathbf{B}_t = \hat{y}\frac{ck_t}{\omega_t}E_t = \hat{y}\frac{n-\beta}{n\beta - 1}E_t, \tag{4.31c}$$

where $n = c\sqrt{\mu_0\epsilon_t}$. The field vectors $\mathbf{H}$ and $\mathbf{D}$ are determined from the vacuum constitutive relations for the incident and reflected waves, and from the moving medium constitutive relations in (2.36) for the transmitted wave:

$$c\mathbf{D}_i = \hat{x}c\epsilon_0 E_i, \tag{4.32a}$$

$$c\mathbf{D}_r = \hat{x}c\epsilon_0 E_r, \tag{4.32b}$$

$$c\mathbf{D}_t = \hat{x}\frac{1}{c\mu_0}\left(p - l\frac{ck_t}{\omega_t}\right)E_t = \hat{x}\frac{1}{c\mu_0}\frac{n(n-\beta)}{1-n\beta}E_t; \tag{4.32c}$$

$$\mathbf{H}_i = -\hat{y}\frac{1}{c\mu_0}E_i, \tag{4.33a}$$

$$\mathbf{H}_r = \hat{y}\frac{1}{c\mu_0}E_r, \tag{4.33b}$$

$$\mathbf{H}_t = \hat{y}\frac{1}{c\mu_0}\left(l + q\frac{ck_t}{\omega_t}\right)E_t$$

$$= -\hat{y}\frac{1}{c\mu_0}nE_t. \tag{4.33c}$$

To calculate amplitudes for the reflected and transmitted waves, we have to use moving boundary conditions 1.7. In our case, we have $E_x - \beta c B_y$ and $H_y - \beta c D_x$ continuous across the boundary. The result is

$$(1+\beta)+(1-\beta)R=\frac{1-\beta^2}{1-n\beta}T, \tag{4.34a}$$

$$-(1+\beta)+(1-\beta)R=\frac{-n(1-\beta^2)}{1-n\beta}T. \tag{4.34b}$$

The solution is immediately obtained for the reflection and transmission coefficients:

$$R=-\frac{1+\beta}{1-\beta}\frac{n-1}{n+1}, \tag{4.35a}$$

$$T=\frac{1-n\beta}{1-\beta}\frac{2}{n+1}. \tag{4.35b}$$

The reflectivity and transmissivity are found to be

$$r=\frac{\hat{z}\cdot(\mathbf{E}_r\times\mathbf{H}_r)}{-\hat{z}\cdot(\mathbf{E}_i\times\mathbf{H}_i)}=|R|^2, \tag{4.36a}$$

$$t=\frac{-\hat{z}\cdot(\mathbf{E}_t\times\mathbf{H}_t)}{-\hat{z}\cdot(\mathbf{E}_i\times\mathbf{H}_i)}=n|T|^2. \tag{4.36b}$$

Apparently, as β increases from -1, r increases and t decreases. Equations 4.35 and 4.36 reduce to the stationary case for normal incidence when $\beta=0$. Similarly, in the stationary case, $R+T\neq1$. Contrary to the stationary case, we also have $r+t\neq1$ when $\beta\neq0$. Is power conservation being violated?

To answer this question, we shall consider the energy and force relations for this case (see Daly and Gruenberg[25]). Conceive a cylinder of unit cross section erected across the boundary with its axis parallel to $\hat{z}$ and containing a portion of the interface. The time-average power is calculated from $\frac{1}{2}\mathrm{Re}(\mathbf{E}\times\mathbf{H}^*)$. The total time-average electromagnetic power flow into the cylinder is given by:

$$\langle P_{\text{elec}}\rangle=\frac{1}{2c\mu_0}(E_i^2-E_r^2-nE_t^2)$$

$$=\frac{4cU_0\beta(n-1)(1-n\beta)}{(1-\beta)^2(1+n)}, \tag{4.37}$$

where

$$U_0 = \frac{1}{2c\mu_0}|E_0|^2. \tag{4.38}$$

Inside the cylinder, there is an increase in the time-average electromagnetic energy, $\frac{1}{4}\mathrm{Re}(\mathbf{E} \cdot \mathbf{D}^* + \mathbf{H} \cdot \mathbf{B}^*)$, as the moving dielectric occupies more free space. The rate of this increase in the stored energy is given by the velocity times the difference between the electromagnetic energy in the dielectric and that in the vacuum.

$$\begin{aligned}\langle P_{\text{stored}} \rangle &= \frac{\beta}{2c\mu_0}\left[n\left(\frac{n-\beta}{1-n\beta} \right) E_t^2 - E_r^2 - E_i^2 \right] \\ &= \frac{2cU_0\beta(n-1)(1-2n\beta+\beta^2)}{(1-\beta)^2(1+n)}.\end{aligned} \tag{4.39}$$

When the medium is stationary, $\langle P_{\text{elec}} \rangle = \langle P_{\text{stored}} \rangle = 0$. When the medium is in motion, mechanical power is required to keep the dielectric moving at a constant velocity. The rate at which mechanical work has to be supplied to the system is given by the difference between $\langle P_{\text{stored}} \rangle$ and $\langle P_{\text{elec}} \rangle$:

$$\begin{aligned}\langle P_{\text{mech}} \rangle &= \langle P_{\text{stored}} \rangle - \langle P_{\text{elec}} \rangle \\ &= \frac{-2cU_0\beta(n-1)(1+\beta)}{(1-\beta)(1+n)}.\end{aligned} \tag{4.40}$$

The negative sign indicates that mechanical work has been done to the system. The force per unit area acting on the dielectric medium is obtained from $\mathbf{F} \cdot \mathbf{v} = \langle P_{\text{mech}} \rangle$. Thus

$$\mathbf{F}_{\text{mech}} = \frac{-\hat{z}2U_0(n-1)(1+\beta)}{(1-\beta)(1+n)}. \tag{4.41}$$

This mechanical force is needed to maintain the medium at constant velocity. We note that the force is in the negative $\hat{z}$ direction; this means that mechanical force must be applied to stop the medium from accelerating toward the wave. The electromagnetic force $\mathbf{F}_{\text{elec}}$ exerted on the medium by the wave is equal to the negative of $\mathbf{F}_{\text{mech}}$.

We may double-check this assertion by using the conservation theorem derived in Problem 1.2 to calculate the electromagnetic force $\mathbf{F}_{\text{elec}}$. When the force density is integrated over the volume of the cylinder, we find that the force per unit area acting on the surface is

$$\mathbf{F}_{\text{elec}} = -\hat{z}\left(\langle T_{zz}\rangle_i + \langle T_{zz}\rangle_r - \langle T_{zz}\rangle_t\right) + v\left(\langle \mathbf{G}\rangle_i + \langle \mathbf{G}\rangle_r - \langle \mathbf{G}\rangle_t\right)$$

$$= \hat{z}\left\{\frac{1}{2c^2\mu_0}\left[-E_i^2 - E_r^2 + n\left(\frac{n-\beta}{1-n\beta}\right)E_t^2\right] + \frac{\beta}{2c^2\mu_0}\left[-E_i^2 + E_r^2 + n\left(\frac{n-\beta}{1-n\beta}\right)^2 E_t^2\right]\right\}$$

$$= \frac{\hat{z}2U_0(n-1)(1+\beta)}{(1-\beta)(1+n)}. \tag{4.42}$$

Clearly, $\mathbf{F}_{\text{elec}}$ and $\mathbf{F}_{\text{mech}}$ are indeed in opposite directions. We conclude that the radiation pressure exerted on a dielectric half-space by a plane wave at normal incidence results in a force attracting the medium toward the wave. This force is there whether the dielectric is stationary or is in motion, as demonstrated in (4.42). A mechanical force counterbalancing $\mathbf{F}_{\text{elec}}$ is needed either to keep the dielectric medium stationary or to maintain its constant velocity when it is in motion.

It is an interesting exercise to examine the case of a perfect conductor, where $E_t = 0$ and $R = -(1+\beta)/(1-\beta)$. We obtain, instead of (4.40),

$$\langle P_{\text{mech}}\rangle = \frac{1}{2c\mu_0}\left[-\beta(E_i^2 + E_r^2) - (E_i^2 - E_r^2)\right] = \frac{2cU_0\beta(1+\beta)}{1-\beta}.$$

It follows that $\mathbf{F}_{\text{mech}} = \hat{z}2U_0(1+\beta)/(1-\beta)$. Note in particular the sign of $\mathbf{F}_{\text{mech}}$, which is now in the positive $\hat{z}$ direction, demonstrating that the wave is exerting force to push the conductor away. Thus the electromagnetic force is attractive when the medium is a dielectric, and repulsive when the medium is a perfect conductor.

4.3 STRATIFIED MEDIA

4.3a Wave Amplitude and Wave Impedance

Consider a stratified medium with boundaries at $x = 0, -d_1, \ldots, -d_n$ (Fig. 4.10). The $(n+1)$th region is semiinfinite and is labeled region t, $t = n+1$.

We let all regions be composed of biaxial media oriented with principal axes parallel to the coordinate axes. The permittivities and the permeabilities in each region l are denoted by $\bar{\epsilon}_l$ and $\bar{\mu}_l$. Consider the simple case of a plane wave incident upon the stratified medium from region 0 and having the plane of incidence parallel to the $x-z$ plane. The problem is much more complicated if the plane of incidence is oriented arbitrarily. The polarization of the incident wave is decomposed into TE and TM components which are treated separately. As before, we concentrate on the TE component; solutions for the TM case can be obtained by duality with the replacements $\mathbf{E}\rightarrow\mathbf{H}$, $\mathbf{H}\rightarrow-\mathbf{E}$, and $\bar{\epsilon}\leftrightarrow\bar{\mu}$. All results reduce to those for an isotropic stratified medium, by letting $\bar{\epsilon}=\epsilon\bar{\mathbf{I}}$ and $\bar{\mu}=\mu\bar{\mathbf{I}}$.

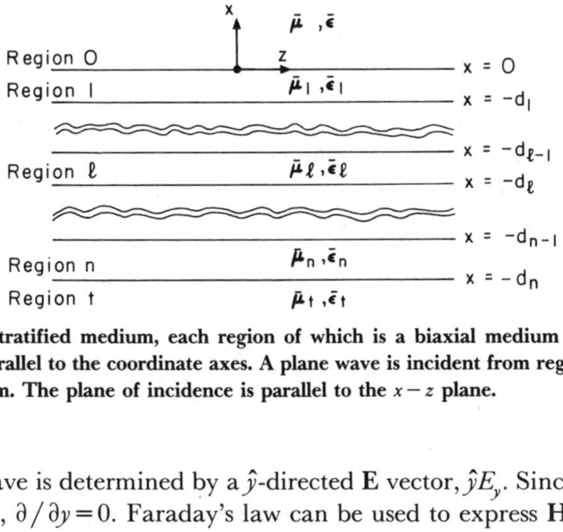

Figure 4.10 A stratified medium, each region of which is a biaxial medium with principal axes oriented parallel to the coordinate axes. A plane wave is incident from region 0 upon the stratified medium. The plane of incidence is parallel to the $x-z$ plane.

The TE wave is determined by a $\hat{y}$-directed $\mathbf{E}$ vector, $\hat{y}E_y$. Since there is no y dependence, $\partial/\partial y=0$. Faraday's law can be used to express $\mathbf{H}$ in terms of E_y:

$$H_x=-\frac{1}{i\omega\mu_x}\frac{\partial}{\partial z}E_y, \tag{4.43a}$$

$$H_z=\frac{1}{i\omega\mu_z}\frac{\partial}{\partial x}E_y. \tag{4.43b}$$

Substituting (4.43) in the source-free Ampère's law, we obtain

$$\left(\frac{\mu_y}{\mu_z}\frac{\partial^2}{\partial x^2}+\frac{\mu_y}{\mu_x}\frac{\partial^2}{\partial z^2}+\omega^2\mu_y\epsilon_y\right)E_y=0. \tag{4.44}$$

This is a wave equation for E_y in all regions.

In region l the TE wave solutions can be written as

$$E_{ly} = \left(A_l e^{ik_{lx}x} + B_l e^{-ik_{lx}x}\right) e^{ik_z z}, \tag{4.45a}$$

$$H_{lx} = -\frac{k_z}{\omega\mu_{lx}} \left(A_l e^{ik_{lx}x} + B_l e^{-ik_{lx}x}\right) e^{ik_z z}, \tag{4.45b}$$

$$H_{lz} = \frac{k_{lx}}{\omega\mu_{lz}} \left(A_l e^{ik_{lx}x} - B_l e^{-ik_{lx}x}\right) e^{ik_z z}. \tag{4.45c}$$

We do not write a subscript l for k_z because, from phase-matching conditions, the k_z in all regions must be the same. Substitution of (4.45a) in wave equation 4.44 yields the dispersion relation in the lth medium:

$$\frac{\mu_{ly}}{\mu_{lz}} k_{lx}^2 + \frac{\mu_{ly}}{\mu_{lx}} k_z^2 = \omega^2 \mu_{ly} \epsilon_{ly}. \tag{4.46}$$

The wave amplitudes A_l and B_l are related to wave amplitudes in neighboring regions by the boundary conditions.

The boundary condition at $x = -d_l$ requires that E_y and H_z be continuous across the boundary, and we have

$$A_l e^{-ik_{lx}d_l} + B_l e^{ik_{lx}d_l} = A_{l+1} e^{-ik_{(l+1)x}d_l} + B_{l+1} e^{ik_{(l+1)x}d_l} \tag{4.47a}$$

$$\frac{k_{lx}}{\omega\mu_{lz}} \left(A_l e^{-ik_{lx}d_l} - B_l e^{ik_{lx}d_l}\right) = \frac{k_{(l+1)x}}{\omega\mu_{(l+1)z}} \left[A_{l+1} e^{-ik_{(l+1)x}d_l} - B_{l+1} e^{ik_{(l+1)x}d_l}\right]. \tag{4.47b}$$

Equations 4.47a and 4.47b are the basic equations for subsequent discussions. The number of unknowns is $2n+2$, which includes the reflection and the transmission coefficients R and T, as well as A_l and B_l, $l = 1, 2, \ldots, n$. There are $n+1$ boundaries, and at each boundary there are two equations such as (4.47). We therefore have $2n+2$ equations to solve for the $2n+2$ unknowns. The equations are linear, and the solution is unique. We can arrange the equations in matrix form with the unknowns forming a $2n+2$ column matrix and the coefficients forming a $(2n+2) \times (2n+2)$ square matrix. The solution is obtained by inverting the square matrix. This procedure is straightforward but very tedious. We shall now describe simpler ways of dealing with the problem.

From (4.45), we see that A_l / B_l is the ratio of the amplitude of the wave propagating in the positive $\hat{x}$ direction to that of the wave propagating in the negative $\hat{x}$ direction. We define a space-dependent complex reflection coefficient R_l such that

$$R_l(x) = \frac{A_l}{B_l} e^{i2k_{lx}x}.$$

On the complex $R_l(x)$ plane (Fig. 4.11), as the phase $\phi = 2k_{lx}x$ increases with x, $R_l(x)$ rotates counterclockwise. If k_{lx} is real, the locus of the tip of $R_l(x)$ is a circle. If k_{lx} is complex, $R_l(x)$ decreases with increasing x.

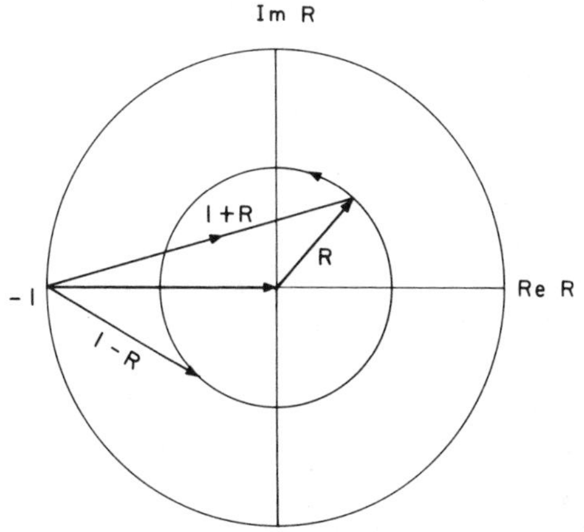

Figure 4.11 Complex R plane. As the phase of the reflection coefficient increases, R rotates counterclockwise and the locus of its tip traces a circle when the medium is lossless.

We can define a wave impedance $Z_l(x)$ in the negative $\hat{x}$ direction by the ratio of (4.45a) to (4.45c):

$$Z_{lx}(x) \equiv -\frac{E_{ly}}{H_{lz}} = \frac{\omega\mu_{lz}}{k_{lx}} \frac{1 + R_l(x)}{1 - R_l(x)}. \tag{4.48a}$$

The wave impedance is a complex number. For a plane wave propagating in free space in the absence of any medium, the wave impedance in the direction of wave propagation is $\eta = \omega\mu_0/k = (\mu_0/\epsilon_0)^{1/2} \approx 377\Omega$.

With the definition of the complex impedance, the ratio of (4.47a) to (4.47b) gives $Z_{lx}(x = -d_l) = Z_{(l+1)x}(x = -d_l)$. Thus at each interface the wave impedances are continuous across the boundary.

On the complex R plane, $Z_{lx}(x)$ can be interpreted as the ratio of the two lengths as shown in Fig. 4.11. The magnitude of $Z_{lx}(x)$ is maximum when R_l

is real and positive, and minimum when R_l is real and negative. We define a dimensionless wave impedance as

$$z_l = \frac{Z_{lx}}{\omega\mu_{lz}/k_{lx}} = \frac{1+R_l}{1-R_l}. \tag{4.48b}$$

For all possible complex values of $R_l(x)$, we can map the corresponding $z_l(x)$ values onto the complex $R_l(x)$ plane. The result is in the form of a Smith chart, which is most frequently used in transmission line theory.

To illustrate the use of the wave impedance concept, we consider a stratified medium composed of $2N+2$ isotropic dielectric layers (corresponding to $2N+2$ boundaries) with alternating high and low permittivities, ϵ_h and ϵ_l; regions $1,3,5,\ldots,2N+1$ are high-permittivity layers, and regions $2,4,6,\ldots,2N$ are low-permittivity layers. Region 0 has permittivity ϵ and permeability μ. The thickness of each layer is a quarter-wavelength inside the dielectric. The transmitted region is $2N+2=t$ and has permittivity ϵ_t. Permeabilities for all layers are equal to μ (Fig. 4.12).

Consider a wave normally incident upon the stratified medium, $k_z=0$, $k_{lx}=\omega(\mu\epsilon_l)^{1/2}$ for all l. The wave impedance of region t, since there is no reflection, is $Z_t=(\mu/\epsilon_t)^{1/2}$. Because of the continuity of wave impedance across the boundary, the impedance across the interface separating regions $2N+1$ and t is $Z_{2N+1}=(\mu/\epsilon_t)^{1/2}$. The relative impedance is $z_{2N+1}=(\mu/\epsilon_t)^{1/2}/(\mu/\epsilon_h)^{1/2}=(\epsilon_h/\epsilon_t)^{1/2}$. In view of (4.48b), the relative impedance at the interface separating regions $2N$ and $2N+1$ is the inverse of z_{2N+1} because R_{2N+1} changes sign after an increase of a quarter-wavelength

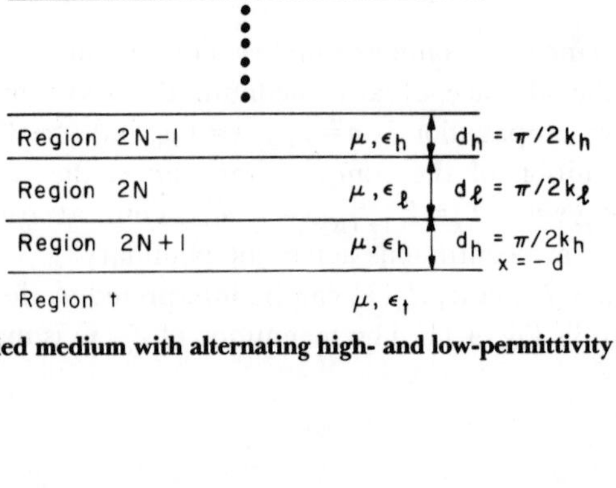

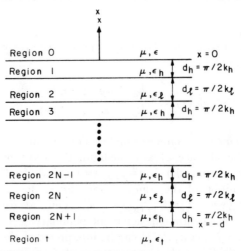

Figure 4.12 **Stratified medium with alternating high- and low-permittivity layers.**

in distance x or, equivalently, a rotation of 180° on the complex R plane. The wave impedance across the interface separating regions $2N$ and $2N+1$ is $Z_{2N}=(\epsilon_t/\epsilon_h)^{1/2}(\mu/\epsilon_h)^{1/2}$. The relative impedance becomes $z_{2N}=(\epsilon_t/\epsilon_h)^{1/2}(\epsilon_l/\epsilon_h)^{1/2}$. After a 180° rotation of R_{2N} on the complex R plane, we find that the wave impedance across the interface separating regions $2N-1$ and $2N$ is $Z_{2N-1}=(\epsilon_h/\epsilon_t)^{1/2}(\epsilon_h/\epsilon_l)^{1/2}(\mu/\epsilon_l)^{1/2}$. The relative impedance is $z_{2N-1}=(\epsilon_h/\epsilon_t)^{1/2}(\epsilon_h/\epsilon_l)^{1/2}(\epsilon_h/\epsilon_l)^{1/2}$. Rotating 180° on the complex plane and multiplying by $(\mu/\epsilon_h)^{1/2}$, we find that the wave impedance across the interface separating regions $2N-1$ and $2N-2$ is $Z_{2N-2}=(\epsilon_t/\epsilon_h)^{1/2}(\epsilon_l/\epsilon_h)(\mu/\epsilon_h)^{1/2}$. Continuing the process and noting the periodicity of the structure, we determine the wave impedance at $x=0$:

$$Z_0=(\epsilon_t/\epsilon_h)^{1/2}(\epsilon_l/\epsilon_h)^N(\mu/\epsilon_h)^{1/2}.$$

The reflection coefficient R at $x=0$ is readily calculated from (4.48a):

$$R_0=\frac{Z_0/(\mu/\epsilon)^{1/2}-1}{Z_0/(\mu/\epsilon)^{1/2}+1}$$

$$=\frac{(\epsilon_t/\epsilon_h)^{1/2}(\epsilon_l/\epsilon_h)^N(\epsilon/\epsilon_h)^{1/2}-1}{(\epsilon_t/\epsilon_h)^{1/2}(\epsilon_l/\epsilon_h)^N(\epsilon/\epsilon_h)^{1/2}+1}. \tag{4.49}$$

We observe that, for a high ϵ_h/ϵ_l ratio and for a larger number of layers, the reflection coefficient R_0 approaches the value -1, and the structure is highly reflective. Such structures are useful at optical frequencies where metallic reflectors are subject to corrosion and tarnishing problems.

4.3b Reflection Coefficients Expressed in Continuous Fractions

A closed-form formula for reflection coefficients can be derived from (4.47). Solving for A_l and B_l from (4.47), we obtain.

$$A_l e^{-ik_{lx}d_l}=\frac{1}{2}\left[1+\frac{\mu_{lz}k_{(l+1)x}}{\mu_{(l+1)z}k_{lx}}\right]\left[A_{l+1}e^{-ik_{(l+1)x}d_l}+R_{l(l+1)}B_{l+1}e^{ik_{(l+1)x}d_l}\right], \tag{4.50a}$$

$$B_l e^{ik_{lx}d_l}=\frac{1}{2}\left[1+\frac{\mu_{lz}k_{(l+1)x}}{\mu_{(l+1)z}k_{lx}}\right]\left[R_{l(l+1)}A_{l+1}e^{-ik_{(l+1)x}d_l}+B_{l+1}e^{ik_{(l+1)x}d_l}\right], \tag{4.50b}$$

where

$$R_{l(l+1)} = \frac{1 - \mu_{lz} k_{(l+1)x} / \mu_{(l+1)z} k_{lx}}{1 + \mu_{lz} k_{(l+1)x} / \mu_{(l+1)z} k_{lx}} \tag{4.51}$$

is the reflection coefficient for TE waves in region l, caused by the boundary separating regions l and $l+1$. Forming the ratio of (4.50a) to (4.50b), we have

$$\frac{A_l}{B_l} e^{-i2k_{lx}d_l} = \frac{1}{R_{l(l+1)}} + \frac{\left[1 - \left(1/R_{l(l+1)}^2\right)\right] e^{-i2k_{(l+1)x}(d_{l+1} - d_l)}}{\left[1/R_{l(l+1)}\right] e^{-i2k_{(l+1)x}(d_{l+1} - d_l)} + \left(A_{l+1}/B_{l+1}\right) e^{-i2k_{(l+1)x}d_{l+1}}}$$

$$= \frac{1}{R_{l(l+1)}} + \frac{\left[1 - \left(1/R_{l(l+1)}^2\right)\right] e^{-i2k_{(l+1)x}(d_{l+1} - d_l)}}{\left[1/R_{l(l+1)}\right] e^{-i2k_{(l+1)x}(d_{l+1} - d_l)}} \Bigg/ + \frac{A_{l+1}}{B_{l+1}} e^{-i2k_{(l+1)x}d_{l+1}} \tag{4.52}$$

The second equality introduces a notation for writing a continuous fraction. Equation 4.52 expresses $(A_l/B_l)e^{-i2k_{lx}d_l}$ in terms of $(A_{l+1}/B_{l+1})e^{-i2k_{(l+1)x}d_{l+1}}$, which is in turn expressed by $(A_{l+2}/B_{l+2})e^{-i2k_{(l+2)x}d_{l+2}}$, and so on until the transmitted region t, where $A_t/B_t = 0$, is reached. The reflection coefficient attributable to the stratified medium is $R = A_0/B_0$. Using this process, we obtain

$$R = \frac{1}{R_{01}} + \frac{\left[1 - \left(1/R_{01}^2\right)\right] e^{-i2k_{1x}d_1}}{\left(1/R_{01}\right) e^{-i2k_{1x}d_1}} \Bigg/ + \frac{1}{R_{12}} + \frac{\left[1 - \left(1/R_{12}^2\right)\right] e^{-i2k_{2x}(d_2 - d_1)}}{\left(1/R_{12}\right) e^{-i2k_{2x}(d_2 - d_1)}} \Bigg/$$

$$+ \cdots + \frac{1}{R_{(n-1)n}} + \frac{\left[1 - \left(1/R_{(n-1)n}^2\right)\right] e^{-i2k_{nx}(d_n - d_{n-1})}}{\left[1/R_{(n-1)n}\right] e^{-i2k_{nx}(d_n - d_{n-1})}} \Bigg/ + R_{nt}. \tag{4.53}$$

This is a closed-form solution for the reflection coefficient expressed in continuous fractions. For a stratified medium with any given layer number (the number of layers is defined to be the number of boundaries), the reflection coefficient is derived from (4.53) by taking terms, starting from the last one, until the subscript l of $R_{l(l+1)}$ becomes zero.

For example, in the case of reflection from a one-layer (half-space) medium, $n=0$, (4.53) is simply

$$R = R_{0t} = \frac{1 - \left(\mu_z k_{tx} / \mu_{tz} k_x\right)}{1 + \left(\mu_z k_{tx} / \mu_{tz} k_x\right)}. \tag{4.54}$$

When the medium is isotropic, this becomes the Fresnel reflection coefficient for TE waves as derived in Section 4.2. When $n=1$, (4.53) gives

$$R=\frac{1}{R_{01}}+\frac{\left[1-(1/R_{01}^2)\right]e^{-i2k_{1x}d_1}}{(1/R_{01})e^{-i2k_{1x}d_1}+R_{1t}}$$

$$=\frac{R_{01}+R_{1t}e^{i2k_{1x}d_1}}{1+R_{01}R_{1t}e^{i2k_{1x}d_1}}. \tag{4.55}$$

This is the reflection coefficient caused by a two-layer stratified medium.

We note that, although our derivation has been carried out for TE waves, the final result as expressed in (4.53) is equally applicable to TM waves. The only difference is that in (4.53) we must now use the Fresnel reflection coefficients for TM waves for $R_{l(l+1)}$, which, in view of duality, is obtained from (4.51) with the replacement $\mu\rightarrow\epsilon$.

4.3c Propagation Matrices

Using (4.50), we can express the wave amplitudes A_l and B_l in region l in terms of A_{l+1} and B_{l+1} in region $l+1$:

$$\begin{bmatrix} A_l e^{-ik_{lx}d_l} \\ B_l e^{ik_{lx}d_l} \end{bmatrix} = \overline{\mathbf{U}}_{l(l+1)}\cdot \begin{bmatrix} A_{l+1} e^{-ik_{(l+1)x}d_{l+1}} \\ B_{l+1} e^{ik_{(l+1)x}d_{l+1}} \end{bmatrix}. \tag{4.56}$$

Here $\overline{\mathbf{U}}_{l(l+1)}$ is called the *backward propagation matrix*:

$$\overline{\mathbf{U}}_{l(l+1)}=\frac{1}{2}\left(1+\frac{\mu_{lz}k_{(l+1)x}}{\mu_{(l+1)z}k_{lx}}\right)$$

$$\cdot\begin{bmatrix} e^{ik_{(l+1)x}(d_{l+1}-d_l)} & R_{l(l+1)}e^{-ik_{(l+1)x}(d_{l+1}-d_l)} \\ R_{l(l+1)}e^{ik_{(l+1)x}(d_{l+1}-d_l)} & e^{-ik_{(l+1)x}(d_{l+1}-d_l)} \end{bmatrix}. \tag{4.57}$$

By the same token, we can express A_{l+1} and B_{l+1} in terms of A_l and B_l. Solving A_{l+1} and B_{l+1} from (4.47), we write

$$\begin{bmatrix} A_{l+1} e^{-ik_{(l+1)x}d_{l+1}} \\ B_{l+1} e^{ik_{(l+1)x}d_{l+1}} \end{bmatrix} = \overline{\mathbf{V}}_{(l+1)l}\cdot \begin{bmatrix} A_l e^{-ik_{lx}d_l} \\ B_l e^{ik_{lx}d_l} \end{bmatrix}. \tag{4.58}$$

Here $\overline{\mathbf{V}}_{(l+1)l}$ is called the *forward propagation matrix*:

$$\overline{\mathbf{V}}_{(l+1)l}=\tfrac{1}{2}\left(1+\frac{\mu_{(l+1)z}k_{lx}}{\mu_{lz}k_{(l+1)x}}\right)$$

$$\cdot\begin{bmatrix} e^{-ik_{(l+1)x}(d_{l+1}-d_l)} & R_{(l+1)l}e^{-ik_{(l+1)x}(d_{l+1}-d_l)} \\ R_{(l+1)l}e^{ik_{(l+1)x}(d_{l+1}-d_l)} & e^{ik_{(l+1)x}(d_{l+1}-d_l)} \end{bmatrix}, \tag{4.59}$$

where $R_{(l+1)l}=-R_{l(l+1)}$ is the reflection coefficient for TE waves in region $l+1$ caused by the boundary separating regions $l+1$ and l. It can be shown that the product of the forward and the backward propagation matrices at a boundary surface yields the identity matrix:

$$\overline{\mathbf{U}}_{l(l+1)}\cdot\overline{\mathbf{V}}_{(l+1)l}=\overline{\mathbf{V}}_{(l+1)l}\cdot\overline{\mathbf{U}}_{l(l+1)}=\overline{\mathbf{I}},$$

where $\overline{\mathbf{I}}$ is the 2×2 identity matrix. This is expected because propagating the wave amplitudes across the boundary and then back again gives the original wave amplitudes. Note again that these results are derived for TE waves, and that we have omitted the superscript TE on the propagation matrices and reflection coefficients for simplification.

The propagation matrices can be used to determine wave amplitudes in any region in terms of those in any other region. Let $l<m$; then

$$\begin{bmatrix} A_l e^{-ik_{lx}d_l} \\ B_l e^{ik_{lx}d_l} \end{bmatrix}=\overline{\mathbf{U}}_{lm}\cdot\begin{bmatrix} A_m e^{-ik_{mx}d_m} \\ B_m e^{ik_{mx}d_m} \end{bmatrix}, \tag{4.60}$$

where $\overline{\mathbf{U}}_{lm}$ is a multiplication of $m-l$ backward propagation matrices:

$$\overline{\mathbf{U}}_{lm}=\overline{\mathbf{U}}_{l(l+1)}\cdot\overline{\mathbf{U}}_{(l+1)(l+2)}\cdots\overline{\mathbf{U}}_{(m-1)m}. \tag{4.61}$$

Similarly, the forward propagation matrices can be used to express wave amplitudes in any region l in terms of those in any region j for $l>j$.

As an example, we calculate the transmission coefficients by using the backward propagation matrices. In using the propagation matrices, we note that d_t is not a physical parameter and we let $d_t=0$. For a one-layer medium,

$$\begin{bmatrix} R \\ 1 \end{bmatrix}=\tfrac{1}{2}\left(1+\frac{\mu_z k_{tx}}{\mu_{tz}k_x}\right)\begin{bmatrix} 1 & R_{0t} \\ R_{0t} & 1 \end{bmatrix}\cdot\begin{bmatrix} 0 \\ T \end{bmatrix}.$$

We have

$$T=\frac{2}{1+(\mu_z k_{tx}/\mu_{tz}k_x)}, \tag{4.62}$$

which reduces to (4.16) for isotropic media.

For a two-layer medium the transmission coefficient is obtained from

$$\begin{bmatrix} R \\ 1 \end{bmatrix} = \tfrac{1}{4}\left(1+\frac{\mu_z k_{1x}}{\mu_{1z}k_x}\right)\left(1+\frac{\mu_{1z}k_{tx}}{\mu_{tz}k_{1x}}\right)\begin{bmatrix} e^{ik_{1x}d_1} & R_{01}e^{-ik_{1x}d_1} \\ R_{01}e^{ik_{1x}d_1} & e^{-ik_{1x}d_1} \end{bmatrix}$$

$$\cdot\begin{bmatrix} e^{-ik_{tx}d_1} & R_{1t}e^{ik_{tx}d_1} \\ R_{1t}e^{-ik_{tx}d_1} & e^{ik_{tx}d_1} \end{bmatrix}\cdot\begin{bmatrix} 0 \\ T \end{bmatrix}.$$

Thus

$$T=\frac{4e^{i(k_{1x}-k_{tx})d_1}}{[1+(\mu_z k_{1x}/\mu_{1z}k_x)][1+(\mu_{1z}k_{tx}/\mu_{tz}k_{1x})](1+R_{01}R_{1t}e^{i2k_{1x}d_1})}. \tag{4.63}$$

The reflectivity and the transmissivity for the two-layer medium, in view of (4.18), are as follows:

$$r=|R|^2, \tag{4.64a}$$

$$t=\frac{\mu_z k_{tx}}{\mu_{tz}k_x}|T|^2. \tag{4.64b}$$

It is straightforward to prove that $r+t=1$, with R and T given by (4.55) and (4.63).

As another example, we determine the transmission and the reflection coefficients for the periodic medium made of alternating high and low permittivities, which was treated in Section 4.3a (Fig. 4.12). The reflection coefficient R has been calculated with the wave impedance approach [shown in (4.49)]. Using the forward propagation matrices, we have

$$\begin{bmatrix} 0 \\ T \end{bmatrix}=\overline{\mathbf{V}}_{th}\cdot(\overline{\mathbf{V}}_{hl}\cdot\overline{\mathbf{V}}_{lh})^N\cdot\overline{\mathbf{V}}_{h0}\cdot\begin{bmatrix} R \\ 1 \end{bmatrix}.$$

Since, in region $m+1$, $k_{(m+1)x}=k_{m+1}$ and $(d_{m+1}-d_m)$ is a quarter-wavelength thick, we have $k_{(m+1)x}(d_{m+1}-d_m)=\pi/2$. Note also the fact that

$\mu_{(m+1)z}k_{mx}/\mu_{mz}k_{(m+1)x}=(\epsilon_m/\epsilon_{m+1})^{1/2}$. The forward propagation matrices, according to (4.59), become

$$\overline{\mathbf{V}}_{h0}=-\frac{i}{2}\begin{bmatrix} 1+(\epsilon/\epsilon_h)^{1/2} & 1-(\epsilon/\epsilon_h)^{1/2} \\ -\left[1-(\epsilon/\epsilon_h)^{1/2}\right] & -\left[1+(\epsilon/\epsilon_h)^{1/2}\right] \end{bmatrix},$$

$$\overline{\mathbf{V}}_{hl}\cdot\overline{\mathbf{V}}_{lh}=-\frac{1}{2}\begin{bmatrix} (\epsilon_l/\epsilon_h)^{1/2}+(\epsilon_h/\epsilon_l)^{1/2} & (\epsilon_l/\epsilon_h)^{1/2}-(\epsilon_h/\epsilon_l)^{1/2} \\ (\epsilon_l/\epsilon_h)^{1/2}-(\epsilon_h/\epsilon_l)^{1/2} & (\epsilon_l/\epsilon_h)^{1/2}+(\epsilon_h/\epsilon_l)^{1/2} \end{bmatrix},$$

$$\overline{\mathbf{V}}_{th}=\frac{1}{2}\begin{bmatrix} \left[1+(\epsilon_h/\epsilon_t)^{1/2}\right]e^{ik_t d} & \left[1-(\epsilon_h/\epsilon_t)^{1/2}\right]e^{ik_t d} \\ \left[1-(\epsilon_h/\epsilon_t)^{1/2}\right]e^{-ik_t d} & \left[1+(\epsilon_h/\epsilon_t)^{1/2}\right]e^{-ik_t d} \end{bmatrix},$$

with d representing the total thickness of the stratified medium. The N product, $(\overline{\mathbf{V}}_{hl}\cdot\overline{\mathbf{V}}_{lh})^N$, can be calculated by noting that

$$\begin{bmatrix} a+b & a-b \\ a-b & a+b \end{bmatrix}^N = 2^{N-1}\begin{bmatrix} a^N+b^N & a^N-b^N \\ a^N-b^N & a^N+b^N \end{bmatrix}.$$

Thus we have

$$\begin{bmatrix} 0 \\ Te^{ik_t d} \end{bmatrix} = \frac{i(-1)^{N+1}}{2}\begin{bmatrix} \left(\frac{\epsilon}{\epsilon_h}\right)^{1/2}\left(\frac{\epsilon_l}{\epsilon_h}\right)^{N/2}+\left(\frac{\epsilon_h}{\epsilon_t}\right)^{1/2}\left(\frac{\epsilon_h}{\epsilon_l}\right)^{N/2} & -\left(\frac{\epsilon}{\epsilon_h}\right)^{1/2}\left(\frac{\epsilon_l}{\epsilon_h}\right)^{N/2}+\left(\frac{\epsilon_h}{\epsilon_t}\right)^{1/2}\left(\frac{\epsilon_h}{\epsilon_l}\right)^{N/2} \\ \left(\frac{\epsilon}{\epsilon_h}\right)^{1/2}\left(\frac{\epsilon_l}{\epsilon_h}\right)^{N/2}-\left(\frac{\epsilon_h}{\epsilon_t}\right)^{1/2}\left(\frac{\epsilon_h}{\epsilon_l}\right)^{N/2} & -\left(\frac{\epsilon}{\epsilon_h}\right)^{1/2}\left(\frac{\epsilon_l}{\epsilon_h}\right)^{N/2}-\left(\frac{\epsilon_h}{\epsilon_t}\right)^{1/2}\left(\frac{\epsilon_h}{\epsilon_l}\right)^{N/2} \end{bmatrix}\cdot\begin{bmatrix} R \\ 1 \end{bmatrix}$$

The first row of the matrix equation yields the solution for R, which is identical to (4.49). The second row yields the solution for T, which is

$$T=\frac{i(-1)^N 2(\epsilon/\epsilon_t)^{1/2}e^{-ik_t d}}{(\epsilon/\epsilon_h)^{1/2}(\epsilon_l/\epsilon_h)^{N/2}+(\epsilon_h/\epsilon_t)^{1/2}(\epsilon_h/\epsilon_l)^{N/2}}. \tag{4.65}$$

In view of (4.18), the reflectivity is $r=|R|^2$, and the transmissivity is $t=(\epsilon_t/\epsilon)^{1/2}|T|^2$. Again it can be shown that $r+t=1$. Note that, although both TE and TM waves become TEM at normal incidence, we should not use (4.23) to calculate t because here R and T are amplitude reflection and transmission coefficients for electric field vectors. If we calculate r and t with R^{TM} and T^{TM}, we use (4.23).

We summarize the procedure for calculating reflection and transmission coefficients for an n-layer stratified medium by using the propagation matrices. With the forward propagation matrices, the field amplitudes in the incident and the transmitted regions are related by

$$\begin{bmatrix} 0 \\ T \end{bmatrix} = \overline{\mathbf{V}}_{t0} \cdot \begin{bmatrix} R \\ 1 \end{bmatrix} = \begin{bmatrix} V_{11} & V_{12} \\ V_{21} & V_{22} \end{bmatrix} \cdot \begin{bmatrix} R \\ 1 \end{bmatrix}, \tag{4.66a}$$

$$\overline{\mathbf{V}}_{t0} = \overline{\mathbf{V}}_{tn} \cdot \overline{\mathbf{V}}_{n(n-1)} \cdot \cdots \cdot \overline{\mathbf{V}}_{10}, \tag{4.66b}$$

where $\overline{\mathbf{V}}_{t0}$ embeds all information about the stratified medium. From (4.66), it can be determined that

$$R = -\frac{V_{12}}{V_{11}}, \tag{4.67a}$$

$$T = \frac{V_{22}V_{11} - V_{12}V_{21}}{V_{11}}. \tag{4.67b}$$

With a similar formula we use backward propagation matrix U_{0t} to find R and T. We have

$$\begin{bmatrix} R \\ 1 \end{bmatrix} = \overline{\mathbf{U}}_{0t} \cdot \begin{bmatrix} 0 \\ T \end{bmatrix} = \begin{bmatrix} U_{11} & U_{12} \\ U_{21} & U_{22} \end{bmatrix} \cdot \begin{bmatrix} 0 \\ T \end{bmatrix}, \tag{4.68a}$$

$$\overline{\mathbf{U}}_{0t} = \overline{\mathbf{U}}_{0t} \cdot \overline{\mathbf{U}}_{12} \cdots \overline{\mathbf{U}}_{nt}, \tag{4.68b}$$

$$R = \frac{U_{12}}{U_{22}}, \tag{4.69a}$$

$$T = \frac{1}{U_{22}}. \tag{4.69b}$$

When either R or T is determined, wave amplitudes in any other layers can be calculated by successive application of the forward and backward propagation matrices.

4.3d Waves in Stratified Bianisotropic Media

Consider a stratified bianisotropic medium composed of n layers of moving uniaxial media with optic axes along the $\hat{z}$ direction. Assume that the plane of incidence is parallel to the direction of motion. We make use of constitutive relations 2.37 and Maxwell's source-free equations. Following a procedure similar to that leading to (4.43) and (4.44), we find the governing equations for the TE and TM waves in a moving uniaxial medium:

$$H_x = -\frac{1}{i\omega\mu}\left(\frac{\partial}{\partial z} - i\omega\xi\right)E_y, \tag{4.70a}$$

$$H_z = \frac{1}{i\omega\mu_z}\frac{\partial}{\partial x}E_y, \tag{4.70b}$$

$$\left[\frac{\mu}{\mu_z}\frac{\partial^2}{\partial x^2} + \left(\frac{\partial}{\partial z} - i\omega\xi\right)^2 + \omega^2\mu\epsilon\right]E_y = 0, \tag{4.70c}$$

for TE waves, and

$$E_x = \frac{1}{i\omega\epsilon}\left(\frac{\partial}{\partial z} - i\omega\xi\right)H_y, \tag{4.71a}$$

$$E_z = -\frac{1}{i\omega\epsilon_z}\frac{\partial}{\partial x}H_y, \tag{4.71b}$$

$$\left[\frac{\epsilon}{\epsilon_z}\frac{\partial^2}{\partial x^2} + \left(\frac{\partial}{\partial z} - i\omega\xi\right)^2 + \omega^2\mu\epsilon\right]H_y = 0 \tag{4.71c}$$

for TM waves. The constitutive parameters ϵ, ϵ_z, μ, μ_z, and ξ are given in Table 2.1. Equations 4.70 and 4.71 are dual of each other; both are similar to (4.43) and (4.44). When the plane of incidence is not parallel to the direction of medium motion, it is more convenient to use E_z and H_z instead of E_y and H_y to classify TE to $\hat{z}$ and TM to $\hat{z}$ waves.

In view of the governing equations given above, we observe that the problem can be solved by following exact procedures as in the last three sections. The reflection coefficients and propagation matrices can be similarly obtained. Instead of dispersion relation 4.46, we now have

$$k_x^2 + \frac{\mu_z (k_z - \omega\xi)^2}{\mu} = \omega^2 \mu_z \epsilon \tag{4.72a}$$

for TE waves, and

$$k_x^2 + \frac{\epsilon_z (k_z - \omega\xi)^2}{\epsilon} = \omega^2 \mu \epsilon_z \tag{4.72b}$$

for TM waves. Equation 4.72a is equivalent to (3.75a) or (3.77a) with $k_y = 0$; equation 4.72b, to (3.76a) or (3.77b) with $k_y = 0$. We observe that the TE and TM waves correspond to the Type I and Type II waves discussed in Chapter 3.

Some conclusions can be drawn from a dispersion analysis. For generality, we assume that the incident region 0 is also a moving medium. Total reflection occurs when the transmitted wave is evanescent. Critical angles can be found by requiring that $k_{tx}^2 \leqslant 0$. We use either the dispersion relation in (3.75a) or that in (3.76a). The condition for total reflection gives

$$(\omega - \chi_t k_z)^2 - \kappa_t \nu_t k_z^2 \leqslant 0. \tag{4.73}$$

To find the corresponding angles of incidence, we note that $k_z = k \sin\theta$, and the dispersion relation in the incident region is

$$(\omega - \chi k_z)^2 - \kappa \nu k_z^2 = \kappa \nu_z k_x^2 \tag{4.74a}$$

for Type I waves or

$$(\omega - \chi k_z)^2 - \kappa \nu k_z^2 = \kappa_z \nu k_x^2 \tag{4.74b}$$

for Type II waves, where $k_x = k \cos\theta$. Phase matching requires that all k_z's be equal. We solve ω from (4.74), substitute in (4.73), and compute θ.

The calculation becomes particularly simple when the moving media are isotropic and the medium velocities are nonrelativistic. Using Table 2.1 and keeping only terms of first order in β, we can approximate $\kappa\nu \approx c^2/n^2$, $\kappa_t \nu_t \approx c^2/n_t^2$, $\chi \approx c\beta(1 - 1/n^2)$, and $\chi_t \approx c\beta_t(1 - 1/n_t^2)$. For moving isotropic media (4.74a) and (4.74b) are the same; we find, for nonrelativistic velocities, $\omega \approx \chi k_z + ck/n$. Substituting in (4.73) yields

$$\sin\theta \geqslant \frac{n_t/n}{1+n_t\beta_t(1-1/n_t^2)-n_t\beta(1-1/n^2)}$$

$$\approx \frac{n_t}{n}\left[1-n_t\beta_t\left(1-\frac{1}{n_t^2}\right)+n_t\beta\left(1-\frac{1}{n^2}\right)\right]. \tag{4.75}$$

The last two terms are due to the motion of the media in regions 0 and t. The faster the medium in region t moves, the smaller the critical angle. Motion of the medium in region 0 induces opposite effects. Note that in any region l, when k_{lx} is imaginary, the waves inside the region experience attenuation in the $\hat{x}$ direction. Thus even the transmitted wave is not evanescent; its amplitude may still be small because of attenuations in the stratified regions.

The reflection coefficients due to the stratified moving medium take the same form as (4.53). For two media moving relative to each other, the Fresnel reflection coefficients are as follows:

$$R_{0t}^{\mathrm{TE}} = \frac{1-\mu_z k_{tx}/\mu_{tz}k_x}{1+\mu_z k_{tx}/\mu_{tz}k_x}, \tag{4.76}$$

$$R_{0t}^{\mathrm{TM}} = \frac{1-\epsilon_z k_{tx}/\epsilon_{tz}k_x}{1+\epsilon_z k_{tx}/\epsilon_{tz}k_x}, \tag{4.77}$$

with the constitutive parameters given by Table 2.1 and the dispersion relations by (4.72).

It is of interest to calculate Brewster angles by setting the reflection coefficients equal to zero. For two isotropic media moving with nonrelativistic velocities, $\omega \approx \chi k_z + ck/n$. Using dispersion relation 4.74 for the isotropic case, we find the Brewster angle θ_b for TE waves:

$$\sin\theta_b^{\mathrm{TE}} = \left[\frac{(n_t/n)^2-(\mu_t'/\mu')^2}{1-(\mu_t'/\mu')^2}\right]^{1/2}$$

$$+\frac{n_t^2/n}{1-(\mu_t'/\mu')^2}\left[\beta\left(1-\frac{1}{n^2}\right)-\beta_t\left(1-\frac{1}{n_t^2}\right)\right]. \tag{4.78}$$

By duality, we find for TM waves

$$\sin\theta_b^{\mathrm{TM}} = \left[\frac{(n_t/n)^2 - (\epsilon_t'/\epsilon')^2}{1-(\epsilon_t'/\epsilon')^2} \right.$$

$$\left. + \frac{n_t^2/n}{1-(\epsilon_t'/\epsilon')^2}\left[\beta\left(1-\frac{1}{n^2}\right) - \beta_t\left(1-\frac{1}{n_t^2}\right)\right]\right]^{1/2}. \tag{4.79}$$

The first terms in (4.78) and (4.79) are the results for two stationary media, with the effects of permeability also taken into account. The first term in (4.79) reduces to the familiar results for TM waves (4.29) when we set $\mu_t = \mu$ and $\epsilon_t'/\epsilon' = n_t^2/n^2$. The velocity-dependent terms in (4.78) and (4.79) account for the effects arising from the motion.

PROBLEMS

4.1. Let a plane wave be incident on a plane boundary from the inside of a negative uniaxial crystal. Consider the special case in which the optic axis is perpendicular to the plane of incidence. Find the range of θ such that there is total internal reflection for the ordinary wave but not for the extraordinary wave.

4.2. The phase change produced on total internal reflection may be utilized to obtain circularly polarized light from linearly polarized light. The scheme, devised by Fresnel, is shown in Fig. P4.2. The essential element is a glass prism ($n = 1.6$), made in the form of a rhomb having an apex angle α. Linearly polarized light whose direction of polarization is at an angle of 45° with respect to the face edge of the rhomb enters normally on one face. What should α be so that the light coming out is circularly polarized?

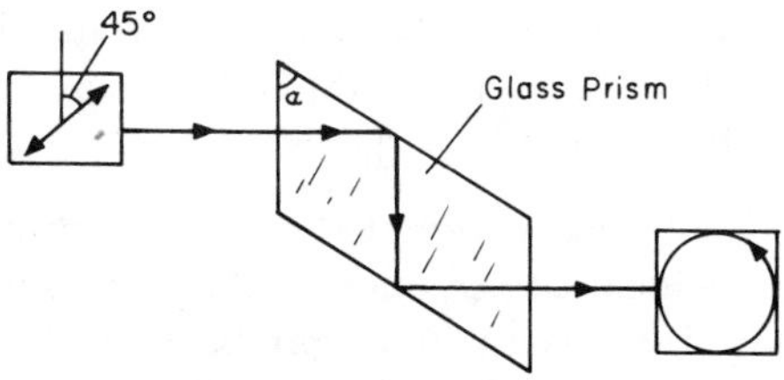

Figure P4.2 Fresnel's device for producing circularly polarized light.

4.3. Linearly polarized waves can be obtained from unpolarized waves by reflection at the Brewster angle. Choose an apex angle α for the rhomb in Problem 4.2 to produce waves with linear polarization.

4.4. A plane wave is incident from free space on a half-space conducting medium. Show that for large $\sigma/\omega\epsilon_0$ the transmitted angle is

$$\theta_t \approx \sin^{-1}\left[\left(\frac{2\omega\epsilon_0}{\sigma}\right)^{1/2}\sin\theta\right],$$

which is a very small angle, and the transmitted wave is almost perpendicular to the boundary.

4.5. For a cold plasma with collisions,

$$\epsilon(\omega) = \epsilon_0\left[1 - \frac{\omega_p^2}{\omega^2+\nu^2} + \frac{i\nu\omega_p^2}{\omega(\omega^2+\nu^2)}\right].$$

Consider a plane wave incident on this medium. Determine $\mathbf{k}_t$ in magnitude and direction as a function of ν.

4.6. A plane wave is incident on a stratified medium with increasing dielectric constant. Use phase matching to determine the direction of the $\mathbf{k}$ vector inside the medium.

4.7. The ionosphere can be modeled as a stratified medium with permittivity $\epsilon = \epsilon_0(1-\omega_p^2/\omega^2)$, where the plasma frequency ω_p is a function of height. Describe how a wave incident upon the ionosphere can be totally reflected.

4.8. Consider a solid-state Fabry-Perot etalon filter made of an eight-layer stratified medium. Regions 1, 3, 5 and 7 have refractive index 1.35 and are a quarter-wavelength thick. Regions 2, 4, and 6 have refractive index 2.3. Regions 2 and 6 are a quarter-wavelength thick, but region 4 is a half-wavelength thick. What are the reflectivity and transmissivity for a plane wave normally incident upon this stratified medium? Explain why the structure can be used for filtering purposes.

4.9. A plane wave is totally reflected when incident upon a glass-air boundary. At these incident angles, another glass is brought very close to the first one so that there is a very small air gap between the two. Calculate the reflection and transmission coefficients as a function of the gap dimension. Show that transmission is now possible.

4.10. Calculate the reflection and transmission coefficients for a three-layer stratified medium. Find transmissivity as a function of incident angle when layer 1 is identical to layer 3 but layer 2 has a smaller permittivity.

4.11. A gas laser is often a tube containing the gas fitted with Brewster angle windows and external mirrors. The output of the laser will be linearly polarized—for what reason and in which direction? Let a solid-state laser be fabricated of rods with ends beveled at the Brewster angle. Sketch at what locations and at what angles you would place the external mirrors. Calculate and indicate on your sketch all appropriate angles, including the bevel angle of the glass rod, which has dielectric constant $\epsilon = 2.5$.

4.12. In microwave remote sensing of the Earth from satellite or aircraft, a radiometer is used to measure the emissivity of the area under observation. The emissivity e is related to reflectivity by $e = 1 - r$. Theoretically, we should be able to detect, for instance, the ice thickness on a lake. Assume that the lake ice permittivity is $\epsilon = 3.2(1 + i0.01)\epsilon_0$ and that the water is a perfect reflector. Calculate the depth in terms of free space wavelength that the radiometer can see through the ice.

4.13. When the incident **k** vector is normal to a plane boundary, a TE wave becomes a TEM wave; a TM wave also becomes a TEM wave. Compare the reflection and transmission coefficients for TE and TM waves at normal incidence. Do both TE and TM results reduce to the same answer? If not, why? Do the reflectivities and transmissivities for TE and TM waves at normal incidence reduce to the same result?

4.14. A plane wave is incident upon a half-space medium with $\epsilon = 10\epsilon_0$, $\mu = \mu_0$, and $\sigma = 10^{-3}$mho/m. The transmitted wave is picked up by a receiver located at a distance $d = 100$ m from the interface. At the frequency of 1 MHz, calculate the ratio of the received power density to the incident power density.

4.15. A plane wave linearly polarized along the $\hat{x}$ direction is propagating along the $\hat{z}$ direction and is normally incident upon an anisotropic plasma slab. Find the reflection and transmission coefficients when the dc magnetic field is along the (*a*) $\hat{z}$ direction, (*b*) $\hat{y}$ direction, and (*c*) $\hat{x}$ direction.

4.16. An extraordinary wave is incident from inside a positive uniaxial medium upon the interface separating free space. The optic axis of the uniaxial medium is inclined 45° with respect to the interface, and the incident angle of the wave is perpendicular to the optic axis. Under what conditions will the transmitted wave be evanescent? When will the reflected wave be a backward wave? Find the reflection and transmission coefficients and the reflectivity and transmissivity.

4.17. Determine the propagation matrices for TM waves in a stratified medium.

4.18. A shock-wave front travels in a gaseous material of permittivity ϵ and the medium is ionized. Behind the shock, the permittivity is identical to that of an isotropic collisionless plasma with plasma frequency ω_p. A plane wave with $\mathbf{E} = \hat{x}E_0 e^{-ikz}$ is propagating in a direction opposite to the shock-wave front velocity $\hat{z}v$. Calculate the reflection and transmission coefficients.

5

Guidance and Resonance

In the study of guided waves, natural modes associated with a particular guidance structure will be our primary interest in this chapter. We restrict ourselves to cylindrical structures with uniform cross sections in the guiding direction, which is taken to be along the z axis. Formulations are generalized to treat guided waves in moving gyrotropic media. The modes are classified with respect to the $\hat{z}$ direction as TE, TM, and hybrid modes. We distinguish three concepts in dealing with guided waves: (*a*) dispersion relations that follow directly from the wave equations, (*b*) guidance conditions imposed by the boundary conditions, and (*c*) cutoff criteria stating conditions under which a guided wave will propagate or attenuate. When the modes have been determined for a guiding structure, they can be used to express source excitations and small perturbations by making use of their orthogonality relations. The derivation of natural modes in cavity resonators follows the same reasoning process. A cavity with a uniform cross section in any direction can be viewed as a waveguide with two ends closed. Perturbation of resonant frequencies caused by a small variation in a cavity is also examined.

5.1 PLANAR WAVEGUIDES

5.1a Parallel-Plate Waveguides

We use a parallel-plate waveguide as the simplest example to illustrate the general characteristics of guided waves, such as propagation, evanescence, cutoff, attenuation, mode expansion, and source excitation. Consider two parallel metallic plates situated at $x=0$ and $x=d$. The width of the waveguide dimension along y is w (Fig. 5.1). The medium between the plates is homogeneous and isotropic. Assume that $w \gg d$ so that fringing fields can be neglected, and we have $\partial/\partial y = 0$. From Maxwell's equations, the x and z components of the field vectors can be expressed in terms of E_y and H_y; both satisfy the wave equation:

$$H_x = -\frac{1}{i\omega\mu}\frac{\partial E_y}{\partial z}, \tag{5.1a}$$

$$H_z = \frac{1}{i\omega\mu}\frac{\partial E_y}{\partial x}, \tag{5.1b}$$

$$\left(\frac{\partial^2}{\partial x^2} + \frac{\partial^2}{\partial z^2} + k^2\right)E_y = 0; \tag{5.1c}$$

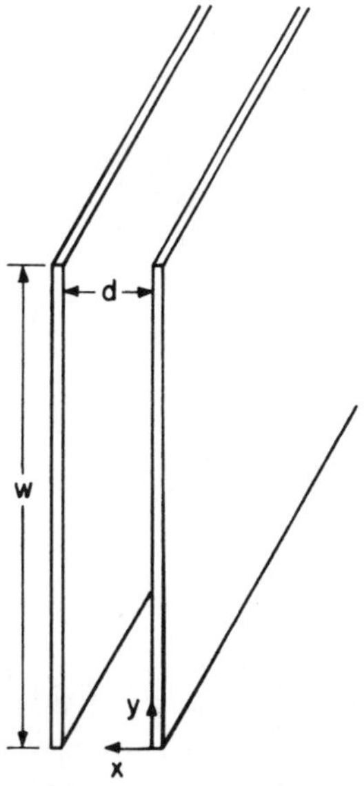

Figure 5.1 Parallel-plate waveguide. The medium between the plates is homogeneous and isotropic with permittivity ϵ and permeability μ.

$$E_x = \frac{1}{i\omega\epsilon}\frac{\partial H_y}{\partial z}, \tag{5.2a}$$

$$E_z = -\frac{1}{i\omega\epsilon}\frac{\partial H_y}{\partial x}, \tag{5.2b}$$

$$\left(\frac{\partial^2}{\partial x^2}+\frac{\partial^2}{\partial z^2}+k^2\right)H_y = 0. \tag{5.2c}$$

Equations 5.1 and 5.2 are equivalent to all of Maxwell's equations for the parallel-plate problem.

Waves are guided along the $\hat{z}$ direction. For time-harmonic excitations, we write the z and t dependence of all field vectors as $e^{ik_z z - i\omega t}$. The z component of the $\mathbf{k}$ vector, k_z, is called the *propagation constant*. With this dependence, we can replace $\partial/\partial z$ in Maxwell's equation with ik_z. The guided waves can be classified as follows with respect to the $\hat{z}$ direction:

1. TEM waves or TEM modes; both E_z and H_z are zero. TEM waves do not possess x or y dependence.

2. TE waves or the TE modes; $E_z=0$. TE modes are derived from (5.1).
3. TM waves or TM modes; $H_z=0$. TM modes are derived from (5.2).

First, we consider TEM waves. In view of the boundary condition that tangential electric fields must vanish at the plate boundaries, the solution is seen to be

$$\mathbf{H}=\hat{y}H_0 e^{ikz}, \tag{5.3a}$$

where H_0 is a constant. By (5.2a),

$$\mathbf{E}=\hat{x}\sqrt{\frac{\mu}{\epsilon}}\; H_0 e^{ikz}. \tag{5.3b}$$

The electric field is perpendicular to the plates, and the magnetic field is parallel to the plates. The propagation constant is $k=\omega(\mu\epsilon)^{1/2}$. The wave will attenuate in the propagation direction if the medium filling the waveguide is lossy; this gives a complex k. Even when the medium is not lossy, the finite conductivity of the plates, although very large, causes the wave to attenuate as it propagates. Attenuation arising from wall loss can be treated with a perturbation approach. By the boundary condition, $\mathbf{J}_s=\hat{n}\times\mathbf{H}$. We find a surface current flowing in the $\hat{z}$ direction. Because the metal is not perfectly conducting, a small tangential electric field is needed to support this surface current sheet. Thus, instead of being perpendicular to the plate boundaries, the electric field is slightly tilted (Fig. 5.2). There is then Poynting's power directed into the plates in addition to Poynting's power propagating in the $\hat{z}$ direction.

In the perturbation approach, we assume that the tangential magnetic fields at the walls $\mathbf{H}_w$ are the same as in the unperturbed case with perfect conductors and that the plate thickness is much greater than the skin depth. For large conductivity σ, the tangential electric field required to support the current in the plate at $x=0$ is related to $\mathbf{H}_w$ by

$$\begin{aligned}\mathbf{E}_w &= \sqrt{\frac{\mu}{\epsilon_w}}\;\hat{x}\times\mathbf{H}_w\\ &\approx\sqrt{\frac{\omega\mu}{i\sigma}}\;\hat{x}\times\mathbf{H}_w.\end{aligned} \tag{5.4}$$

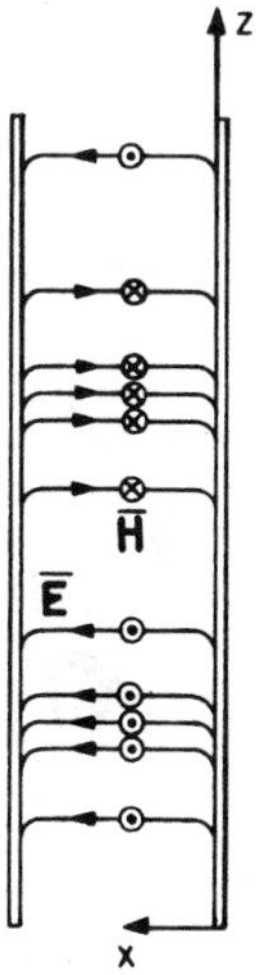

Figure 5.2 Parallel plates with finite conductivity, which requires a small tangential electric field to support the surface currents.

The time-average power density dissipated into the wall at $x=0$ is

$$\langle S_d \rangle = -\hat{x}\cdot\frac{1}{2}\,\mathrm{Re}\,\mathbf{E}_w \times \mathbf{H}_w^*$$

$$= \frac{1}{2}\sqrt{\frac{\omega\mu}{2\sigma}}\;|\mathbf{H}_w|^2, \tag{5.5}$$

which is also equal to that dissipated into the upper wall at $x=d$. Equation 5.5 is a perturbation formula applicable to general cases with $\mathbf{H}_w$ interpreted as the tangential magnetic field at the conducting surface.

The power dissipation per unit length along the guiding direction for both plates is

$$P_d = \int_0^w 2\langle S_d \rangle\, dy = w\sqrt{\frac{\omega\mu}{2\sigma}}\;|\mathbf{H}_w|^2 \quad \text{Watts/m}. \tag{5.6}$$

This power dissipation causes the wave to attenuate in the $\hat{z}$ direction. Let the attenuation rate be α, so that the z dependence of the fields in (5.3) becomes $e^{ikz-\alpha z}$. The time-average power density propagating in the $\hat{z}$ direction is

$$\langle \mathbf{S}_f \rangle = \hat{z}\,\frac{1}{2}\sqrt{\frac{\mu}{\epsilon}}\;|H_0|^2 e^{-2\alpha z}. \tag{5.7}$$

The total power flowing in the $\hat{z}$ direction is

$$P_f = \int_0^w dy \int_o^d dx\, \hat{z}\cdot\langle \mathbf{S}_f \rangle$$

$$= \frac{wd}{2}\sqrt{\frac{\mu}{\epsilon}}\;|H_0|^2 e^{-2\alpha z} \quad \text{Watts.} \tag{5.8}$$

The decrease of P_f with z is attributed to the power dissipation, P_d. Thus

$$P_d = -\frac{dP_f}{dz} = 2\alpha P_f$$

and

$$\alpha = \frac{P_d}{2P_f} \quad m^{-1}. \tag{5.9}$$

The perturbation formula for the attenuation rate α as given by (5.9) is also applicable to general cases.

For parallel-plate waveguides, insertion of (5.6) and (5.8) into (5.9) gives

$$\alpha = \frac{1}{d}\sqrt{\frac{\omega\epsilon}{2\sigma}} . \tag{5.10}$$

As expected, the larger σ, the smaller is α. The attenuation is also smaller when the plate separation is large.

Next, we consider TE modes. The governing equations are (5.1), and the boundary condition requires that the E_y be zero on both boundary surfaces. The solution is

$$E_y = E_0 \sin k_x x e^{ik_z z} \tag{5.11a}$$

and

$$k_x = \frac{m\pi}{d} ; \tag{5.11b}$$

where m is any integer. Equation 5.11b is the *guidance condition*. Substituting (5.11a) into (5.1c), we find that the propagation constant k_z satisfies the *dispersion relation*

$$k_z = \left[k^2 - \left(\frac{m\pi}{d}\right)^2\right]^{1/2} . \tag{5.12}$$

We see that k_z will be imaginary if $k < m\pi/d$. The wave number k_c at which $k_z = 0$ is called the *cutoff wave number*. Thus $k_c d = m\pi$, which corresponds to a cutoff frequency $\omega_c = m\pi/d(\mu\epsilon)^{1/2}$, and a cutoff wavelength $\lambda_c = 2d/m$. According to the cutoff wave number, TE waves are classified as TE_m modes. When k_z is imaginary, the wave attenuates exponentially along the $\hat{z}$ direction. For a TE_m mode to propagate in the waveguide, the excitation wave number must be larger than $m\pi/d$. Note that, if k_z is real for a TE_m mode, then, for all TE_l modes with $l < m$, it is also real. Thus, for a given wave number k such that $m\pi < kd < (m+1)\pi$, there is a total of m admissible TE modes in the waveguide. There is no TE_0 mode because, for $m = 0$, $E_y = 0$. The lowest-order mode is TE_1, which can be excited when $\pi < kd = 2\pi d/\lambda$, namely, with the excitation wavelength smaller than $2d$. For kd between π and 2π only a single TE mode can propagate.

Field solution 5.11a admits a simple physical interpretation. Writing in terms of exponential functions instead of sine, we have

$$E_y = \frac{E_0}{2i}\left[e^{i(m\pi/d)x + ik_z z} - e^{-i(m\pi/d)x + ik_z z}\right],$$

which represents a superposition of two plane waves. We may view the guided wave as resulting from a plane wave bouncing back and forth between the two plates and propagating in the $\hat{z}$ direction with the propagation constant k_z (Fig. 5.3). The guidance condition states that in the direction perpendicular to the plates the bouncing waves interfere constructively, $2k_x d = 2m\pi$.

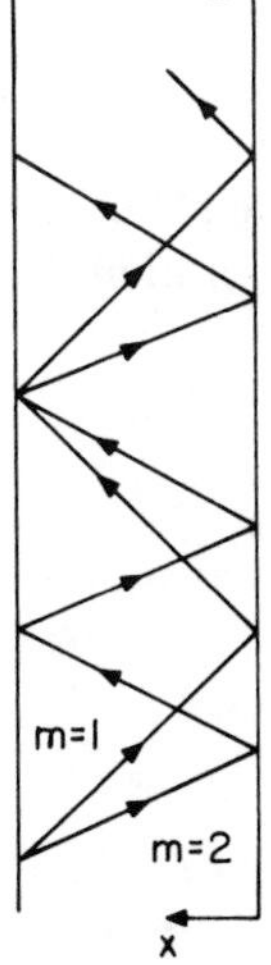

Figure 5.3 Interpretation of a mode as bouncing waves.

Finally, we consider the TM modes. The governing equations are (5.2). To satisfy the boundary conditions, the solution is found to be

$$H_y = H_0 \cos k_x x e^{ik_z z} \tag{5.13a}$$

and the guidance condition is

$$k_x = \frac{m\pi}{d}. \tag{5.13b}$$

The other field components are determined by (5.2a) and (5.2b). Solution 5.13a is not the dual of (5.11a) for E_y because the boundary conditions for H_y and E_y are different. The dispersion relation is $k_z^2 = k^2 - (m\pi/d)^2$, and the cutoff wave numbers for the TM_m modes are $k_c = m\pi/d$. Both the dispersion relation and the cutoff wave numbers are identical to those for the TE_m modes. The TM_m and the TE_m modes are said to be degenerate. In particular, note that now we can have $m=0$. The TM_0 mode, in fact, is the TEM wave discussed in the text after (5.3). We see that the TEM mode has no cutoff wave number and it can propagate at all frequencies.

Guided waves are excited by external sources. A source can be expressed in terms of an expansion of the waveguide modes. Assume that the source is a current sheet located at $z=0$. This current sheet generates propagating, as well as evanescent, guided modes in both positive $\hat{z}$ and negative $\hat{z}$ directions. The boundary conditions at $z=0$ require that (a) E_x and E_y be continuous, (b) discontinuity in H_x be equal to a current sheet flowing in the $\hat{y}$ direction, and (c) discontinuity in H_y be equal to a current sheet flowing in the $\hat{x}$ direction. Let the current sheet be a line source located at $x=a$ and flowing in the $\hat{y}$ direction. We write

$$\mathbf{J}_s = \hat{y} I_0 \delta(x-a).$$

According to the boundary conditions, only TE waves are excited. Let the solution for $z \geqslant 0$ be denoted by the superscript plus and the solution for $z \leqslant 0$ by the superscript minus. We write

$$E_y^+ = \sum_{m=1}^{\infty} E_m \sin \frac{m\pi x}{d} e^{ik_z z}, \qquad z \geqslant 0,$$

$$E_y^- = \sum_{m=1}^{\infty} E_m \sin \frac{m\pi x}{d} e^{-ik_z z}, \qquad z \leqslant 0.$$

The amplitudes E_m are determined by the nature of the source. Because of symmetry and the boundary condition of continuity of E_y at $z=0$, the amplitudes E_m in regions $z<0$ and $z>0$ are equal. The magnetic field components required in matching the boundary condition at $z=0$ are obtained from (5.1a):

$$H_x^+ = \sum_{m=1}^{\infty} -\frac{k_z}{\omega\mu} E_m \sin\frac{m\pi x}{d} e^{ik_z z}, \qquad z \geqslant 0,$$

$$H_x^- = \sum_{m=1}^{\infty} \frac{k_z}{\omega\mu} E_m \sin\frac{m\pi x}{d} e^{-ik_z z}, \qquad z \leqslant 0,$$

and at $z=0$ we have

$$\begin{aligned} I_0 \delta(x-a) &= (H_x^+ - H_x^-)_{z=0} \\ &= \sum_{m=1}^{\infty} -\frac{2k_z}{\omega\mu} E_m \sin\frac{m\pi x}{d}. \end{aligned}$$

Using the orthogonality properties of the sinusoidal function, we multiply both sides by $\sin(m\pi x/d)$ and integrate from 0 to d. The mode amplitude E_m is determined as

$$\begin{aligned} E_m &= -\frac{\omega\mu}{k_z d} I_0 \sin\frac{m\pi a}{d} \\ &= \frac{-\omega\mu I_0}{\sqrt{k^2 d^2 - m^2\pi^2}} \sin\frac{m\pi a}{d}. \end{aligned}$$

For the TE_1 mode, E_1 is maximum when $a=d/2$. This is because E_y is also maximum at $x=d/2$ and the coupling of source energy into the TE_1 mode is largest. The time-average Poynting's power propagating along the waveguide in the $\hat{z}$ direction at $z=0$ is given by

$$\begin{aligned} P &= \frac{1}{2}\,\mathrm{Re} \int_0^d dx \int_0^w dy \left(\sum_{m=1}^{\infty} E_m \sin\frac{m\pi x}{d} \right) \left(\sum_{m=1}^{\infty} \frac{k_z}{\omega\mu} E_m \sin\frac{m\pi x}{d} \right)^* \\ &= \frac{1}{2}\,\mathrm{Re}\left[\frac{wd}{2} \sum_{m=1}^{\infty} |E_m|^2 \left(\frac{k_z^*}{\omega\mu} \right) \right]. \end{aligned}$$

It is important to observe that the total power is a summation of the power associated with each individual mode. There is no coupling among the various modes. The guided modes are said to be orthogonal to one another. If the mth mode is a propagating mode, k_z is real and for that mode

$$P = \frac{wd}{4}\frac{1}{\omega\mu}\sqrt{k^2 - \left(\frac{m\pi}{d}\right)^2}\,|E_m|^2. \tag{5.14}$$

If the mth mode is evanescent, k_z is imaginary and no time average power is propagating down the waveguide.

Assume that the isotropic medium is also nondispersive; then we find that the phase delay for TE_m and TM_m modes along $\hat{z}$ is $\tau_p = k_z/\omega$, which is smaller than the phase delay of a plane wave in the unbounded medium. But the group delay, $\tau_g = \partial k_z/\partial\omega = \omega\mu\epsilon/k_z$, is larger than the group delay of a plane wave in the unbounded medium. Consequently, as compared with a plane wave in the unbounded medium, the phase velocity v_p becomes larger and the group velocity v_g becomes smaller. It can be shown that $\tau_p\tau_g = \mu\epsilon$ and $v_p v_g = 1/\mu\epsilon$.

Calculation of attenuation rates caused by wall loss for the TE and TM modes is similar to that for the TEM mode. The results are

$$\alpha_{\mathrm{TE}} = \frac{1}{d}\left(\frac{\omega\epsilon}{2\sigma}\right)^{1/2}\frac{2k_c^2/k^2}{\left(1 - k_c^2/k^2\right)^{1/2}} \tag{5.15a}$$

for the TE modes, and

$$\alpha_{\mathrm{TM}} = \frac{1}{d}\left(\frac{\omega\epsilon}{2\sigma}\right)^{1/2}\frac{2}{\left(1 - k_c^2/k^2\right)^{1/2}} \tag{5.15b}$$

for the TM modes. In Fig. 5.4 the attenuation constants for TEM, TE, and TM modes are plotted. Note that for the TEM mode α increases as the square root of the wave number. For both TE and TM modes, α is very large near cutoff. For TE modes, α decreases monotonically with increasing frequency.

In conclusion, we should appreciate that three mechanisms cause a guided wave to attenuate along the propagation direction:

1. Evanescence of nonpropagating modes. For nonpropagating modes, the propagation constant k_z is imaginary, $k_z = i[(m\pi/d)^2 - k^2]^{1/2}$, and the wave decays exponentially along $\hat{z}$.

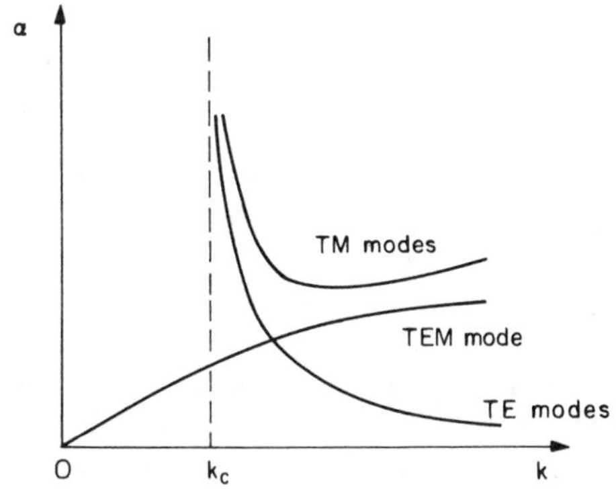

Figure 5.4 Attenuation constant α inside a parallel-plate waveguide.

2. Attenuation attributable to loss caused by material filling the waveguide. Note that in dispersion relation 5.12a $k=\omega(\mu\epsilon)^{1/2}$. Suppose that the material filling the waveguide is slightly lossy so that $\epsilon=\epsilon'+i\epsilon''$. By (5.12a), we find that the propagation constant k_z becomes complex. The imaginary part of k_z causes the wave to decay exponentially along $\hat{z}$.

3. Attenuation caused by wall loss. The results are obtained by a perturbation approach and expressed in (5.10) and (5.15).

5.1b Slab Waveguides

Consider a single-slab waveguide with boundaries at $x=0$ and $x=-d$. The permittivities and permeabilities for the three regions are μ, $\epsilon(0<x)$; μ_1, $\epsilon_1(-d<x<0)$; and μ_t, $\epsilon_t(x<-d)$. The reflection coefficient for the two-layer medium is determined in (4.55). For guided waves, there is no incident wave. Setting the denominator of the reflection coefficient equal to zero, we find the guidance condition to be

$$R_{01}R_{1t}e^{i2k_{1x}d}+1=0. \tag{5.16}$$

If regions 0 and t are perfect conductors, (5.16) gives $\sin k_{1x}d=0$, which is identical to the guidance condition for the parallel-plate waveguides. Remembering that $R_{01}=-R_{10}$, we write condition 5.16 in the following form:

$$R_{10}R_{1t}e^{i2k_{1x}d}=1. \tag{5.17}$$

Thus the magnitudes for both R_{10} and R_{1t} must be equal to unity, under the assumption that all media are lossless. This can happen only at total reflection from top and bottom boundaries when viewed from inside the slab.

The guidance condition may be stated as follows: the sum of the phase shifts at the top and the bottom boundaries plus the phase shift caused by traversing the waveguide in the $\hat{x}$ direction must be equal to an integer number of 2π:

$$2\phi_{10}+2\phi_{1t}+2k_{1x}d=2m\pi, \tag{5.18}$$

where the phase shifts ϕ_{10} and ϕ_{1t} are defined for TE waves in (4.19), and for TM waves in (4.24).

Compare this condition with that of a pair of conducting parallel plates. The guidance condition can be illustrated by a bouncing-wave approach similar to that illustrated in Fig. 5.3. The waves are totally reflected at both boundaries. The media in regions 0 and t must be less dense than the slab medium. Of all totally reflected waves, only those satisfying condition 5.18 are allowed. In accordance with the integer number m in (5.18), we may classify guided waves in TE_m and TM_m modes.

The TE solutions for the slab waveguide may be written as

$$E_y=E_0\cos\phi_{10}e^{-\alpha_x x+ik_z z}, \qquad 0\leqslant x, \tag{5.19a}$$

$$E_y=E_0\cos(k_{1x}x-\phi_{10})e^{ik_z z}, \qquad -d\leqslant x\leqslant 0, \tag{5.19b}$$

$$E_y=E_0\cos(\phi_{1t}-m\pi)e^{\alpha_{tx}(x+d)+ik_z z}, \qquad x\leqslant -d, \tag{5.19c}$$

where

$$\alpha_x=\sqrt{k_z^2-k^2}\ , \tag{5.20a}$$

$$\alpha_{tx}=\sqrt{k_z^2-k_t^2}\ , \tag{5.20b}$$

and k_{1x} is determined by guidance condition 5.18. According to (4.19),

$$\phi_{10}=-\tan^{-1}\frac{\mu_1\alpha_x}{\mu k_{1x}}, \tag{5.21a}$$

$$\phi_{1t}=-\tan^{-1}\frac{\mu_1\alpha_{tx}}{\mu_t k_{1x}}. \tag{5.21b}$$

Obviously, (5.19) satisfies wave equation 5.1c. The magnetic fields are determined from (5.1a) and (5.1b). In view of (5.21), it can be verified that all boundary conditions are satisfied. A similar set of solutions for the TM waves can be written by duality.

Cutoff wave numbers are determined differently from the parallel-plate case. For parallel plates, cutoff occurs when $k_z^2 \leqslant 0$; the wave begins to decay in the propagation direction. In slab waveguides, $k_z^2 = \alpha_x^2 + k^2 = \alpha_{tx}^2 + k_t^2$ will never be zero for real α_x and α_{tx}. Cutoff occurs when either $\alpha_x^2 \leqslant 0$ or $\alpha_{tx}^2 \leqslant 0$; according to either (5.19a) or (5.19c), the wave outside the slab is no longer evanescent in the $\hat{x}$ direction. The surface waves become plane waves, and energy guided inside the slab begins to radiate into its surroundings. Thus for the slab waveguide the cutoff condition is $k_z = k$ or $k_z = k_t$.

Consider the simple case of a slab waveguide with region t identical to region 0, $\phi_{10} = \phi_{1t}$. At cutoff, $\phi_{10} = \phi_{1t} = 0$ for both TE and TM waves. By the guidance condition and the dispersion relation,

$$k_c d = \frac{m\pi}{\sqrt{1 - n^2/n_1^2}}, \tag{5.22}$$

where $k_c = \omega_c \sqrt{\mu_1 \epsilon_1}$, with ω_c denoting the cutoff angular frequency, and n_1 and n are refractive indices for media 1 and 0. Comparison of (5.22) with (5.12) shows that $k_c d$ for the slab waveguide is larger than the corresponding $k_c d$ for a parallel-plate waveguide, especially when n is close to n_1. It is interesting to observe that both TE_0 and TM_0 modes exist and their cutoff wave numbers are zero. This should not be surprising, however, because as the slab medium properties approach those of the surrounding medium, the 0th-order modes approach TEM waves in an unbounded medium.

A graphical approach is useful in determining the propagation constant k_z for a symmetric slab waveguide. By (5.20a),

$$(k_z d)^2 = (\alpha_x d)^2 + (kd)^2. \tag{5.23}$$

To find $\alpha_x d$, we note from (5.18) and (5.21) that for TE modes

$$\alpha_x \frac{d}{2} = \frac{\mu}{\mu_1} k_{1x} \frac{d}{2} \tan\left(\frac{k_{1x} d - m\pi}{2}\right). \tag{5.24a}$$

Note also that

$$\left(\alpha_x \frac{d}{2}\right)^2 + \left(k_{1x} \frac{d}{2}\right)^2 = (k_z^2 - k^2)\left(\frac{d}{2}\right)^2 + (k_1^2 - k_z^2)\left(\frac{d}{2}\right)^2 = (n_1^2 - n^2)\left(k_0 \frac{d}{2}\right)^2, \tag{5.24b}$$

where $k_0 = \omega/c$. Values of $\alpha_x d/2$ are determined by solving (5.24). Equation 5.24b, plotted on the two-dimensional plane of positive $\alpha_x d/2$ and $k_{1x} d/2$, describes a circular arc with radius equal to $(k_0 d/2)(n_1^2 - n^2)^{1/2}$. Equation 5.24a is plotted in Fig. 5.5 for m even or odd. The intersections among the

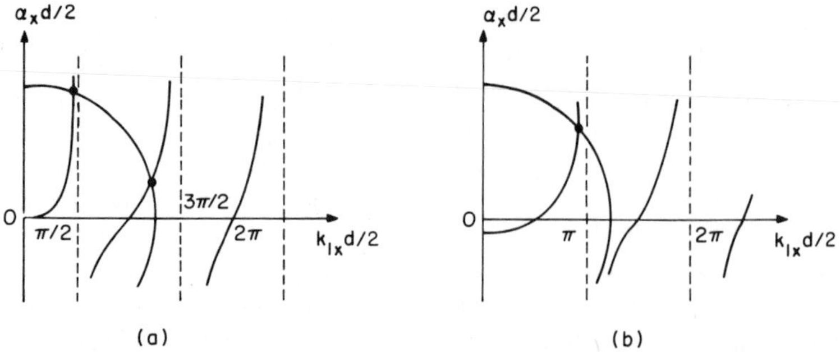

Figure 5.5 Graphical solution of $\alpha_x d/2$ and $k_{1x}d/2$. (*a*) For even modes. (*b*) For odd modes.

families of curves represented by (5.24a) and (5.24b) give the values for $\alpha_x d/2$, which in turn determine the propagation constant k_z by using (5.23). We observe that the larger the applied wave number k_0, the more intersections, and hence more modes can be guided. At $m=0$, the TE_0 mode exists no matter how small $k_0 d$ becomes. Cutoff occurs when $\alpha_x d=0$ or $k_{1x}d=m\pi$, in which case (5.24b) becomes (5.22).

Next, we consider the case in which medium t in this example is a perfect conductor. The guidance conditions for the TE and TM waves are now different. For TM waves, $\phi_{1t}^{TM}=0$ and we obtain the guidance condition from (5.18):

$$-2\tan^{-1}\frac{\epsilon_1\alpha_x}{\epsilon k_{1x}}+2k_{1x}d=2m\pi. \tag{5.25}$$

At cutoff, $\alpha_x=0$ and $k_z=k$. The cutoff wave numbers for TM_m modes are identical to the one given by (5.22). As in the former case, the TM_0 mode has zero cutoff wave number.

The field solutions for the TM modes can be written by observation:

$$H_y=H_0\cos(k_{1x}d)e^{-\alpha x+ik_z z}, \qquad 0\leqslant x,$$

$$H_y=H_0\cos(k_{1x}x+k_{1x}d)e^{ik_z z}, \qquad -d\leqslant x\leqslant 0.$$

The electric fields are derived from (5.2a) and (5.2b). It can be checked that all boundary conditions are satisfied.

By the same token, we write the solution for TE modes:

$$E_y=E_0\sin(k_{1x}d)e^{-\alpha x+ik_z z}, \qquad 0\leqslant x,$$

$$E_y=E_0\sin(k_{1x}x+k_{1x}d)e^{ik_z z}, \qquad -d\leqslant x\leqslant 0.$$

We cannot apply duality to write TE fields from TM fields because the boundary conditions for E_y and H_y at the perfect conductor are different.

The guidance conditions for TE waves are again obtained from (5.18) by noting that $\phi_{1t}^{\mathrm{TE}} = -\pi$:

$$-2\tan^{-1}\frac{\mu_1\alpha_x}{\mu k_{1x}} - \pi + 2k_{1x}d = 2m\pi.$$

At cutoff, $\alpha_x = 0$ and $k_z = k$. The cutoff wave numbers for TE_m modes are now $k_c d = (2m+1)\pi/2(1-n^2/n_1^2)^{1/2}$. Comparing this expression with (5.22), we notice that the TE_0 mode no longer has zero cutoff wave number.

5.1c Guided Waves in a Moving Dielectric Slab

The procedures illustrated in Section 5.1b apply to the treatment of moving isotropic slab waveguides. Now consider TE modes guided by a slab medium moving between two identical stationary isotropic media. Let the guidance direction be parallel to the direction of motion of the slab. The guidance condition is similar to (5.18) with the phase shifts determined differently. The field solutions also have the same form as (5.19). We follow the graphical approach to determine the cutoff wave numbers and the propagation constant k_z.

The transverse wave number k_{1x} inside the waveguide, in view of (4.72) or (4.74), may be cast into the following form:

$$k_{1x}^2 = pk_0^2 - 2lk_z k_0 - qk_z^2, \tag{5.26}$$

where $k_0 = \omega/c$ and $k_z = k_0(n^2 + \alpha_x^2/k_0^2)^{1/2}$. Equation 5.24a, which follows from the guidance condition, remains unchanged. Equation 5.24b, however, now becomes

$$\begin{aligned}(\alpha_x d)^2 + (k_{1x}d)^2 &= (k_z^2 - k^2)d^2 + (pk_0^2 - 2lk_0k_z - qk_z^2)d^2 \\ &= (k_0 d)^2\left\{\frac{n_1^2-\beta^2}{1-\beta^2} - n^2 - 2\beta\frac{n_1^2-1}{1-\beta^2}\sqrt{n^2+\left(\frac{\alpha_x}{k_0}\right)^2}\right. \\ &\qquad \left. + \beta^2\frac{n_1^2-1}{1-\beta^2}\left[n^2+\left(\frac{\alpha_x}{k_0}\right)^2\right]\right\}.\end{aligned}$$

The shape of the curve described by this equation is a function of the slab velocity. At $\beta=0$, the curve is a circular arc, the expected result of a stationary slab. When the isotropic medium surrounding the slab is free space, $n=1$, and this equation becomes

$$(\alpha_x d)^2+(k_{1x}d)^2=(k_0 d)^2\left(\frac{n_1^2-1}{1-\beta^2}\right)\left[1-\beta\sqrt{1+\left(\frac{\alpha_x}{k_0}\right)^2}\,\right]^2,$$

a result that can be derived by applying the Lorentz transformations directly.[113] We observe that at cutoff $\alpha_x=0$ and

$$k_0 d=\frac{\sqrt{(1+\beta)/(1-\beta)}\; m\pi}{\sqrt{n_1^2-1}};$$

the cutoff wave number increases as β increases. In general, as β increases, α_x decreases, and so does k_z. Thus the guided waves possess higher cutoff frequencies and propagate at larger phase velocities. A parallel analysis results in similar conclusions for the TM waves.

Next, we consider an isotropic medium moving between two perfectly conducting parallel plates. The guidance condition for both TE and TM modes is

$$\frac{m\pi}{d}=k_{1x}=\gamma\sqrt{(n_1^2-\beta^2)k_0^2-2\beta(n_1^2-1)k_0k_z-(1-n_1^2\beta^2)k_z^2}$$

The second equality follows from (5.26). Note that $m\pi/d$ is the cutoff wave number for the mth mode when the medium is stationary. Solving for k_z, we obtain from this equation

$$k_z=k_0\frac{\left\{-\beta(n_1^2-1)\pm(1-\beta^2)\sqrt{n_1^2-[(1-n_1^2\beta^2)(1-\beta^2)](m\pi/k_0d)^2}\right\}}{1-n_1^2\beta^2} \tag{5.27}$$

Observe that cutoff occurs when k_z becomes imaginary. When the velocity of the medium exceeds Čerenkov velocity so that $n_1\beta>1$, k_z will always be real for real n_1 and no cutoff will occur. The propagation constant k_z is always positive when $\beta(n_1^2-1) \geqslant [n_1^2(1-\beta^2)^2-(1-\beta^2)(1-n_1^2\beta^2)(m\pi/k_0 d)]^{1/2}$ or, equivalently, $k_0 d \geqslant m\pi[(1-\beta^2)/(n_1^2-\beta^2)]^{1/2}$. Thus, at sufficiently high frequencies, the phase velocities of all guided waves are in the positive $\hat{z}$ direction. Phase velocities in both directions become possible when $k_0 d < m\pi[(1-\beta^2)/(n_1^2-\beta^2)]^{1/2}$.

In the low-velocity regime when $n_1 \beta < 1$, we find that cutoff occurs as $(k_0 d)^2 \leqslant (m\pi)^2(1-n_1^2\beta^2)/(1-\beta^2)n_1^2$. Comparing this with the stationary case in which $k_c d = m\pi/n_1^2$, we see that motion of the medium always lowers the cutoff wave number. For frequencies above cutoff but with the square root term smaller than the first term in the equation, the phase velocities of the guided waves are all in the negative $\hat{z}$ direction. Phase velocities in both directions become possible when $k_0 d \geqslant m\pi[(1-\beta^2)/(n_1^2-\beta^2)]^{1/2}$.

It is of interest to investigate the power flow carried by each mode in the waveguide. Consider the TE modes with

$$E_y = E_m \sin \frac{m\pi x}{d}\, e^{ik_z z}.$$

The magnetic field components are determined from Maxwell's equations and constitutive relations 2.36 for the moving medium:

$$\mathbf{B} = \frac{1}{i\omega}\nabla\times\mathbf{E} = \left(-\hat{x}\frac{k_z}{\omega}E_m \sin\frac{m\pi x}{d} - \hat{z}i\frac{m\pi}{\omega d}E_m\cos\frac{m\pi x}{d}\right)e^{ik_z z},$$

$$\mathbf{H} = \frac{1}{c\mu'}\left[\hat{x}\,(-lE_y + qc\beta_x) + \hat{z}\,(c\beta_z)\right]$$

$$= \hat{x}\frac{1}{c\mu'}\left(-l - \frac{qck_z}{\omega}\right)E_y - \hat{z}i\frac{m\pi}{\omega\mu' d}E_m\cos\frac{m\pi x}{d}\,e^{ik_z z}.$$

The power flow in the $\hat{z}$ direction after integrating over the waveguide cross section is found to be, assuming real k_z,

$$P_z = -\int_0^d dx\,\frac{1}{2}\,\mathrm{Re}\,E_y H_x^* = \mathrm{Re}\sum_{m=1}^{\infty}\frac{d}{2c\mu'}\left(l + \frac{qk_z}{k_0}\right)|E_m|^2$$

$$= \mathrm{Re}\sum_{m=1}^{\infty}\frac{d}{2c\mu'}\left[\pm\sqrt{n_1^2 - \left(\frac{1-n_1^2\beta^2}{1-\beta^2}\right)\left(\frac{m\pi}{k_0 d}\right)^2}\,\right]|E_m|^2.$$

Thus each individual mode carries its own power; the total power is the sum of all individual components. At $\beta = 0$ this result reduces to the case of a stationary medium. Above Čerenkov velocity, the square root always gives real values and all modes carry time-average power. Below Čerenkov velocity, modes below cutoff will not carry time-average power. In all cases, the $\pm$ sign in front of the square root indicates that power can propagate in both positive and negative $\hat{z}$ directions, as opposed to the phase velocities, which in some cases can be in only one direction. In those velocity ranges, guided backward waves can be generated in the moving medium.

5.2 CYLINDRICAL WAVEGUIDES

5.2a Formulation

In treating guided waves along the $\hat{z}$ direction, the z and t dependence of all field vectors is written as $e^{ik_z z - i\omega t}$, where k_z is the propagation constant. With this dependence, we can replace $\partial/\partial z$ in Maxwell's equations by ik_z. Because of its unique position in guided wave theory, we use the $\hat{z}$ direction to characterize guided modes. From Maxwell's equations, we can express all field components transverse to $\hat{z}$ in terms of the longitudinal field components parallel to the z axis. When all vectors are separated into their transverse and longitudinal components, Maxwell's two curl equations for isotropic media become

$$(\nabla_s + \hat{z}ik_z) \times (\mathbf{E}_s + \mathbf{E}_z) = i\omega\mu(\mathbf{H}_s + \mathbf{H}_z), \tag{5.28a}$$

$$(\nabla_s + \hat{z}ik_z) \times (\mathbf{H}_s + \mathbf{H}_z) = -i\omega\epsilon(\mathbf{E}_s + \mathbf{E}_z), \tag{5.28b}$$

where the subscript s denotes transverse components. In terms of the transverse and longitudinal components, we have

$$i\omega\mu\mathbf{H}_s = \nabla_s \times \mathbf{E}_z + ik_z\hat{z} \times \mathbf{E}_s, \tag{5.29a}$$

$$-i\omega\epsilon\mathbf{E}_s = \nabla_s \times \mathbf{H}_z + ik_z\hat{z} \times \mathbf{H}_s, \tag{5.29b}$$

$$\nabla_s \times \mathbf{E}_s = i\omega\mu\mathbf{H}_z, \tag{5.29c}$$

$$\nabla_s \times \mathbf{H}_s = -i\omega\epsilon\mathbf{E}_z. \tag{5.29d}$$

From (5.29a) and (5.29b), we can express $\mathbf{E}_s$ and $\mathbf{H}_s$ in terms of E_z and H_z:

$$\mathbf{E}_s = \frac{i}{k^2 - k_z^2}(k_z \nabla_s E_z + \omega\mu \nabla_s \times \mathbf{H}_z), \tag{5.30a}$$

$$\mathbf{H}_s = \frac{i}{k^2 - k_z^2}(k_z \nabla_s H_z - \omega\epsilon \nabla_s \times \mathbf{E}_z), \tag{5.30b}$$

where $k^2 = \omega^2\mu\epsilon$ is the dispersion relation for isotropic media. Substituting (5.30) in (5.29c) and (5.29d), we obtain two wave equations for E_z and H_z:

$$(\nabla_s^2 + k^2 - k_z^2)E_z = 0, \tag{5.31a}$$

$$(\nabla_s^2 + k^2 - k_z^2)H_z = 0. \tag{5.31b}$$

These are homogeneous Helmholtz equations for E_z and H_z. When the longitudinal components are solved from (5.31) and the transverse components determined from (5.30), we can proceed to match the appropriate boundary conditions imposed by the guiding structure.

We classify the guided-wave modes in the following manner:

1. TE modes. All electric fields are transverse, $E_z = 0$. Field solutions that satisfy all boundary conditions are derived from H_z alone.

2. TM modes. All magnetic fields are transverse, $H_z = 0$. Field solutions that satisfy all boundary conditions are derived from E_z alone.

3. Hybrid modes. Both E_z and H_z are required to satisfy Maxwell's equations and the boundary conditions. If the boundary conditions do not require the presence of both E_z and H_z, the guided waves will be either TE or TM waves.

In addition to the TE, TM, and hybrid modes, there are TEM (transverse electric and magnetic) modes for which both E_z and H_z are zero. It is observed from (5.29c) and (5.29d) that for this set of modes the transverse fields are curl-free. The transverse fields can thus be derived from potential functions that satisfy the Laplace equation instead of the Helmholtz equation.

5.2b Rectangular Waveguides

Consider a metallic rectangular waveguide with dimensions a along the x axis and b along the y axis (Fig. 5.6). For TE modes

$$H_z = H_{mn} \cos\frac{m\pi x}{a} \cos\frac{n\pi y}{b} e^{ik_z z}, \tag{5.32}$$

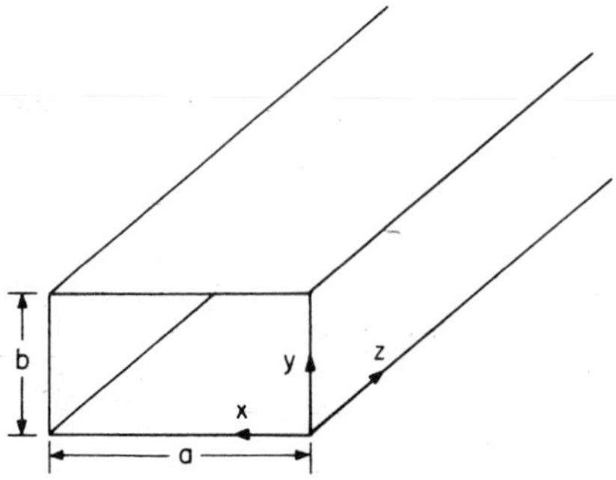

Figure 5.6 Metallic rectangular waveguide.

and for TM modes

$$E_z = E_{mn} \sin \frac{m\pi x}{a} \sin \frac{n\pi y}{b} e^{ik_z z}. \tag{5.33}$$

The dispersion relations for both the TE and the TM modes, in view of wave equation 5.31, are

$$k_z^2 = k^2 - \left(\frac{m\pi}{a}\right)^2 - \left(\frac{n\pi}{b}\right)^2. \tag{5.34}$$

The guided waves are cut off when $k_z^2 \leqslant 0$, and (5.34) yields

$$k_c a = \sqrt{(m\pi)^2 + \left(n\pi \frac{a}{b}\right)^2} . \tag{5.35}$$

Inserting (5.32) and (5.33) in (5.30), we find that the transverse fields for both TE and TM modes have the same spatial dependence:

$$E_x = \frac{i}{k^2 - k_z^2} \left[k_z \left(\frac{m\pi}{a} E_{mn} \right) - \omega\mu \left(\frac{n\pi}{b} H_{mn} \right) \right] \cos \frac{m\pi x}{a} \sin \frac{n\pi y}{b} e^{ik_z z}, \tag{5.36a}$$

$$E_y = \frac{i}{k^2 - k_z^2} \left[k_z \left(\frac{n\pi}{b} E_{mn} \right) + \omega\mu \left(\frac{m\pi}{a} H_{mn} \right) \right] \sin \frac{m\pi x}{a} \cos \frac{n\pi y}{b} e^{ik_z z}, \tag{5.36b}$$

$$H_x = \frac{i}{k^2 - k_z^2} \left[k_z \left(- \frac{m\pi}{a} H_{mn} \right) - \omega\epsilon \left(\frac{n\pi}{b} E_{mn} \right) \right] \sin \frac{m\pi x}{a} \cos \frac{n\pi y}{b} e^{ik_z z}, \tag{5.36c}$$

$$H_y=\frac{i}{k^2-k_z^2}\left[k_z\left(-\frac{n\pi}{b}H_{mn}\right)+\omega\epsilon\left(\frac{m\pi}{a}E_{mn}\right)\right]\cos\frac{m\pi x}{a}\sin\frac{n\pi y}{b}e^{ik_z z}. \quad (5.36\text{d})$$

It can be seen that tangential **E** fields at the conducting walls all vanish.

We must note that for TE_{mn} modes $E_{mn}=0$, and for TM_{mn} modes $H_{mn}=0$. The time-average Poynting's power flowing in the waveguide, evaluated at $z=0$, is

$$P=\frac{1}{2}\operatorname{Re}\left\{\sum_{m=1}^{\infty}\frac{\omega\mu k_z}{k^2-k_z^2}\frac{ab}{2}|H_{m0}|^2+\sum_{n=1}^{\infty}\frac{\omega\mu k_z}{k^2-k_z^2}\frac{ab}{2}|H_{0n}|^2\right.$$

$$\left.+\sum_{m,n=1}^{\infty}\left(\frac{\omega\mu k_z}{k^2-k_z^2}\frac{ab}{4}|H_{mn}|^2+\frac{\omega\epsilon k_z}{k^2-k_z^2}\frac{ab}{4}|E_{mn}|^2\right)\right\}. \quad (5.37)$$

This is simply a summation of the power carried by the individual TE and TM modes. All TE and TM modes are orthogonal to one another. In a lossless waveguide, when a mode is not propagating, k_z is imaginary, and no power flow is associated with the mode.

The attenuation constants caused by imperfectly conducting walls are evaluated in terms of k_c/k:

$$\alpha_{\text{TE}_{m0}}=\frac{1+(2b/a)(k_c/k)^2}{b\sqrt{2\sigma/\omega\epsilon}\sqrt{1-(k_c/k)^2}}, \quad (5.38\text{a})$$

$$\alpha_{\text{TE}_{mn}}=\frac{2}{b\sqrt{2\sigma/\omega\epsilon}\sqrt{1-(k_c/k)^2}}\left\{\left(1+\frac{b}{a}\right)\left(\frac{k_c}{k}\right)^2\right.$$

$$\left.+\left[1-\left(\frac{k_c}{k}\right)^2\right]\frac{m^2(b/a)^2+n^2b/a}{m^2(b/a)^2+n^2}\right\}, \quad (5.38\text{b})$$

$$\alpha_{\text{TM}_{mn}}=\frac{2}{b\sqrt{2\sigma/\omega\epsilon}\sqrt{1-(k_c/k)^2}}\frac{m^2(b/a)^3+n^2}{m^2(b/a)^2+n^2}. \quad (5.38\text{c})$$

These results are obtained by integrating (5.5) over the wall area and using (5.9). As $b\to\infty$, (5.38a) reduces to (5.15a) for TE waves in a parallel-plate waveguide.

In a rectangular waveguide with $b<a$, the TE_{10} mode has the lowest cutoff frequency and is called the *fundamental* or the *dominant* mode. This mode is excited for $k>\pi/a$, independent of dimension b. It has the electric field in the $\hat{y}$ direction and the magnetic field in the $\hat{x}$ and $\hat{z}$ directions. As in the parallel-plate case, the mode can be viewed as plane waves bouncing between walls at $x=0$ and $x=a$. The walls at $y=0$ and $y=b$ provide resting places for the charges induced by E_y. The cutoff frequency for higher-order modes depends largely on dimension b. If $b<a/2$, the next higher-order mode will be TE_{20}.

5.2c Circular Metallic Waveguides

In cylindrical coordinates, the wave equation for E_z and H_z becomes

$$\left[\frac{1}{\rho}\frac{\partial}{\partial\rho}\left(\rho\frac{\partial}{\partial\rho}\right)+\frac{1}{\rho^2}\frac{\partial^2}{\partial\phi^2}+k_\rho^2\right]\left\{\begin{matrix}E_z\\H_z\end{matrix}\right\}=0, \tag{5.39}$$

where $k_\rho^2=k^2-k_z^2$. Solutions to the wave equation are Bessel functions multiplied by sinusoidal functions. The sinusoidals can be combinations of $\sin m\phi$, $\cos m\phi$, or $e^{\pm im\phi}$. Substituting in (5.39) and making the transformation $\xi=k_\rho\rho$, we have the Bessel equation

$$\left[\frac{1}{\xi}\frac{d}{d\xi}\left(\xi\frac{d}{d\xi}\right)+\left(1-\frac{m^2}{\xi^2}\right)\right]B(\xi)=0. \tag{5.40}$$

This has solutions in the form of Bessel function $J_m(\xi)$, Neumann function $N_m(\xi)$, Hankel function of the first kind $H_m^{(1)}(\xi)$, or Hankel function of the second kind $H_m^{(2)}(\xi)$. The asymptotic values are listed in Table 5.1. As $\xi\to\infty$, J_m behaves as cosine, N_m as sine, $H_m^{(1)}$ as $e^{i\xi}$, and $H_m^{(2)}$ as $e^{-i\xi}$. At $\xi\to 0$, all but $J_m(\xi)$ become singular. The two independent solutions to (5.40) can be regarded as J_m and N_m, and the Hankel functions are defined in terms of them:

$$H_m^{(1)}(\xi)=J_m(\xi)+iN_m(\xi), \tag{5.41a}$$

$$H_m^{(2)}(\xi)=J_m(\xi)-iN_m(\xi). \tag{5.41b}$$

Let $B_m(\xi)$ represent $J_m(\xi)$, $N_m(\xi)$, $H_m^{(1)}(\xi)$, or $H_m^{(2)}(\xi)$. The recurrence formu-

Table 5.1
Bessel Functions and Their Asymptotic Values

B_m \ ξ	$\xi \to 0$		$\xi \to \infty$
	$m=0$	$\mathrm{Re}\, m>0$	
$J_m(\xi)$	1	$\dfrac{(\xi/2)^m}{\Gamma(m+1)}$	$\sqrt{\dfrac{2}{\pi\xi}}\cos\left(\xi-\dfrac{m\pi}{2}-\dfrac{\pi}{4}\right)$
$N_m(\xi)$	$\left(\dfrac{2}{\pi}\right)\ln\xi$	$-\dfrac{\Gamma(m)}{\pi}\left(\dfrac{2}{\xi}\right)^m$	$\sqrt{\dfrac{2}{\pi\xi}}\sin\left(\xi-\dfrac{m\pi}{2}-\dfrac{\pi}{4}\right)$
$H_m^{(1)}(\xi)$	$i\dfrac{2}{\pi}\ln\xi$	$-\dfrac{i\Gamma(m)}{\pi}\left(\dfrac{2}{\xi}\right)^m$	$\sqrt{\dfrac{2}{\pi\xi}}\, e^{i(\xi-m\pi/2-\pi/4)}$
$H_m^{(2)}(\xi)$	$-i\dfrac{2}{\pi}\ln\xi$	$\dfrac{i\Gamma(m)}{\pi}\left(\dfrac{2}{\xi}\right)^m$	$\sqrt{\dfrac{2}{\pi\xi}}\, e^{-i(\xi-m\pi/2-\pi/4)}$

las for the Bessel functions $B_m(\xi)$ are as follows:

$$B'_m(\xi)=B_{m-1}(\xi)-\frac{m}{\xi}B_m(\xi)$$

$$=-B_{m+1}(\xi)+\frac{m}{\xi}B_m(\xi). \tag{5.42}$$

In Fig. 5.7, J_m, J'_m, and N_m are plotted for $m=0$, 1, and 2.

Consider a circular metallic waveguide with radius a (Fig. 5.8). The boundary condition requires that E_z and E_ϕ vanish at $\rho=a$. The solution must not be singular at $\rho=0$. We find for TM modes

$$E_z=E_{mn}J_m(k_\rho\rho)\begin{Bmatrix}\sin m\phi\\ \cos m\phi\end{Bmatrix}e^{ik_z z}, \tag{5.43}$$

$$J_m(k_\rho a)=0. \tag{5.44}$$

Equation 5.44 is the guidance condition, which determines the cutoff wave numbers as

$$k_c a=\xi_{mn}, \tag{5.45}$$

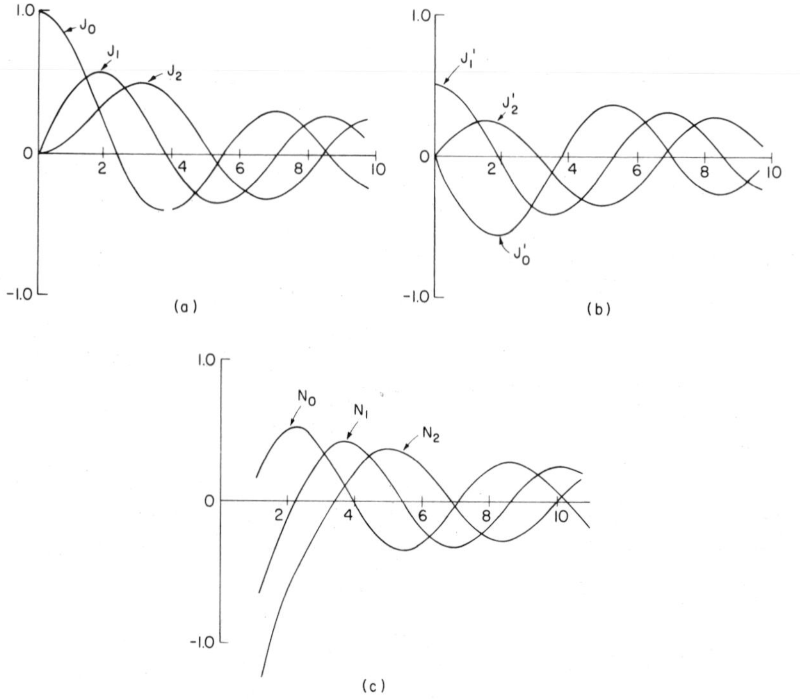

Figure 5.7 **(*a*) The first three Bessel functions of order 0, 1, 2. (*b*) Derivatives of the first three Bessel functions of order 0, 1, 2. (*c*) The first three Neumann functions of order 0, 1, 2.**

where ξ_{mn} is the nth root of the mth-order Bessel function, $J_m(\xi_{mn})=0$. From Fig. 5.7a, we see that $k_c a$ is 2.4 for the TM_{01} mode, and 3.8 for the TM_{11} mode.

For TE modes, we have

$$H_z = H_{mn} J_m(k_\rho \rho) \begin{Bmatrix} \sin m\phi \\ \cos m\phi \end{Bmatrix} e^{ik_z z}. \tag{5.46}$$

In view of (5.30a) and the boundary condition of vanishing E_ϕ at $\rho = a$, we find the guidance condition

$$J_m'(k_\rho a)=0. \tag{5.47}$$

The cutoff wave numbers are determined as

$$k_c a = \xi_{mn}', \tag{5.48}$$

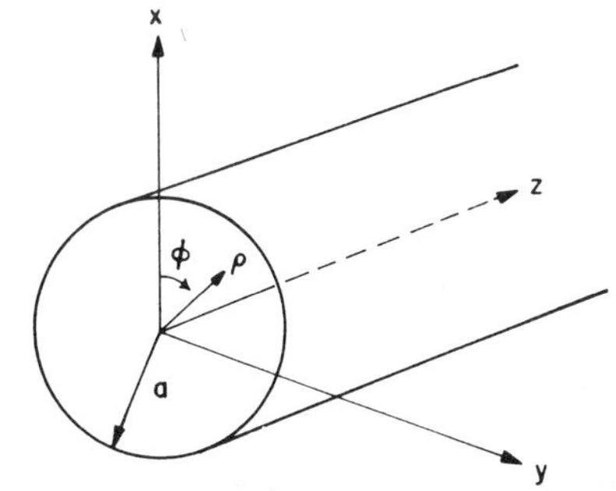

Figure 5.8 **Metallic waveguide with circular cross section.**

where ξ'_{mn} is the nth root of the derivative of the mth-order Bessel function, $J'_m(\xi'_{mn})=0$. From Fig. 5.7b, we see that $k_c a$ is 1.8 for the TE_{11} mode, 3.05 for the TE_{21} mode, and 3.8 for the TE_{01} mode. All TM_{1n} and TE_{0n} modes are degenerate because $J'_0 = -J_1$.

Arranging the cutoff wave number k_c times the waveguide radius a in increasing order, we list the first few guided modes:

$k_c a$:	1.8		2.4		3.05		3.8		4.2		5.1	
Modes:		TE_{11}		TM_{01}		TE_{21}		TE_{01}, TM_{11}		TE_{31}		TM_{21}

In practice, the metallic wall is not a perfect conductor and the wall loss causes attenuation. Using the procedure illustrated in (5.5)–(5.10), we calculate the attenuation constant:

$$\alpha_{TM} = \frac{(\omega\epsilon/2\sigma)^{1/2}}{a\left[1-(k_c/k)^2\right]^{1/2}}, \tag{5.49a}$$

$$\alpha_{TE} = \frac{(\omega\epsilon/2\sigma)^{1/2}}{a\left[1-(k_c/k)^2\right]^{1/2}}\left[\left(\frac{k_c}{k}\right)^2 + \frac{m^2}{(\xi'_{mn})^2 - m^2}\right], \tag{5.49b}$$

where ξ'_{mn} is the root of $J'_m(\xi')$. For the three modes plotted in Fig. 5.9, we see that the attenuation constant of the TE_{01} mode decreases monotonically with frequency. In long-distance transmission, where attenuation caused by wall loss is of primary concern, the TE_{01} mode is usually preferred. The field

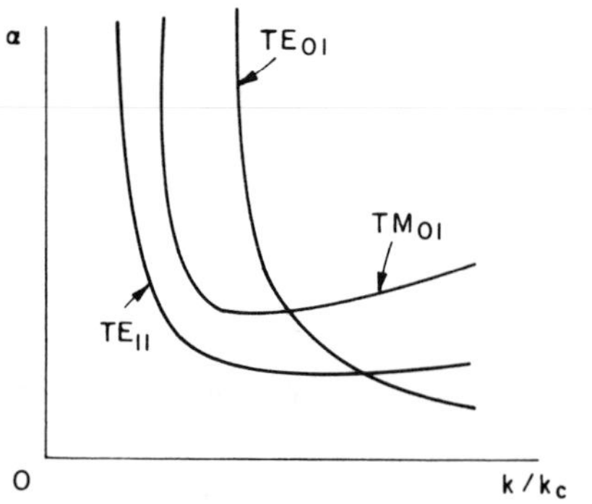

Figure 5.9 Attenuation arising from wall loss in a circular waveguide.

solutions for the TE_{01} mode take the following forms:

$$H_z = H_{mn} J_m\left(\frac{\xi'_{mn}\rho}{a}\right) e^{ik_z z},$$

$$H_\rho = \frac{ik_z}{k_\rho} H_{mn} J'_m\left(\frac{\xi'_{mn}\rho}{a}\right) e^{ik_z z},$$

$$E_\phi = \frac{i\omega\mu}{k_\rho} H_{mn} J'_m\left(\frac{\xi'_{mn}\rho}{a}\right) e^{ik_z z}.$$

The surface currents on the walls are circumferential. Note that, when this mode is excited, other modes, such as TM_{01}, TE_{21}, and TM_{21}, may propagate. To discourage these other modes, thin conducting sheets may be placed on radial planes to damp modes with E_ρ or E_z. We can also place closely spaced parallel conducting rings along the waveguide wall to encourage the circumferential current and to discourage longitudinal currents. By filling the waveguide with gas and heating the waveguide wall to have high-density cold gas flowing in the center, we can concentrate the field at the center and thus decrease wall loss.

We shall now consider guided waves in a coaxial line made of a concentric conducting cylinder of radius a inside a metallic waveguide of radius b. Coaxial lines are usually used as transmission lines. The fundamental mode is TEM. We have

$$\mathbf{E} = \hat{\rho}\,\frac{E_0}{k\rho}\,e^{ikz}, \tag{5.50a}$$

$$\mathbf{H} = \hat{\phi}\sqrt{\frac{\epsilon}{\mu}}\,\frac{E_0}{k\rho}\,e^{ikz}, \tag{5.50b}$$

which has no cutoff. Both **E** and **H** in (5.50) can be expressed in terms of potential functions satisfying the Laplace equation.

The higher-order modes are governed by (5.30) and (5.31). Boundary conditions require that E_z and E_ϕ vanish at $\rho=a$ and $\rho=b$. The origin $\rho=0$ is now excluded from the region of interest. We must write the solution to wave equation 5.31 as a superposition of two independent solutions to the Bessel equation. We find

$$E_z=\left[H_n^{(1)}(k_\rho a)H_n^{(2)}(k_\rho \rho)-H_n^{(2)}(k_\rho a)H_n^{(1)}(k_\rho \rho)\right]\begin{Bmatrix}\sin m\phi\\ \cos m\phi\end{Bmatrix}e^{ik_z z} \quad (5.51a)$$

for TM modes, and

$$H_z=\left[H_n^{(1)\prime}(k_\rho a)H_n^{(2)}(k_\rho \rho)-H_n^{(2)\prime}(k_\rho a)H_n^{(1)}(k_\rho \rho)\right]\begin{Bmatrix}\sin m\phi\\ \cos m\phi\end{Bmatrix}e^{ik_z z} \quad (5.51b)$$

for TE modes. It is easy to verify that with these solutions the boundary condition at $\rho=a$ is satisfied. To satisfy the boundary condition at $\rho=b$, we obtain the guidance condition

$$H_n^{(1)}(k_\rho a)H_n^{(2)}(k_\rho b)-H_n^{(2)}(k_\rho a)H_n^{(1)}(k_\rho b)=0$$

for TM modes, and

$$H_n^{(1)\prime}(k_\rho a)H_n^{(2)\prime}(k_\rho b)-H_n^{(2)\prime}(k_\rho a)H_n^{(1)\prime}(k_\rho b)=0$$

for TE modes. In terms of the Bessel and Neumann functions, we have

$$J_n(k_\rho a)N_n(k_\rho b)-J_n(k_\rho b)N_n(k_\rho a)=0 \quad (5.52a)$$

for TM modes, and

$$J_n'(k_\rho a)N_n'(k_\rho b)-J_n'(k_\rho b)N_n'(k_\rho a)=0 \quad (5.52b)$$

for TE modes. The TM and TE modes can be labeled with two subscripts; the first subscript refers to the order of the Hankel functions, and the second to the position of the roots of (5.52) arranged according to increasing magnitude. Note that the longitudinal field components E_z and H_z can also be written in terms of summations of Bessel functions J_m and Neumann functions N_m. Written in terms of the two Hankel functions, the guided wave may be viewed as a superposition of two bouncing cylindrical waves, similar to the bouncing plane waves in parallel-plate waveguides.

5.2d Circular Dielectric Waveguides

Consider a circular waveguide made of isotropic medium μ_1 and ϵ_1 with radius a, embedded in another isotropic medium μ and ϵ. The boundary conditions require that tangential **E** and **H** be continuous at $\rho = a$. As we shall see, decomposition of fields into TE and TM modes will, in general, no longer be possible. The boundary conditions cannot be satisfied without using both H_z and E_z; thus the modes are hybrid. Inside the waveguide, the solutions are Bessel functions:

$$E_z = A_m J_m(k_{1\rho}\rho)\cos m\phi e^{ik_z z}, \tag{5.53a}$$

$$H_z = B_m J_m(k_{1\rho}\rho)\sin m\phi e^{ik_z z}, \tag{5.53b}$$

$$k_{1\rho} = \sqrt{k_1^2 - k_z^2}\ . \tag{5.53c}$$

In order for waves to be guided inside the waveguide, the field outside must be evanescent in the $\hat{\rho}$ direction. The proper choice of the solution will be Hankel functions with imaginary arguments, known as the modified Hankel functions. Let $k_\rho = i\alpha_\rho$. We write for $\rho \geqslant a$

$$E_z = C_m H_m^{(1)}(i\alpha_\rho\rho)\cos m\phi e^{ik_z z}, \tag{5.54a}$$

$$H_z = D_m H_m^{(1)}(i\alpha_\rho\rho)\sin m\phi e^{ik_z z}, \tag{5.54b}$$

$$\alpha_\rho = \sqrt{k_z^2 - k^2}\ . \tag{5.54c}$$

To match the boundary conditions, the transverse ϕ components need to be determined from (5.30).

The boundary conditions of continuity of tangential fields at $\rho = a$ yield

$$A_m J_m(k_{1\rho}a) = C_m H_m^{(1)}(i\alpha_\rho a), \tag{5.55a}$$

$$\frac{\omega\epsilon_1}{k_{1\rho}a} A_m J_m'(k_{1\rho}a) + \frac{mk_z}{k_{1\rho}^2 a^2} B_m J_m(k_{1\rho}a)$$

$$= -i\frac{\omega\epsilon}{\alpha_\rho a} C_m H_m^{(1)\prime}(i\alpha_\rho a) - \frac{mk_z}{\alpha_\rho^2 a^2} D_m H_m^{(1)}(i\alpha_\rho a); \tag{5.55b}$$

$$B_m J_m(k_{1\rho}a) = D_m H_m^{(1)}(i\alpha_\rho a), \tag{5.56a}$$

$$\frac{mk_z}{k_{1\rho}^2 a^2} A_m J_m(k_{1\rho}a) + \frac{\omega\mu_1}{k_{1\rho}a} B_m J_m'(k_{1\rho}a)$$

$$= -\frac{mk_z}{\alpha_\rho^2 a^2} C_m H_m^{(1)}(i\alpha_\rho a) - i\frac{\omega\mu}{\alpha_\rho a} D_m H_m^{(1)'}(i\alpha_\rho a). \qquad (5.56b)$$

We observe that TE and TM modes are possible only if $m=0$. In this case, (5.56a) and (5.56b) give the guidance condition for the TE_{0p} modes:

$$\frac{k_{1\rho}a}{\mu_1}\frac{J_0(k_{1\rho}a)}{J_0'(k_{1\rho}a)} = i\frac{\alpha_\rho a}{\mu}\frac{H_0^{(1)}(k\alpha_\rho a)}{H_0^{(1)'}(i\alpha_\rho a)}. \qquad (5.57)$$

Equations 5.55a and 5.55b give the guidance condition for the TM_{0p} modes:

$$\frac{k_{1\rho}a}{\epsilon_1}\frac{J_0(k_{1\rho}a)}{J_0'(k_{1\rho}a)} = i\frac{\alpha_\rho a}{\epsilon}\frac{H_0^{(1)}(i\alpha_\rho a)}{H_0^{(1)'}(i\alpha_\rho a)}. \qquad (5.58)$$

The subscript p on TE_{0p} or TM_{0p} indicates the mode number arranged in order of increasing cutoff frequency.

When $m \neq 0$, separation into TE and TM modes is no longer possible; the $m \neq 0$ modes are hybrid. Eliminating A_m, B_m, C_m, and D_m from (5.55) and (5.56), we find

$$J_m^2 H_m^{(1)^2}\left[\left(\frac{\omega\mu_1}{k_{1\rho}a}\frac{J_m'}{J_m} + \frac{i\omega\mu}{\alpha_\rho a}\frac{H_m^{(1)'}}{H_m^{(1)}}\right)\left(\frac{\omega\epsilon_1}{k_{1\rho}a}\frac{J_m'}{J_m} + \frac{i\omega\epsilon}{\alpha_\rho a}\frac{H_m^{(1)'}}{H_m^{(1)}}\right)\right.$$

$$\left. - m^2k_z^2\left(\frac{1}{k_{1\rho}^2 a^2} + \frac{1}{\alpha_\rho^2 a^2}\right)^2\right] = 0. \qquad (5.59)$$

Remember that the arguments for the Bessel and the Hankel functions are, respectively, $k_{1\rho}a$ and $\alpha_\rho a$. Equation 5.59 is the guidance condition for the hybrid modes.

Cutoff occurs when the field outside ceases to be described by Hankel functions with imaginary arguments and requires those with real arguments, a condition that characterizes radiation. This cutoff criterion is identical to the one used in slab waveguides. Near cutoff, $\alpha_\rho \to 0$ and $k_z \to k$. In view of the asymptotic values for the Hankel functions as $\alpha_\rho \to 0$, (5.57) and (5.58) yield identical equations for TE and TM modes at cutoff:

$$J_0\left(k_c a\sqrt{1-\frac{n^2}{n_1^2}}\,\right)=0, \tag{5.60}$$

where $k_c=n_1\omega_c/c$. We therefore have $k_c a(1-n^2/n_1^2)^{1/2}=2.40$, 5.52, 8.65, and so on. For the hybrid modes, we have either $J_m[k_c a(1-n^2/n_1^2)^{1/2}]=0$ or the bracket term in (5.59) equal to zero. When

$$J_m\left(k_c a\sqrt{1-n^2/n_1^2}\,\right)=0, \qquad k_c a\neq 0, \tag{5.61}$$

the modes are called EH_{mp} modes. We can show that for $m\neq 1$ equation 5.59 does not admit the solution $k_x a=0$ as $\alpha_\rho a$ and $k_{1\rho}a$ both approach zero.

When the bracket terms in (5.59) vanish the modes are called HE_{mp} modes. We use the fact that $k_z^2=k_1^2-k_{1\rho}^2=k^2+\alpha_\rho^2$ and recurrence relations 5.42 to obtain

$$\left(\frac{\omega\mu_1 J_m'}{k_{1\rho}aJ_m}+i\frac{\omega\mu H_{m-1}^{(1)}}{\alpha_\rho a H_m^{(1)}}-\frac{m\omega\mu}{\alpha_\rho^2a^2}\right)\left(\frac{\omega\epsilon_1 J_m'}{k_{1\rho}aJ_m}+i\frac{\omega\epsilon H_{m-1}^{(1)}}{\alpha_\rho a H_m^{(1)}}-\frac{m\omega\epsilon}{\alpha_\rho^2a^2}\right)$$
$$-m^2\left(\frac{k_1^2}{k_{1\rho}^2a^2}+\frac{k^2}{\alpha_\rho^2a^2}\right)\left(\frac{1}{k_{1\rho}^2a^2}+\frac{1}{\alpha_\rho^2a^2}\right)=0. \tag{5.62}$$

Next we observe that for $m>1$ the asymptotic formula for $H_{m-1}^{(1)}/H_m^{(1)}$ as $\alpha_\rho a\rightarrow 0$ gives

$$\frac{H_{m-1}^{(1)}(i\alpha_\rho a)}{H_m^{(1)}(i\alpha_\rho a)}\approx\frac{i\alpha_\rho a}{2(m-1)}. \tag{5.63}$$

Equation 5.62, keeping terms of the order of $1/\alpha_\rho^2a^2$, becomes

$$\frac{\mu\epsilon}{m-1}-\frac{\mu\epsilon_1+\mu_1\epsilon}{k_{1\rho}a}\frac{J_m'(k_{1\rho}a)}{J_m(k_{1\rho}a)}-m\frac{\mu_1\epsilon_1+\mu\epsilon}{k_{1\rho}^2a^2}=0. \tag{5.64}$$

This equation determines cutoff wave numbers for HE_{mp} modes with $m>1$. For $m=1$ and $\alpha_\rho a\rightarrow 0$, we have

$$\frac{H_0^{(1)}(i\alpha_\rho a)}{H_1^{(1)}(i\alpha_\rho a)}\approx i\alpha_\rho a\ln(i\alpha_\rho a).$$

Equation 5.62 becomes

$$\lim_{\alpha_p a \to 0}\left[\frac{\mu\epsilon_1+\mu_1\epsilon}{k_{1\rho}a}\frac{J_1'(k_{1\rho}a)}{J_1(k_{1\rho}a)}+\frac{\mu_1\epsilon_1+\mu\epsilon}{k_{1\rho}^2a^2}-2\mu\epsilon\ln(i\alpha_\rho a)\right]=0.$$

Thus the cutoff wave numbers are determined by

$$J_1\left(k_c a\sqrt{1-\frac{n^2}{n_1^2}}\right)=0. \tag{5.65}$$

This set of cutoff wave numbers determines the HE_{1p} modes. It is important to observe that the first root of J_1 is zero, implying that the cutoff wave number for the HE_{11} mode is also zero. We can appreciate that, as $\mu_1 \to \mu$ and $\epsilon_1 \to \epsilon$, the HE_{11} mode approaches a TEM wave for which there is no cutoff. The next higher cutoff frequency occurs at $(\omega_c/c)a = 2.4/(n_1^2-n^2)^{1/2}$. By choosing a small radius a and a refractive index n slightly smaller than n_1, the operating frequency range for the fundamental HE_{11} mode can be made very large before any higher-order mode can propagate.

5.2e Guided Waves in Moving Media

For guided waves in anisotropic and bianisotropic media, the wave equations for E_z and H_z are usually coupled. We illustrate this with the general case of a bianisotropic medium realized by a moving gyrotropic medium. The gyrotropic medium in its rest frame has tensor permittivity 2.32:

$$\bar{\epsilon}'=\begin{bmatrix}\epsilon' & -i\epsilon_g' & 0\\ i\epsilon_g' & \epsilon' & 0\\ 0 & 0 & \epsilon_z'\end{bmatrix}. \tag{5.66}$$

In the laboratory frame the medium is moving along the $\hat{z}$ direction and the constitutive matrix is given by (2.33). Using (1.17b), we transform to **EH** representation and obtain the following constitutive relations:

$$\mathbf{D}=\bar{\epsilon}_s\cdot\mathbf{E}_s+\epsilon_z'\mathbf{E}_z+\bar{\xi}_s\cdot\mathbf{H}_s,$$

$$\mathbf{B}=\bar{\mu}_s\cdot\mathbf{H}_s+\mu'\mathbf{H}_z-\bar{\xi}_s\cdot\mathbf{E}_s,$$

where

$$\bar{\epsilon}_s = \epsilon' \begin{bmatrix} a & -ia_g \\ ia_g & a \end{bmatrix}$$

$$= \frac{(1-\beta^2)\epsilon'}{n^2\left[(1-n^2\beta^2)^2 - n_g^4\beta^4\right]} \begin{bmatrix} n^2(1-n^2\beta^2)+n_g^4\beta^2 & -in_g^2 \\ in_g^2 & n^2(1-n^2\beta^2)+n_g^4\beta^2 \end{bmatrix}, \tag{5.67a}$$

$$\bar{\mu}_s = \mu' \begin{bmatrix} b & -ib_g \\ ib_g & b \end{bmatrix} = \frac{(1-\beta^2)\mu'}{(1-n^2\beta^2)^2 - n_g^4\beta^4} \begin{bmatrix} 1-n^2\beta^2 & -in_g^2\beta^2 \\ in_g^2\beta^2 & 1-n^2\beta^2 \end{bmatrix}, \tag{5.67b}$$

$$\bar{\xi}_s = \frac{1}{c} \begin{bmatrix} -i\xi_g & -\xi \\ \xi & -i\xi_g \end{bmatrix} = \frac{\beta}{c} \begin{bmatrix} -in^2a_g & -\gamma^2(n^2a-b) \\ \gamma^2(n^2a-b) & -in^2a_g \end{bmatrix} \tag{5.67c}$$

At $\beta=0$, (5.67a) reduces to (5.66), $\bar{\mu}_s=\mu'\bar{\mathbf{I}}$, $\bar{\xi}_s=0$, and we have the gyrotropic medium at rest.

Following a procedure similar to that used in deriving (5.30) and (5.31), we can express the transverse components in terms of the longitudinal components E_z and H_z and derive wave equations for E_z and H_z. Analogous to (5.29), we have

$$\nabla_s \times \mathbf{E}_z = i\omega\bar{\mu}_s \cdot \mathbf{H}_s - \bar{\mathbf{d}} \cdot \mathbf{E}_s, \tag{5.68a}$$

$$\nabla_s \times \mathbf{H}_z = -i\omega\bar{\epsilon}_s \cdot \mathbf{E}_s - \bar{\mathbf{d}} \cdot \mathbf{H}_s, \tag{5.68b}$$

$$\nabla_s \times \mathbf{E}_s = i\omega\mu'\mathbf{H}_z, \tag{5.69a}$$

$$\nabla_s \times \mathbf{H}_s = -i\omega\epsilon_z'\mathbf{E}_z, \tag{5.69b}$$

where

$$\bar{\mathbf{d}} = \begin{bmatrix} d_g & -id \\ id & d_g \end{bmatrix} = \begin{bmatrix} \omega\xi_g/c & -i(k_z + \omega\xi/c) \\ i(k_z + \omega\xi/c) & \omega\xi_g/c \end{bmatrix}. \tag{5.70}$$

In terms of E_z and H_z, the transverse components are

$$\mathbf{E}_s = \left(\bar{\mathbf{I}} - \omega^2 \bar{\mathbf{d}}^{-1} \cdot \bar{\boldsymbol{\mu}}_s \cdot \bar{\mathbf{d}}^{-1} \cdot \bar{\boldsymbol{\epsilon}}_s\right)^{-1}$$

$$\cdot \left[-\bar{\mathbf{d}}^{-1} \cdot (\nabla \times \mathbf{E}_z) - i\omega \bar{\mathbf{d}}^{-1} \cdot \bar{\boldsymbol{\mu}}_s \cdot \bar{\mathbf{d}}^{-1} \cdot (\nabla_s \times \mathbf{H}_z) \right], \tag{5.71a}$$

$$\mathbf{H}_s = \left(\bar{\mathbf{I}} - \omega^2 \bar{\mathbf{d}}^{-1} \cdot \bar{\boldsymbol{\epsilon}}_s \cdot \bar{\mathbf{d}}^{-1} \cdot \bar{\boldsymbol{\mu}}_s\right)^{-1}$$

$$\cdot \left[-\bar{\mathbf{d}}^{-1} \cdot (\nabla_s \times \mathbf{H}_z) + i\omega \bar{\mathbf{d}}^{-1} \cdot \bar{\boldsymbol{\epsilon}}_s \cdot \bar{\mathbf{d}}^{-1} \cdot (\nabla_s \times \mathbf{E}_z) \right]. \tag{5.71b}$$

After considerable algebraic manipulations, the wave equations for the longitudinal field components are determined to be (see Problem 5.9)

$$\left(\nabla_s^2 + \frac{\epsilon_z'}{\epsilon'} k^2 e\right) E_z = i\omega\mu' h_g H_z, \tag{5.72a}$$

$$(\nabla_s^2 + k^2 h) H_z = -i\omega\epsilon_z' e_g E_z, \tag{5.72b}$$

where

$$e = \frac{1}{b}\left[b^2 - b_g^2 + \frac{(bd - b_g d_g)^2}{d_g^2 - k^2 ab} \right], \tag{5.73a}$$

$$h = \frac{1}{a}\left[a^2 - a_g^2 + \frac{(ad - a_g d_g)^2}{d_g^2 - k^2 ab} \right], \tag{5.73b}$$

$$e_g = \frac{1}{a}\left[ad_g - a_g d - \frac{(ad - a_g d_g)(dd_g - k^2 ab_g)}{d_g^2 - k^2 ab} \right], \tag{5.73c}$$

$$h_g = \frac{1}{b}\left[bd_g - b_g d - \frac{(bd - b_g d_g)(dd_g - k^2 a_g b)}{d_g^2 - k^2 ab}\right], \tag{5.73d}$$

where $k = \omega(\mu'\epsilon')^{1/2}$. The two equations in (5.72) for E_z and H_z are coupled. Thus the guided wave modes are hybrid. Compare with preceding sections on isotropic media, where wave equations 5.31 are decoupled and hybrid modes are consequences of boundary conditions.

Solutions to the coupled equations in (5.72) can be facilitated by transforming into decoupled homogeneous Helmholtz equations. We define

$$\psi_j = E_z - i\alpha_j H_z, \qquad j = 1,2. \tag{5.74}$$

Multiply (5.72d) by $i\alpha_j$, and subtract the result from (5.72a). Using (5.74) to eliminate E_z and requiring that the coefficient for H_z be zero, we obtain the following second-order equations for α_j:

$$\omega\epsilon_z' e_j \alpha_j^2 - k^2\left(h - \frac{\epsilon_z'}{\epsilon'} e\right)\alpha_j - \omega\mu' h_g = 0. \tag{5.75}$$

The two roots for α_j from (5.75) are the values for α_1 and α_2. With (5.75) satisfied, (5.72a) and (5.72b) combine to yield a single second-order, two-dimensional, scalar, homogeneous Helmholtz equation:

$$\nabla_s^2 \psi_j + q_j^2 \psi_j = 0, \tag{5.76}$$

where

$$q_j^2 = \frac{\epsilon_z'}{\epsilon'} k^2 e + \alpha_j \omega\epsilon_z' e_g = k^2 h + \frac{1}{\alpha_j}\omega\mu' h_g. \tag{5.77}$$

For waveguides of rectangular or circular cross section, ψ_j can be determined from (5.76) in an appropriate coordinate system. With ψ_1 and ψ_2 found, we obtain from (5.74)

$$E_z = \frac{1}{\alpha_2 - \alpha_1}(\alpha_2\psi_1 - \alpha_1\psi_2), \tag{5.78a}$$

$$H_z = \frac{-i}{\alpha_2 - \alpha_1}(\psi_1 - \psi_2). \tag{5.78b}$$

The transverse field components can then be derived from (5.71) and be made to satisfy the boundary conditions. We see that both E_z and H_z exist in a gyrotropic medium moving along the z-axis. Even when the gyrotropic medium is stationary, the two wave equations for E_z and H_z are still coupled and the modes are hybrid. The two wave equations will be decoupled if the medium is uniaxial, whether it is stationary or in motion, for then $\epsilon_g = 0$ and all parameters with subscript g will vanish.

5.3 CAVITY RESONATORS

5.3a Rectangular Cavities

A resonator with uniform cross section in the $\hat{z}$ direction can be viewed as a waveguide with both ends closed. Instead of guided waves propagating along the z axis, the waves are standing in the $\hat{z}$ direction. The standing wave can be viewed as a superposition of a guided wave in the $+\hat{z}$ direction and a guided wave in the $-\hat{z}$ direction. The formulation for waveguides in Section 5.1a is applicable to resonators if ik_z is replaced by $\partial/\partial z$. Equations 5.30 and 5.31 become

$$\mathbf{E}_s = \frac{1}{k^2 - k_z^2}\left[\nabla_s \frac{\partial}{\partial z} E_z + i\omega\mu \nabla_s \times \mathbf{H}_z\right], \tag{5.79a}$$

$$\mathbf{H}_s = \frac{1}{k^2 - k_z^2}\left[\nabla_s \frac{\partial}{\partial z} H_z - i\omega\epsilon \nabla_s \times \mathbf{E}_z\right], \tag{5.79b}$$

and

$$(\nabla^2 + k^2)E_z = 0, \tag{5.80a}$$

$$(\nabla^2 + k^2)H_z = 0. \tag{5.80b}$$

The Laplacian operator ∇^2 in (5.80) is now a three-dimensional operator.

Consider a metallic rectangular cavity as shown in Fig. 5.10. It is a waveguide closed with metallic walls at $z=0$ and $z=d$. To satisfy the boundary conditions, we find for TM modes

$$E_z = E_{mnp} \sin\frac{m\pi x}{a} \sin\frac{n\pi y}{b} \cos\frac{p\pi z}{d}, \tag{5.81a}$$

$$E_x = -\frac{E_{mnp}}{(m\pi/a)^2 + (n\pi/b)^2} \frac{m\pi}{a} \frac{p\pi}{d} \cos\frac{m\pi x}{a} \sin\frac{n\pi y}{b} \sin\frac{p\pi z}{d}, \tag{5.81b}$$

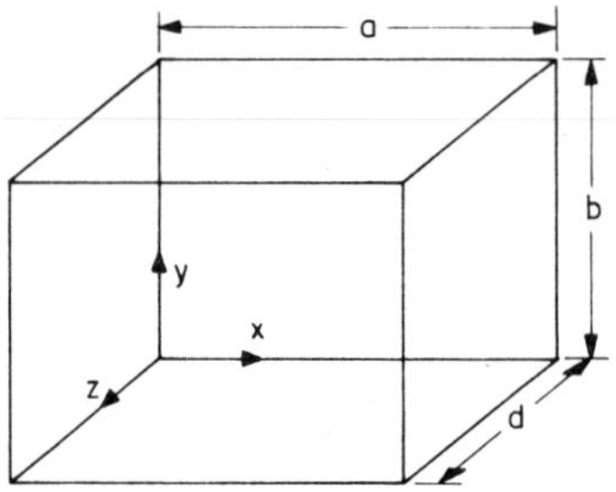

Figure 5.10 Rectangular cavity configuration.

$$E_y = -\frac{E_{mnp}}{(m\pi/a)^2 + (n\pi/b)^2} \frac{n\pi}{b} \frac{p\pi}{d} \sin\frac{m\pi x}{a} \cos\frac{n\pi y}{b} \sin\frac{p\pi z}{d}, \tag{5.81c}$$

$$H_x = \frac{i\omega\epsilon E_{mnp}}{(m\pi/a)^2 + (n\pi/b)^2} \frac{n\pi}{b} \sin\frac{m\pi x}{a} \cos\frac{n\pi y}{b} \cos\frac{p\pi z}{d}, \tag{5.81d}$$

$$H_y = \frac{i\omega\epsilon E_{mnp}}{(m\pi/a)^2 + (n\pi/b)^2} \frac{m\pi}{a} \cos\frac{m\pi x}{a} \sin\frac{n\pi y}{b} \cos\frac{p\pi z}{d}; \tag{5.81e}$$

and for TE modes

$$H_z = H_{mnp} \cos\frac{m\pi x}{a} \cos\frac{n\pi y}{b} \sin\frac{p\pi z}{d}, \tag{5.82a}$$

$$H_x = -\frac{H_{mnp}}{(m\pi/a)^2 + (n\pi/b)^2} \frac{m\pi}{a} \frac{p\pi}{d} \sin\frac{m\pi x}{a} \cos\frac{n\pi y}{b} \cos\frac{p\pi z}{d}, \tag{5.82b}$$

$$H_y = -\frac{H_{mnp}}{(m\pi/a)^2 + (n\pi/b)^2} \frac{n\pi}{b} \frac{p\pi}{d} \cos\frac{m\pi x}{a} \sin\frac{n\pi y}{b} \cos\frac{p\pi z}{d}, \tag{5.82c}$$

$$E_x = -\frac{i\omega\mu H_{mnp}}{(m\pi/a)^2 + (n\pi/b)^2} \frac{n\pi}{b} \cos\frac{m\pi x}{a} \sin\frac{n\pi y}{b} \sin\frac{p\pi z}{d}, \tag{5.82d}$$

$$E_y = \frac{i\omega\mu H_{mnp}}{(m\pi/a)^2 + (n\pi/b)^2} \frac{m\pi}{a} \sin\frac{m\pi x}{a} \cos\frac{n\pi y}{b} \sin\frac{p\pi z}{d}. \tag{5.82e}$$

Substituting (5.81a) and (5.82a) in (5.80), we obtain the same dispersion relation for the TM and TE modes.

$$k_r^2 = \left(\frac{m\pi}{a}\right)^2 + \left(\frac{n\pi}{b}\right)^2 + \left(\frac{p\pi}{d}\right)^2$$

gives the resonant wave number k_r. The resonant wave numbers for TM_{mnp} modes and TE_{mnp} modes are identical. It is interesting to observe that TM_{mn0} modes correspond to waveguide modes at cutoff, where $k_z = 0$.

When the resonator dimensions are such that $a \geqslant b \geqslant d$, the lowest resonant wave number is found to be

$$k_r = \sqrt{\left(\frac{\pi}{a}\right)^2 + \left(\frac{\pi}{b}\right)^2} \tag{5.83}$$

with $m = n = 1$ and $p = 0$. The mode inside the resonator is TM_{110}. The field distribution is illustrated in Fig. 5.11. We see that the electric fields are perpendicular to the plate boundaries at $z = 0$ and $z = d$ and concentrate at the center of the cavity so that the tangential **E** field vanishes at the boundaries $x = 0$, a and $y = 0$, b. This field can also be viewed as a dominant waveguide mode propagating in the $\hat{y}$ direction and reflected at the walls

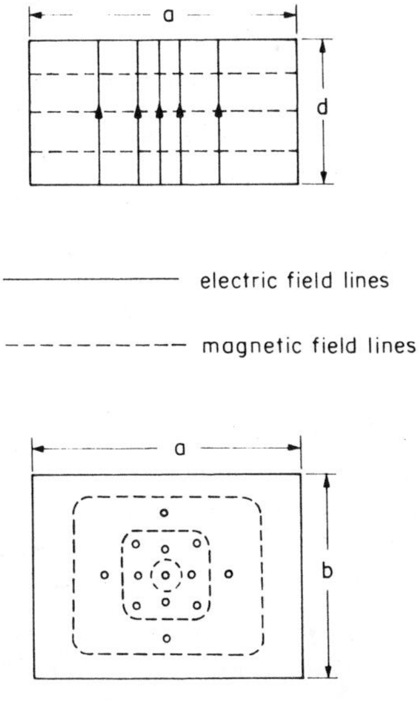

Figure 5.11 Field distributions inside a rectangular cavity. Solid lines: electric field. Broken lines: magnetic field.

$y=0$ and $y=b$ to form a standing wave. If the labels of the coordinate axes y and z are interchanged, this mode may also be called a TE_{101} mode.

In all cavity resonators, the quality factor Q expresses the ratio of energy storage to power dissipation. Let U be the energy stored in a resonator and P_d be the power dissipation in the resonator. We define

$$Q=\frac{\omega_0 U}{P_d},$$

where ω_0 is the resonant angular frequency. Under the assumption of a lossless medium, we calculate

$$U=\frac{1}{2}\operatorname{Re}\left\{\int_0^d dz\int_0^b dy\int_0^a dx\left[\frac{\epsilon}{2}|E|^2+\frac{\mu}{2}|H|^2\right]\right\}=\frac{\epsilon abd}{8}E_{110}^2$$

for the dominant TM_{110} mode in the rectangular cavity. Integrating (5.5) over the cavity walls, we obtain

$$P_d=\frac{1}{2}\sqrt{\frac{\omega_0\mu}{2\sigma}}\left\{2\int_0^d dz\int_0^a dx|H_x|_{y=0}^2+2\int_0^d dz\int_0^b dy|H_y|_{x=0}^2\right.$$

$$\left.+2\int_0^a dx\int_0^b dy\left[|H_x|^2+|H_y|^2\right]_{z=0}\right\}$$

$$=\frac{1}{2}\sqrt{\frac{\omega_0\mu}{2\sigma}}\left[\frac{ad}{b^2}+\frac{bd}{a^2}+\frac{1}{2}\left(\frac{b}{a}+\frac{a}{b}\right)\right]\frac{\pi^2\omega_0^2\epsilon^2E_{110}^2}{\left(\pi^2/a^2+\pi^2/b^2\right)^2}.$$

Therefore

$$Q=\sqrt{\frac{2\sigma}{\omega_0\epsilon}}\;\frac{\pi d\left(a^2+b^2\right)^{3/2}}{2\left[ab\left(a^2+b^2\right)+2d\left(a^3+b^3\right)\right]}. \tag{5.84}$$

In this derivation, we used the fact that $\omega_0\sqrt{\mu\epsilon}=\sqrt{\pi^2/a^2+\pi^2/b^2}$. For a cubic cavity with $a=b=d=2$ cm, the resonant frequency, according to (5.83), is approximately 10GHz; and the quality factor is $Q\approx 10^4$ when the cavity is made of copper walls with air inside. Other sources of loss, such as the material filling the cavity, surface irregularities of the cavity walls, and coupling between other systems, all contribute toward power dissipation P_d and thereby decrease Q.

5.3b Circular Cavities

Consider a circular cavity with height d and radius a with $d < a$ (Fig. 5.12). The fundamental mode is TM_{010}, which corresponds to the waveguide mode TM_{01} at cutoff. The fields inside the cavity are

$$E_z = E_0 J_0(k\rho), \tag{5.85a}$$

$$H_\phi = -i\sqrt{\frac{\epsilon}{\mu}}\, E_0 J_1(k\rho). \tag{5.85b}$$

The resonant wave number is

$$k_r a = 2.405. \tag{5.85c}$$

The time-average energy stored in the cavity is calculated as

$$U = \frac{1}{2}\int_0^a 2\pi\rho\, d\rho\left[\frac{\epsilon}{2}|E_z|^2 + \frac{\mu}{2}|H_\phi|^2\right]d = E_0^2\frac{\pi\epsilon d}{2}a^2 J_1^2(ka).$$

The integral formula for Bessel functions,

$$\int \rho\, d\rho B_m^2(k\rho) = \frac{\rho^2}{2}\left[B_m'^2(k\rho) + \left(1 - \frac{m^2}{k^2\rho^2}\right)B_m^2(k\rho)\right] \tag{5.86}$$

is used, and also $J_0(ka) = 0$. The power dissipation caused by wall loss is

$$P_d = \frac{1}{2}\sqrt{\frac{\omega_0\mu}{2\sigma}}\, E_0^2\left[2\pi a d\frac{\epsilon}{\mu}J_1^2(ka) + 2\int_0^a 2\pi\rho\frac{\epsilon}{\mu}J_1^2(k\rho)\, d\rho\right]$$

$$= \sqrt{\frac{\omega_0\mu}{2\sigma}}\, E_0^2\frac{\epsilon}{\mu}\pi a(d + a)J_1^2(ka). \tag{5.87}$$

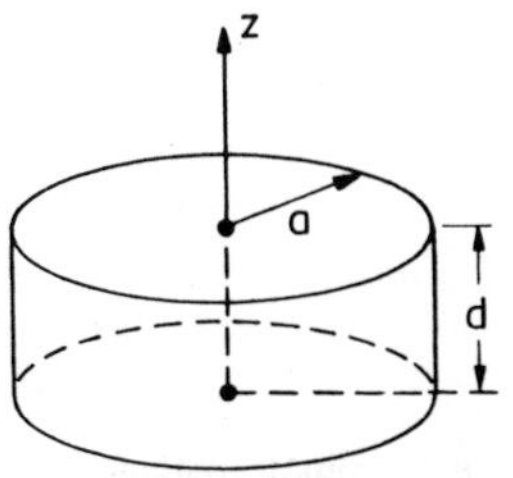

Figure 5.12 Circular cavity configuration.

The first term is due to loss on the side wall, and the second term to loss on the walls at $z=0$ and $z=d$. The quality factor becomes

$$Q=\frac{\omega_0 U}{P_d}=\sqrt{\frac{2\sigma}{\omega_0\epsilon}}\,\frac{2.405}{2(1+a/d)}, \tag{5.88}$$

where we made use of (5.85c). In the mode designation TM_{010} the three subscripts correspond to ϕ, ρ, and z variations, respectively. The TE_{011} mode, for instance, is the waveguide mode TE_{01} forming a standing wave in the $\hat{z}$ direction.

5.3c Spherical Cavities

For a spherical cavity, the formulation in Section 5.1a breaks down because there is no uniform cross section in any direction. We consider Maxwell's equations in spherical coordinates (Fig. 5.13), and treat the case with ϕ symmetry, $\partial/\partial\phi=0$. Instead of decomposing a general field into TM and TE to $\hat{z}$ components, we decompose into TM and TE to $\hat{r}$ components. For the TM waves, Maxwell's equations give

$$\frac{\partial}{\partial r}(rE_\theta)-\frac{\partial E_r}{\partial\theta}=i\omega\mu rH_\phi, \tag{5.89a}$$

$$\frac{1}{r\sin\theta}\frac{\partial}{\partial\theta}(H_\phi\sin\theta)=-i\omega\epsilon E_r, \tag{5.89b}$$

$$-\frac{\partial}{\partial r}(rH_\phi)=-i\omega\epsilon rE_\theta. \tag{5.89c}$$

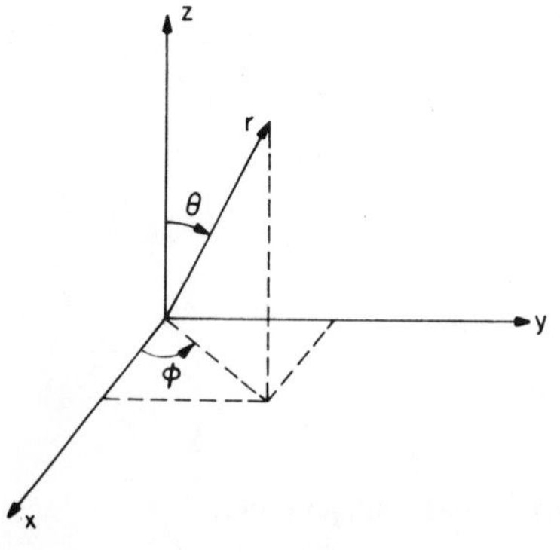

Figure 5.13 Spherical coordinate system.

Inserting (5.89b) and (5.89c) into (5.89a) yields an equation for H_ϕ:

$$\frac{1}{r}\frac{\partial^2}{\partial r^2}(rH_\phi)+\frac{1}{r^2\sin\theta}\frac{\partial}{\partial\theta}\left(\sin\theta\frac{\partial H_\phi}{\partial\theta}\right)-\frac{1}{r^2\sin^2\theta}H_\phi+k^2H_\phi=0. \quad (5.90)$$

A similar equation, which is the dual of (5.90), can be obtained for TE waves. Before treating (5.89) and (5.90), we study general solutions to Helmholtz wave equations in spherical coordinates.

From source-free Maxwell's equations in isotropic media, the wave equation for **E** and **H** is readily derived:

$$(\nabla^2+k^2)\left\{\begin{matrix}\mathbf{E}\\ \mathbf{H}\end{matrix}\right\}=0.$$

Wave equation 5.90 can be derived directly from this equation. Let $W(r,\theta,\phi)$ denote any rectangular component of **E** or **H**. Then the wave equation in spherical coordinates takes the form

$$\frac{1}{r}\frac{\partial^2}{\partial r^2}(rW)+\frac{1}{r^2\sin\theta}\frac{\partial}{\partial\theta}\left(\sin\theta\frac{\partial W}{\partial r}\right)+\frac{1}{r^2\sin^2\theta}\frac{\partial^2 W}{\partial\phi^2}+k^2W=0. \quad (5.91)$$

The solution to this wave equation is obtained by separation of variables and is expressed in terms of spherical Bessel functions $b_n(kr)$, associated Legendre polynomials $L_m^n(\cos\theta)$, and harmonic functions $e^{\pm im\phi}$:

$$\begin{aligned}W&=R(r)\Theta(\theta)\Phi(\phi)\\&=b_n(kr)L_n^m(\cos\theta)e^{\pm im\phi}.\end{aligned} \quad (5.92)$$

The special functions satisfy the following differential equations:

$$r\frac{d^2}{dr^2}(rR)+\left[(kr)^2-n(n+1)\right]R=0, \quad (5.93a)$$

$$\frac{1}{\sin\theta}\frac{d}{d\theta}\left(\sin\theta\frac{d\Theta}{d\theta}\right)+\left[n(n+1)-\frac{m^2}{\sin^2\theta}\right]\Theta=0, \quad (5.93b)$$

$$\frac{d^2\Phi}{d\phi^2}+m^2\Phi=0. \quad (5.93c)$$

The spherical Bessel functions are related to the ordinary Bessel functions by

$$b_n(kr)=\sqrt{\frac{\pi}{2kr}}\;B_{n+1/2}(kr). \quad (5.94)$$

If n is an integer, $B_{n+1/2}$ reduces to simple sinusoids and powers of r. For instance,

$$J_{1/2}(kr)=\sqrt{\frac{2}{\pi kr}}\ \sin kr, \tag{5.95a}$$

$$J_{3/2}(kr)=\sqrt{\frac{2}{\pi kr}}\ \left(\frac{\sin kr}{kr}-\cos kr\right), \tag{5.95b}$$

$$N_{1/2}(kr)=-\sqrt{\frac{2}{\pi kr}}\ \cos kr, \tag{5.96a}$$

$$N_{3/2}(kr)=-\sqrt{\frac{2}{\pi kr}}\ \left(\sin kr+\frac{\cos kr}{kr}\right). \tag{5.96b}$$

The spherical Hankel functions of the first kind take the form

$$h_0^{(1)}(kr)=\sqrt{\frac{\pi}{2kr}}\ \left[J_{1/2}(kr)+iN_{1/2}(kr)\right]$$
$$=\frac{e^{ikr}}{ikr}, \tag{5.97a}$$

$$h_1^{(1)}(kr)=\sqrt{\frac{\pi}{2kr}}\ \left[J_{3/2}(kr)+iN_{3/2}(kr)\right]$$
$$=-\left(1+\frac{i}{kr}\right)\frac{e^{ikr}}{kr}. \tag{5.97b}$$

The spherical Hankel function of the second kind is the complex conjugate of $h_n^{(1)}$.

The first few orders of the associated Legendre polynomials of degree 1 take these forms:

$$P_0^1(\cos\theta)=0, \tag{5.98a}$$

$$P_1^1(\cos\theta)=\sin\theta, \tag{5.98b}$$

$$P_2^1(\cos\theta)=3\sin\theta\cos\theta. \tag{5.98c}$$

It is a general property that all of the associated Legendre polynomials $P_n^1(\cos\theta)$ are zero at $\theta=0$ and π; at $\theta=\pi/2$, they are zero if n is even and maximum if n is odd. For the H_ϕ component,

$$\mathbf{H}=\hat{\phi}H_\phi=(-\hat{x}\sin\theta+\hat{y}\cos\theta)H_\phi.$$

Substituting in (5.91) yields

$$\left(\nabla^2+k^2-\frac{1}{r^2\sin^2\theta}\right)H_\phi=0. \tag{5.99}$$

The effect of the last term on the solution is to increase the associated Legendre polynomial by one more degree in m. Note that (5.90) follows directly from (5.99).

In view of (5.91) and its solutions in (5.92), we see that the solutions for H_ϕ in (5.90) take the form

$$H_\phi=b_n(kr)L_n^1(\cos\theta). \tag{5.100}$$

There is no ϕ dependence. For a spherical cavity with radius a, the spherical Bessel function is used because the origin is included. For the lowest TM mode we let $n=1$. We use three subscripts on TM to denote the variations around r, ϕ, and θ, respectively. The TM_{101} mode has the field solutions

$$H_\phi=H_0\sin\theta\sqrt{\frac{\pi}{2kr}}\,J_{3/2}(kr)$$
$$=H_0\frac{\sin\theta}{kr}\left(\frac{\sin kr}{kr}-\cos kr\right),$$

$$E_r=i2H_0\sqrt{\frac{\mu}{\epsilon}}\,\frac{\cos\theta}{k^2r^2}\left(\frac{\sin kr}{kr}-\cos kr\right),$$

$$E_\theta=-iH_0\sqrt{\frac{\mu}{\epsilon}}\,\frac{\sin\theta}{k^2r^2}\left(\frac{k^2r^2-1}{kr}\sin kr+\cos kr\right).$$

The boundary condition of vanishing E_θ at $r=a$ gives

$$\tan ka=\frac{ka}{1-k^2a^2}.$$

Solving this transcendental equation yields $ka\approx 2.74$, which gives the resonant wave number of the cavity.

5.3d Cavity Perturbation

The resonant frequency of a cavity changes when a small perturbation is applied to either the cavity wall or the medium inside the cavity. First, we consider an inward perturbation of the cavity wall (Fig. 5.14). The unperturbed fields have resonant frequency ω_0 and satisfy Maxwell's equations

$$\nabla \times \mathbf{E}_0 = i\omega_0 \mu \mathbf{H}_0, \tag{5.101a}$$

$$\nabla \times \mathbf{H}_0 = -i\omega_0 \epsilon \mathbf{E}_0. \tag{5.101b}$$

With the perturbation, the resonant frequency becomes ω and the fields satisfy Maxwell's equations:

$$\nabla \times \mathbf{E} = i\omega\mu\mathbf{H}, \tag{5.102a}$$

$$\nabla \times \mathbf{H} = -i\omega\epsilon\mathbf{E}. \tag{5.102b}$$

The task is to calculate the deviation of ω from ω_0. We dot-multiply the complex conjugate of (5.101a) by $\mathbf{H}$ and subtract (5.102b) dot-multiplied by $\mathbf{E}_0^*$. The result is

$$\nabla \cdot (\mathbf{E}_0^* \times \mathbf{H}) = -i\omega_0 \mu \mathbf{H} \cdot \mathbf{H}_0^* + i\omega\epsilon \mathbf{E} \cdot \mathbf{E}_0^*. \tag{5.103a}$$

Next, we dot-multiply (5.102a) by $\mathbf{H}_0^*$ and subtract the complex conjugate of (5.101b) dot-multiplied by $\mathbf{E}$. The result is

$$\nabla \cdot (\mathbf{E} \times \mathbf{H}_0^*) = i\omega\mu\mathbf{H} \cdot \mathbf{H}_0^* - i\omega_0\epsilon\mathbf{E} \cdot \mathbf{E}_0^*. \tag{5.103b}$$

Integrating the sum of (5.103a) and (5.103b) over the unperturbed volume $V_0 = V + \Delta V$, we find

$$\oiint_{\Delta S} d\mathbf{S} \cdot \mathbf{E} \times \mathbf{H}_0^* = i(\omega - \omega_0) \iiint_{V_0} dV (\epsilon \mathbf{E} \cdot \mathbf{E}_0^* + \mu \mathbf{H} \cdot \mathbf{H}_0^*).$$

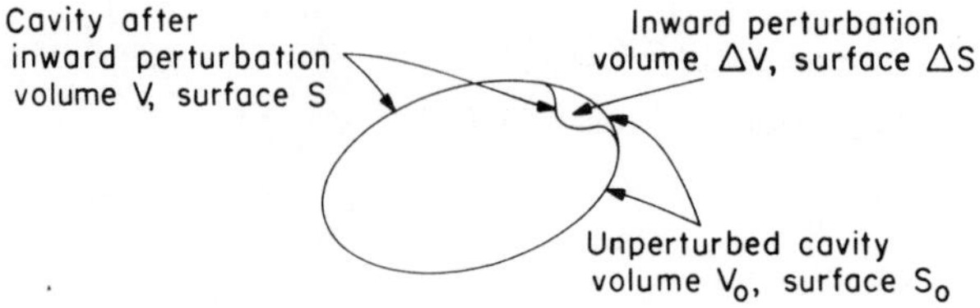

Figure 5.14 Inward perturbation of cavity walls. The unperturbed volume V_0 is related to the volume V after perturbation by $V_0 = V + \Delta V$.

In these calculations, we use the fact that tangential **E** vanishes on the perturbed cavity surface and tangential $\mathbf{E}_0$ vanishes on the unperturbed cavity surface. Note that the surface integration extends over the small perturbed surface ΔS, while the volume integration extends over the unperturbed volume. We obtain the exact equation

$$\omega-\omega_0=-i\frac{\oiint_{\Delta S} d\mathbf{S}\cdot\mathbf{E}\times\mathbf{H}_0^*}{\iiint_{V_0} dV(\epsilon\mathbf{E}\cdot\mathbf{E}_0^*+\mu\mathbf{H}\cdot\mathbf{H}_0^*)} \tag{5.104}$$

Now we assume that the perturbation is so small that we can replace **E** and **H** on the right-hand side of (5.104) by their unperturbed values $\mathbf{E}_0$ and $\mathbf{H}_0$ to obtain approximate values for $\omega-\omega_0$:

$$\omega-\omega_0\approx-i\frac{\oiint_{\Delta S} d\mathbf{S}\cdot\mathbf{E}_0\times\mathbf{H}_0^*}{\iiint_{V_0} dV[\epsilon|\mathbf{E}_0|^2+\mu|\mathbf{H}_0|^2]}$$

$$=\omega_0\frac{\iiint_{\Delta V} dV[\mu|\mathbf{H}_0|^2-\epsilon|\mathbf{E}_0|^2]}{\iiint_{V_0} dV[\mu|\mathbf{H}_0|^2+\epsilon|\mathbf{E}_0|^2]}=\omega_0\frac{\Delta W_m-\Delta W_e}{W_m+W_e}. \tag{5.105}$$

The denominator is the unperturbed total energy storage in the cavity. The numerator is the difference between the magnetic energy and the electric energy removed by the inward perturbation. Thus, if the inward perturbation is produced at a place of large magnetic field, the resonant frequency is raised; if it is produced at a place of large electric field, the resonant frequency is lowered. An opposite effect occurs for an outward perturbation.

Next, we investigate the resonant frequency change caused by material perturbation inside the cavity. Let the unperturbed medium be isotropic. To be more general, we include anisotropy in the perturbation. Maxwell's equations before and after perturbation are

$$\nabla\times\mathbf{E}_0=i\omega_0\mu\mathbf{H}_0, \tag{5.106a}$$

$$\nabla\times\mathbf{H}_0=-i\omega_0\epsilon\mathbf{E}_0, \tag{5.106b}$$

and

$$\nabla\times\mathbf{E}=i\omega\mu\mathbf{H}+i\omega\Delta\bar{\mu}\cdot\mathbf{H}, \tag{5.107a}$$

$$\nabla\times\mathbf{H}=-i\omega\epsilon\mathbf{E}-i\omega\Delta\bar{\epsilon}\cdot\mathbf{E}. \tag{5.107b}$$

We dot-multiply the complex conjugate of (5.106a) by **H** and subtract (5.107b) dot-multiplied by $\mathbf{E}_0^*$. The result is

$$\nabla \cdot (\mathbf{E}_0^* \times \mathbf{H}) = -i\omega_0 \mu \mathbf{H}_0^* \cdot \mathbf{E} + i\omega\epsilon \mathbf{E} \cdot \mathbf{E}_0^* + i\omega(\Delta\bar{\epsilon} \cdot \mathbf{E}) \cdot \mathbf{E}_0^*.$$

Similar operation on (5.106b) and (5.107a) gives

$$\nabla \cdot (\mathbf{E} \times \mathbf{H}_0^*) = i\omega\mu \mathbf{H} \cdot \mathbf{H}_0^* + i\omega(\Delta\bar{\mu} \cdot \mathbf{H}) \cdot \mathbf{H}_0^* - i\omega_0\epsilon \mathbf{E} \cdot \mathbf{E}_0^*.$$

Integrating the sum of these two equations over the cavity volume and making use of the boundary condition that both $\hat{n} \times \mathbf{E} = 0$ and $\hat{n} \times \mathbf{E}_0 = 0$ on the cavity surface, we obtain

$$\frac{\omega - \omega_0}{\omega} = \frac{-\iiint_V dV[(\Delta\bar{\mu} \cdot \mathbf{H}) \cdot \mathbf{H}_0^* + (\Delta\bar{\epsilon} \cdot \mathbf{E}) \cdot \mathbf{E}_0^*]}{\iiint_V dV(\mu \mathbf{H} \cdot \mathbf{H}_0^* + \epsilon \mathbf{E} \cdot \mathbf{E}_0^*)}. \tag{5.108}$$

This is also an exact formula. We write the perturbation formula as

$$\frac{\omega - \omega_0}{\omega_0} \approx -\frac{\Delta W_m + \Delta W_e}{W_m + W_e}. \tag{5.109}$$

The denominator expresses the unperturbed total energy inside the cavity, and the numerator corresponds to the increase in magnetic and electric energies caused by the material perturbation. Thus any increase in the permeability and the permittivity of the material inside the cavity decreases the resonant frequency. Recall that the resonant wave number for the dominant mode in a circular resonator is $k_r a = 2.405$. As μ or ϵ increases, the resonant frequency ω_0 decreases. The material, of course, need not be uniformly perturbed throughout the cavity. The calculation of ΔW_m and ΔW_e corresponding to $\Delta\bar{\mu}$ and $\Delta\bar{\epsilon}$ extends only over the region where perturbation occurs.

PROBLEMS

5.1. A parallel-plate waveguide is closed at $z = -d$ and excited by a line source placed at $z = 0$ and a distance h above the bottom plate. Write the fields produced by this source inside the waveguide as a superposition of the guided modes and determine their mode amplitudes.

5.2. A plane slab of polystyrene ($\epsilon = 2.56\epsilon_0$) is 1 cm thick. What are the cutoff frequencies of the first six modes? At a frequency of 30 GHz, what are the propagating guided modes? Repeat the problem when the slab waveguide has air on one side and another dielectric medium with $\epsilon_t = 2\epsilon_0$ on the other side.

5.3. Consider a rectangular waveguide with dimensions $1\text{cm} \times 0.5\text{cm}$. What are the cutoff frequencies for the first five modes? If the waveguide is excited at 20GHz, what are the propagation constants for the first five modes? If the waveguide is excited at 50GHz, how many modes will propagate?

5.4. The modes in a parallel-plate waveguide are orthogonal to each other because they are sinusoidal functions. In general we can prove that waveguide modes are orthogonal by showing that

$$\int\int_S dS(\mathbf{E}_m \times \mathbf{H}_n^*)\cdot\hat{z} = 0 \qquad \text{for } m \neq n,$$

where $\mathbf{E}_m$ is the $\mathbf{E}$ field for the mth mode, and $\mathbf{H}_n$ is the $\mathbf{H}$ field for the nth mode. The integral is carried out over the cross section of the waveguide. Show that the TE modes of a slab waveguide are orthogonal by verifying that $\int_{-\infty}^{\infty} E_{ym} H_{xn}^* \, dx = 0$. Note that the integration range is from $-\infty$ to $+\infty$; in the calculations field solutions both inside and outside the slab are needed.

5.5. Consider modes in a parallel-plate waveguide filled with an anisotropic medium. Let the medium in the waveguide be a cold plasma with a dc magnetic field $\mathbf{B}_0 = \hat{y}B_0$. Are the TE modes affected by the anisotropy of the plasma? Derive the expressions for TM modes.

5.6. In a parallel-plate waveguide, the region $z < 0$ is free space and the region $z > 0$ is filled with a dielectric medium having permittivity ϵ. How would you write the modes for both regions? For a TM mode incident upon the dielectric from the $z < 0$ region, show that there is no reflected wave when $f/f_c = (1 + \epsilon_0/\epsilon)^{1/2}$. Compare this result with the case of a wave incident upon a half-space medium at the Brewster angle.

5.7. A rectangular waveguide (Fig. 5.6) is excited by a probe at the position $x = d$. Assume that the probe extends from $y = 0$ to $y = b$, and approximate the current on the probe by

$$J_s(x,y) = I_0 \delta(x - d)\cos qy.$$

What are the mode amplitudes as excited by this probe? To achieve maximum excitation for the TE_{10} mode, where should the probe be placed?

5.8. Consider that the rectangular waveguide in Fig. 5.6 is partially filled with a dielectric medium ϵ_1 from $y=0$ to $y=d$, and a dielectric medium ϵ_2 from $y=d$ to $y=b$. Show that the guided waves inside, in general, are hybrid modes. Letting $\epsilon_1 \approx \epsilon_2$, show that the cutoff frequency for the dominant mode is

$$\omega_c \approx \frac{\pi}{a}\left[\frac{\epsilon_1(b-d)+\epsilon_2 d}{\mu\epsilon_1\epsilon_2 b}\right]^{1/2}.$$

5.9. To derive (5.72), first show that $\nabla \cdot \mathbf{D}=0$ yields, in view of (5.69),

$$\omega\epsilon' a\nabla_s \cdot \mathbf{E}_s - id_g\nabla_s \cdot \mathbf{H}_s + k^2 a_g H_z + i\omega\epsilon_z' dE_z = 0.$$

Similarly, $\nabla \cdot \mathbf{B}=0$ yields

$$\omega\mu' b\nabla_s \cdot \mathbf{H}_s + id_g\nabla_s \cdot \mathbf{E}_s - \frac{\epsilon_z'}{\epsilon'}k^2 b_g E_z + i\omega\mu' dH_z = 0.$$

Solve these two equations for $\nabla_s \cdot \mathbf{E}_s$ and $\nabla_s \cdot \mathbf{H}_s$. Take the two dimensional curl of (5.68) and introduce the values for $\nabla_s \cdot \mathbf{E}_s$ and $\nabla_s \cdot \mathbf{H}_s$. Show that (5.72a) and (5.72b) follow from (5.68a) and (5.68b), respectively.

5.10. A cylindrical metallic waveguide with a gap at $z=0$ is excited by a voltage source so that $E_z(a,\phi,z)=V\delta(z)$, where a is the radius of the waveguide. Write the field inside the waveguide as

$$E_z = \int_{-\infty}^{\infty} dk_z g(k_z) J_0(k_\rho \rho) e^{ik_z z}.$$

Show that $g(k_z)=V/[2\pi J_0(k_\rho a)]$. Determine E_z by contour integration.

5.11. In a fiberglass waveguide having center core with a radius of the order of a micron and cladding with a radius of the order of 100 μm, the HE_{11} mode operating range can be extended to the visible range if the refractive indices $n_1 = c(\mu_1\epsilon_1)^{1/2}$ and $n=c(\mu\epsilon)^{1/2}$ are also very close. Because the cladding is very thick in comparison to the core, the wave guidance by an optical fiber can be treated with the dielectric waveguide model as discussed in Section 5.2d. Find the value of $(n_1^2-n^2)^{1/2}$, called the *numerical aperture*, so that the cutoff frequency for the next higher-order mode will be 6×10^{14} Hz. When fiberglass is used as the transmission medium in communication, it not only provides great bandwidth and large channel capacity but also physical compactness and flexibility. Compare the result with the slab and with the metallic waveguides.

5.12. Consider a coaxial line with inner radius a and outer radius b. Assume that $b=a(1+\delta)$ and $\delta \ll 1$. The fundamental mode in this waveguide is TEM, which has zero cutoff wave number. What is the cutoff wave number for the next higher-order mode? The question can be answered by observing that, when δ is very small, the guiding space approaches that between two parallel plates. The field inside the parallel-plate waveguide satisfies periodicity conditions. Using this model, show that the answer to the problem is $k_c \approx 1/a$. Note the similarity to the cutoff wave number for the TE_{20} mode in a rectangular waveguide (Fig. 5.6) with width $2\pi a$ and height δ. Confirm the answer by evaluating (5.52). Note that the cutoff wave number for TE_1 and TM_1 modes in a parallel-plate waveguide is $k_c = \pi/\delta$, which is much larger than $1/a$.

5.13. Apply the formulation in Section 5.2e to a moving uniaxial medium inside a cylindrical metallic waveguide with a rectangular cross section. Show that the forms of the results for guidance conditions and cutoff phenomena are identical to those for a parallel-plate waveguide containing the same moving medium. Find the total Poynting's power attributable to all of the waveguide modes (see Du and Compton[31]).

5.14. Consider a rectangular cavity with a square base and height a. A thin slab of dielectric with thickness d is placed on the base. Determine the change in resonant frequency by using perturbation formula 5.109. Show that, if we assume unperturbed values for the fields, the result is

$$\frac{\omega-\omega_0}{\omega_0} \approx -\tfrac{1}{2}(\epsilon_r - 1)\frac{d}{a}.$$

5.15. The field in a spherical cavity with radius a is given by

$$H_\phi = H_0 j_1\left(2.744\frac{r}{a}\right)\sin\theta.$$

Assume that the cavity is perturbed by a concentric dielectric with radius b and permittivity ϵ. Prove that

$$\frac{\omega-\omega_0}{\omega_0} \approx -0.291\frac{\epsilon_r - 1}{\epsilon_r + 2}\left(2.744\frac{b}{a}\right)^3,$$

by using the quasi-static field inside the dielectric sphere; this is approximately $[3/(2+\epsilon_r)]E_0$, where $\epsilon_r = \epsilon/\epsilon_0$ and E_0 is the unperturbed field outside the dielectric.

5.16. Derive a material perturbation formula for propagation constants in a waveguide. Show that the following formula is exact:

$$k_z - k_{0z} = \omega \frac{\iint_S (\Delta\epsilon \mathbf{E} \cdot \mathbf{E}_0^* + \Delta\mu \mathbf{H} \cdot \mathbf{H}_0^*)\, dS}{\iint_S (\mathbf{E}_0^* \times \mathbf{H} + \mathbf{E} \times \mathbf{H}_0^*) \cdot \hat{z}\, dS},$$

where k_{0z} denotes the unperturbed propagation constant. The field vectors in this formula have no dependence on z. The fields can be approximated by their unperturbed values outside the material and the quasi-static values inside the material.

Consider a circular waveguide of radius a containing a concentric rod of radius b. Show that the change in propagation constant is given by

$$k_z - k_{0z} = k \frac{2.146}{\sqrt{1-(k_c/k)^2}} \frac{\epsilon_r - 1}{\epsilon_r + 1} \left(\frac{b}{a}\right)^2.$$

5.17. A Fabry-Perot resonator is composed of two parallel reflectors within which a standing wave is formed. The field inside is a TEM wave standing between the two plates. Show that the resonant wave numbers are $k_r d = m\pi$. Strictly speaking, because of the finite transverse dimension, a TEM wave is diffracted. Considering diffraction in the cavity, we assume that the transverse field distribution takes a Gaussian form. At $z=0$, $\mathbf{E} = \hat{x}E_0 e^{-y^2/w_0^2}$. At $z>0$,the electric field can be written as a superposition of plane waves with $\mathbf{k} = \hat{z}k_z + \hat{y}k_y$ and $k_y \ll k_z$:

$$\mathbf{E} = \hat{x} \int_{-\infty}^{\infty} dk_y\, E_g e^{ik_z z + ik_y y},$$

where

$$k_z = \sqrt{k^2 - k_y^2} \approx k\left(1 - \frac{1}{2}\frac{k_y^2}{k^2}\right).$$

The amplitude E_g is determined by the field at $z=0$. Determine E_g by using an inverse Fourier transformation. Calculate and show that

$$\mathbf{E} \approx \hat{x} \frac{E_0}{\sqrt{1 + iz/z_F}} e^{ik_z z} e^{(iz/z_F - 1)y^2/w^2},$$

where $z_F = kw_0^2/2$, and $w(z) = w_0\sqrt{1 + z^2/z_F^2}$. Show that, for a given w_0, the locus of w versus z is a hyperbola. The phase front formed by normals to the family of the hyperbolas is curved. The radius of curvature can be determined approximately by $R(z) \approx w(z)/[dw(z)/dz] = z(1 + z_F^2/z^2)$. Draw $w(z)$ and show that the focal points for the right-hand phase front and the left-hand phase front coincide when $z = z_F$. When the mirrors have radius of curvature $R(z_F) = 2z_F$ and are placed at a distance of $d = 2z_F$ apart, the configuration is confocal. Determine modes inside an optical cavity made of confocal mirrors.

6

Radiation and Scattering

The most fundamental radiating element is a Hertzian dipole. We shall study this dipole in detail. The interpretation of Čerenkov radiation, produced by charged particles moving in a medium with velocities exceeding that of light in the medium, marked a great triumph for macroscopic Maxwell's theory. We shall also study this radiation process. To study radiation and scattering caused by a Hertzian dipole in the presence of a stratified medium, we make use of the saddle-point method, which is introduced by first discussing the asymptotic behavior of Hankel functions. Problems involving the scattering of plane waves are then discussed. To illustrate solutions in the rectangular, cylindrical, and spherical coordinate systems, we select examples that can be solved in closed forms.

6.1 RADIATION

6.1a Dyadic Green's Functions

When studying the radiation of electromagnetic waves we are interested in calculating field quantities with given source distributions. A dyadic Green's function expresses a field in terms of its source. The electric field **E** is a vector, and the current **J** is also a vector. The Green's function that operates on **J** to give **E** is then in the form of a dyad:

$$\mathbf{E}(\mathbf{r}) = \int d^3r' \overline{\mathbf{G}}(\mathbf{r},\mathbf{r}') \cdot \mathbf{J}(\mathbf{r}'). \tag{6.1}$$

The differential equation that governs the dyadic Green's function $\overline{\mathbf{G}}(\mathbf{r},\mathbf{r}')$ is a wave equation similar to that satisfied by **E**.

Consider a homogeneous unbounded isotropic medium. Maxwell's equations are as follows:

$$\nabla \times \mathbf{E} = i\omega\mu\mathbf{H}, \tag{6.2a}$$

$$\nabla \times \mathbf{H} = -i\omega\epsilon\mathbf{E} + \mathbf{J}, \tag{6.2b}$$

$$\nabla \cdot \mathbf{H} = 0, \tag{6.2c}$$

$$\nabla \cdot \mathbf{E} = \frac{\rho}{\epsilon}. \tag{6.2d}$$

A wave equation for **E** is easily derived by eliminating **H** from (6.2a) and (6.2b):

$$\nabla \times \nabla \times \mathbf{E}(\mathbf{r}) - k^2\mathbf{E}(\mathbf{r}) = i\omega\mu\mathbf{J}(\mathbf{r}), \tag{6.3}$$

where $k^2 = \omega^2 \mu\epsilon$. Substituting (6.1) in (6.3), we obtain

$$\nabla \times \nabla \times \overline{\mathbf{G}}(\mathbf{r},\mathbf{r}') - k^2 \overline{\mathbf{G}}(\mathbf{r},\mathbf{r}') = i\omega\mu \overline{\mathbf{I}}\,\delta(\mathbf{r}-\mathbf{r}') \tag{6.4}$$

by noting that

$$\mathbf{J}(\mathbf{r}) = \int d^3r'\,\mathbf{J}(\mathbf{r}')\,\delta(\mathbf{r}-\mathbf{r}').$$

The dyadic Green's function can be determined from a scalar Green's function $g(\mathbf{r},\mathbf{r}')$, defined by

$$\overline{\mathbf{G}}(\mathbf{r},\mathbf{r}') = i\omega\mu\left(\overline{\mathbf{I}} + \frac{1}{k^2}\nabla\nabla\right) g(\mathbf{r},\mathbf{r}'). \tag{6.5}$$

Inserting (6.5) into (6.4), we obtain the differential equation satisfied by $g(\mathbf{r},\mathbf{r}')$:

$$(\nabla^2 + k^2) g(\mathbf{r},\mathbf{r}') = -\delta(\mathbf{r}-\mathbf{r}'). \tag{6.6}$$

This is an inhomogeneous Helmholtz wave equation.

To solve (6.6), we translate the coordinate axes so that $\mathbf{r}' = 0$ and (6.6) is spherically symmetric with respect to the new origin. The three-dimensional delta function $\delta(\mathbf{r})$ may be expressed as

$$\delta(\mathbf{r}) = \delta(x)\delta(y)\delta(z) \qquad \text{(rectangular coordinates)}, \tag{6.7a}$$

$$= \frac{\delta(\rho)\delta(z)}{2\pi\rho} \qquad \text{(cylindrical coordinates)}, \tag{6.7b}$$

$$= \frac{\delta(r)}{4\pi r^2} \qquad \text{(spherical coordinates)}. \tag{6.7c}$$

These equalities are proved by noting that

$$\int \delta(\rho)\,d\rho = 1 = \int \delta(x)\delta(y)\,dx\,dy = \int \delta(x)\delta(y)\,2\pi\rho\,d\rho$$

and

$$\int \delta(r)\,dr = 1 = \int \delta(x)\delta(y)\delta(z)\,dx\,dy\,dz = \int \delta(x)\delta(y)\delta(z)\,4\pi r^2\,dr.$$

In the spherical coordinate system (6.6) becomes

$$(\nabla^2 + k^2) g(r) = \frac{-\delta(r)}{4\pi r^2}. \tag{6.8}$$

For $r \neq 0$, the wave equation is homogeneous. The solution must represent an outgoing wave. We have

$$g(r) = C\frac{e^{ikr}}{r}. \tag{6.9}$$

To determine the constant C, we integrate (6.8) over an infinitesimal volume enclosing the origin and let $r \to 0$. In view of (6.9), we obtain $C = 1/4\pi$. The solution for the scalar Green's function $g(\mathbf{r}-\mathbf{r}')$, after transformation to its original coordinate system, is

$$g(\mathbf{r}-\mathbf{r}') = \frac{e^{ik|\mathbf{r}-\mathbf{r}'|}}{4\pi|\mathbf{r}-\mathbf{r}'|}. \tag{6.10}$$

This represents a wave whose constant phase front is a sphere. The scalar Green's function is the response caused by a point source situated at $\mathbf{r}'$.

Substituting (6.10) in (6.5), we obtain the dyadic Green's function and calculate the electric field from (6.1).

$$\mathbf{E}(\mathbf{r}) = i\omega\mu\left(\bar{\mathbf{I}} + \frac{\nabla\nabla}{k^2}\right)\cdot \int d^3r' \frac{e^{ik|\mathbf{r}-\mathbf{r}'|}}{4\pi|\mathbf{r}-\mathbf{r}'|}\mathbf{J}(\mathbf{r}'). \tag{6.11}$$

The dyadic operator can be taken out of the integral because it operates on $\mathbf{r}$. When we are interested in radiation fields generated by a finite, localized source distribution and when the observation point (or the field point) is very far away from the source, the position vector $\mathbf{r}$ and the vector from the source point to the field point $\mathbf{r}-\mathbf{r}'$ are almost parallel. Thus, for $r \gg r'$,

$$|\mathbf{r}-\mathbf{r}'| \approx r - \mathbf{r}'\cdot\hat{r}, \tag{6.12}$$

where $\hat{r}$ is a unit vector along $\mathbf{r}$. Assuming that $kr \gg 1$, we can define a radiation vector $\mathbf{R}$:

$$\mathbf{R} = \int d^3r' \frac{e^{ik|\mathbf{r}-\mathbf{r}'|}}{4\pi|\mathbf{r}-\mathbf{r}'|}\mathbf{J}(\mathbf{r}') \approx \frac{e^{ikr}}{4\pi r}\int d^3r'\mathbf{J}(\mathbf{r}')e^{-i\mathbf{k}\cdot\mathbf{r}'} \tag{6.13}$$

in view of approximation 6.12. Note that we neglect the second term in (6.12) in the denominator but not in the exponent. The $\mathbf{k}$ vector is in the direction of $\mathbf{r}$. We can replace ∇ with $i\mathbf{k}$ and write the electric field in the radiation zone:

$$\mathbf{E}(\mathbf{r}) \approx \frac{i}{\omega\epsilon}(k^2\bar{\mathbf{I}} - \mathbf{k}\mathbf{k})\cdot\mathbf{R} = i\omega\mu\left(\hat{\theta}R_\theta + \hat{\phi}R_\phi\right). \tag{6.14a}$$

The magnetic field strength **H** is determined by (6.2a), as soon as **E** is found. In the radiation zone

$$\mathbf{H}(\mathbf{r}) = ik\left(\hat{\phi}R_\theta - \hat{\theta}R_\phi\right). \tag{6.14b}$$

With a given source $\mathbf{J}(\mathbf{r}')$, the task is to calculate the radiation vector in (6.13). Then the field vectors in the radiation zone are immediately determined by (6.14).

6.1b Hertzian Dipole

Consider an infinitesimal dipole antenna composed of a current element with time-harmonic dependence $e^{-i\omega t}$. Let the dipole be situated at the origin and pointed in the $\hat{z}$ direction (Fig. 6.1). The infinitesimal dipole antenna can be viewed as comprising two opposite charges, $+q$ and $-q$, separated by a distance l so that, in the limit $q\to\infty$, $l\to 0$, $p=ql$ is finite. The dipole oscillates periodically with angular frequency ω. The current of the antenna is $I=-i\omega q$. Thus the dipole moment of the antenna is $Il=-i\omega p$. This antenna model is called a *Hertzian dipole*. We write the current density **J** for the dipole antenna as

$$\mathbf{J} = \hat{z}Il\,\delta(\mathbf{r}'). \tag{6.15}$$

Substituting in (6.11), we find that

$$\begin{aligned}\mathbf{E}(\mathbf{r}) &= i\omega\mu\left(\bar{\mathbf{I}} + \frac{\nabla\nabla}{k^2}\right)\cdot\frac{\hat{z}Ile^{ikr}}{4\pi r} \\ &= \frac{i\omega\mu Ile^{ikr}}{4\pi r}\left[\hat{r}\left(\frac{1}{k^2r^2} - \frac{i}{kr}\right)2\cos\theta + \hat{\theta}\left(\frac{1}{k^2r^2} - \frac{i}{kr} - 1\right)\sin\theta\right].\end{aligned} \tag{6.16a}$$

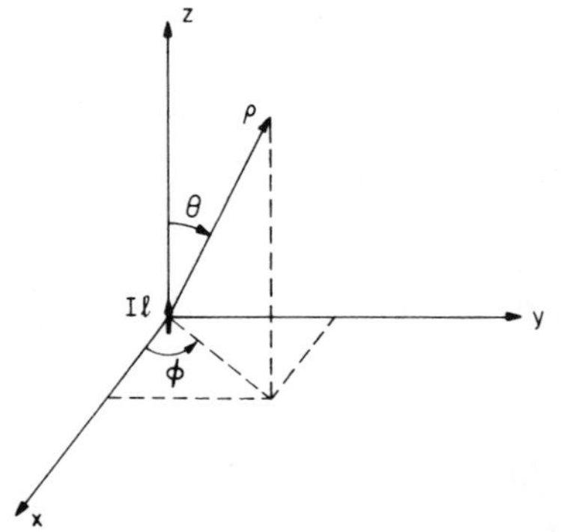

Figure 6.1 Dipole with dipole moment Il located at the origin and pointed in the $\hat{z}$ direction.

In these calculations, we note that $\partial(e^{ikr}/r)/\partial z = (ik - 1/r)\cos\theta e^{ikr}/r$. The solution in (6.16) is valid for the near field, as well as for the radiation field. The associated magnetic field is calculated from Faraday's law:

$$\mathbf{H}(\mathbf{r}) = \frac{1}{i\omega\mu}\nabla\times\mathbf{E} = \hat{\phi}\frac{-ikIle^{ikr}}{4\pi r}\left(\frac{i}{kr}+1\right)\sin\theta. \tag{6.16b}$$

We observe that the electric field has no circumferential component. The magnetic field has only a ϕ component, and the field lines encircle the dipole.

For radiation fields, $kr \gg 1$, and only terms of order $1/r$ are important. Equation 6.16a becomes

$$\mathbf{E} = \hat{\theta}\,\frac{-i\omega\mu Ile^{ikr}}{4\pi r}\sin\theta. \tag{6.17a}$$

This result also follows directly from (6.14a), by noting that (6.13) gives

$$\mathbf{R} = \hat{z}\,\frac{Ile^{ikr}}{4\pi r}.$$

The magnetic field in the radiation zone is determined from either (6.16b) or (6.14b):

$$\mathbf{H} = \hat{\phi}\frac{-ikIle^{ikr}}{4\pi r}\sin\theta = \hat{\phi}\frac{1}{\eta}E_\theta, \tag{6.17b}$$

where $\eta = \sqrt{\mu/\epsilon}$ is the wave impedance. We see that the **E** and **H** fields are perpendicular. At a particular point in space, the field is a plane wave linearly polarized in the plane containing the dipole and the position vector. If we enclose the dipole in a sphere with radius much larger than the wavelength λ, the **E** field is tangent to the sphere. Its magnitude is maximum at $\theta = \pi/2$ and diminsihes as $\sin\theta$ goes to zero at $\theta = 0$. The radiation field pattern is a plot of the magnitude of **E**, $|\mathbf{E}|$, at a constant $r \gg \lambda$. On a plane containing the dipole axis z, the radiation pattern consists of two circles on each side of the dipole axis. Since E_θ is proportional to $\sin\theta$ and independent of ϕ, the field pattern in three-dimensional space is formed by rotating the circles around the axis of the dipole.

The radiation power pattern is a plot of the time-average Poynting's power density $\langle\mathbf{S}\rangle$ at a constant radius r:

$$\langle\mathbf{S}\rangle = \frac{1}{2}\mathrm{Re}(\mathbf{E}\times\mathbf{H}^*) = \hat{r}\,\frac{1}{2}\sqrt{\frac{\mu}{\epsilon}}\left(\frac{kIl}{4\pi r}\right)^2\sin^2\theta. \tag{6.18}$$

Since $\langle \mathbf{S} \rangle$ is proportional to $\sin^2\theta$, the power pattern is flatter than the field pattern. The total time-average power radiated is equal to the integration of $\hat{r}\cdot\langle \mathbf{S} \rangle$ over a large sphere of radius r. The differential area of integration is $r^2 \sin\theta\, d\theta\, d\phi$. The r dependence in the final result drops out. Since $\int_0^\pi \sin^3\theta\, d\theta = 4/3$, we obtain the total radiated power:

$$P = \eta \frac{k^2 (Il)^2}{12\pi}. \tag{6.19}$$

The gain $g(\theta)$ of an antenna in the direction θ is defined as the ratio of the actual radiated power in this direction to the average of the total power of the antenna radiated in all directions:

$$g(\theta) = \frac{\langle S_r(\theta) \rangle}{P/4\pi r^2} = \frac{3}{2}\sin^2\theta. \tag{6.20}$$

Thus the maximum gain of a dipole is 1.5, which occurs at $\theta = \pi/2$. The gain is zero in the polar direction, where $\theta = 0$.

From the circuit point of view, a dipole antenna is an element that dissipates energy. The impedance seen by the source at the terminal is the input impedance of the antenna, which is the ratio of the complex terminal voltage to the complex terminal current. The radiation resistance of the dipole antenna can be obtained from (6.19):

$$R_r = \frac{2P_r}{I^2} = \eta \frac{(kl)^2}{6\pi}. \tag{6.21}$$

It is an equivalent resistance looking from the source terminal. The dissipated power is equal to that radiated into the space surrounding the dipole antenna.

In the static limit, either the frequency is very low or the observation point is very close to the antenna so that $kr \ll 1$, only terms of the order $1/r^3$ in (6.16) are important. We find that

$$\mathbf{E} = \frac{p}{4\pi\epsilon r^3}(\hat{r} 2\cos\theta + \hat{\theta}\sin\theta), \tag{6.22}$$

which is the solution to a static electric dipole.

We now take advantage of these results to treat *Rayleigh scattering*, that is, the scattering of electromagnetic waves by particles much smaller than a wavelength. Consider a spherical particle with radius a and dielectric constant ϵ_t, situated at the origin of a coordinate system. A plane wave polarized in the $\hat{z}$ direction is incident upon the particle, $\mathbf{E}=\hat{z}E_0e^{ikx}$. Because the particle is very small, the scattered field is essentially that resulting from a point source. The $\hat{z}$-directed electric field induces a dipole moment, and the particle reradiates as a dipole antenna. The solution takes the form of (6.16). The dipole moment Il is determined by E_0 and ϵ_t.

Very near the origin, $kr \ll 1$, the scattered field is electric in nature:

$$\mathbf{E} \approx \frac{ikIl}{4\pi r}\sqrt{\frac{\mu}{\epsilon}}\,\frac{1}{(kr)^2}(\hat{r}2\cos\theta+\hat{\theta}\sin\theta).$$

The field inside is uniform and is in the same direction as the incident one:

$$\mathbf{E}=\hat{z}E_i, \qquad r \leqslant a.$$

On the boundary surface, $r=a$, the boundary conditions require that the tangential $\mathbf{E}$ field and the normal $\mathbf{D}$ be continuous. Since $\hat{z}=\hat{r}\cos\theta-\hat{\theta}\sin\theta$, we have

$$-E_0+\frac{ikIl}{4\pi a}\sqrt{\frac{\mu}{\epsilon}}\,\frac{1}{(ka)^2}=-E_i,$$

$$\epsilon E_0+\frac{i\epsilon kIl}{2\pi a}\sqrt{\frac{\mu}{\epsilon}}\,\frac{1}{(ka)^2}=\epsilon_t E_i.$$

When these two equations are solved for Il, the result is

$$Il=-i4\pi ka^3E_0\sqrt{\frac{\epsilon}{\mu}}\,\frac{\epsilon_t-\epsilon}{\epsilon_t+2\epsilon}.$$

We shall now study the scattered field as $kr \gg 1$. Equation 6.17 gives

$$E_\theta=-\frac{\epsilon_t-\epsilon}{\epsilon_t+2\epsilon}k^2a^2E_0\frac{a}{r}e^{ikr}\sin\theta, \qquad (6.23a)$$

$$H_\phi=\sqrt{\frac{\epsilon}{\mu}}\,E_\theta. \qquad (6.23b)$$

The total scattered power from the sphere is

$$P_s = \frac{1}{2}\int_0^{\pi}\int_0^{2\pi} r^2 \sin\theta \, d\theta \, d\phi \, E_\theta H_\phi^*,$$

$$= \frac{4\pi}{3}\sqrt{\frac{\epsilon}{\mu}}\left(\frac{\epsilon_t - \epsilon}{\epsilon_t + 2\epsilon} k^2 a^3 E_0\right)^2. \tag{6.24}$$

The scattering cross section is calculated as

$$\sigma_s = \frac{P_s}{\frac{1}{2}\sqrt{\epsilon/\mu}\,|E_0|^2} = \frac{8\pi}{3}\left(\frac{\epsilon_t - \epsilon}{\epsilon_t + 2\epsilon}\right)^2 k^4 a^6. \tag{6.25}$$

Thus the total scattered power is proportional to the fourth power of the wave number; high-frequency waves are scattered more than lower ones. The scattered power is also proportional to the sixth power of the radius. But as the radius of the sphere becomes larger, these approximations become inaccurate, and other approaches must be used to solve the problem. The scattering process is then called *Mie scattering*. As a matter of fact, the scattering of an electromagnetic plane wave by a sphere of arbitrary size can be solved exactly in closed form.

6.1c Linear Antennas and Array Antennas

In the Hertzian dipole model, the source element has a constant current I. In practice, antennas are made of conductors. For a linear antenna, we use the model of a perfectly conducting wire of finite length and infinitesimal radius. By boundary conditions, currents at the two ends of the conductor are zero. For a very short physical dipole, we assume that the current distributes itself in a triangular shape. We assume also that the dipole is center fed, and let the amplitude at the feeding point be I_0. The average current is $I_0/2$, and the previous results for a Hertzian dipole apply if we replace Il with $I_0 l/2$. The field strengths are decreased by a factor of 2 and the power magnitude is decreased by a factor of 4. In particular, with reference to the input current I_0, the radiation resistance in (6.21) will be down by a factor of 4.

When the antenna is not very short, the detailed distribution of currents on the wire becomes important. The exact form of current distribution cannot be found without solving the complete boundary value problem. In computing radiation fields, we may assume a current distribution and use Green's function to calculate the fields. Let the length of an antenna be l, and let it be center fed. Since the current at the two ends of the wire must be zero, we assume that the current distribution is

$$\mathbf{J}(z)=\hat{z}I_0\sin\left[k\left(\frac{l}{2}-|z|\right)\right]\delta(x)\delta(y). \tag{6.26}$$

The radiation vector is determined from (6.13):

$$\begin{aligned}\mathbf{R}&=\hat{z}\frac{e^{ikr}}{4\pi r}\int_{-l/2}^{l/2}dz'\,I_0\sin\left[k\left(\frac{l}{2}-|z'|\right)\right]e^{-ikz'\cos\theta}\\&=\hat{z}\frac{2I_0}{k\sin^2\theta}\left[\cos\left(k\frac{l}{2}\cos\theta\right)-\cos\left(k\frac{l}{2}\right)\right]\frac{e^{ikr}}{4\pi r}.\end{aligned} \tag{6.27}$$

The radiation vector is $\hat{z}$-directed and has $\hat{r}$ and $\hat{\theta}$ components. In view of (6.14), we obtain

$$\mathbf{E}=\hat{\theta}\left(-\frac{i\eta I_0 e^{ikr}}{2\pi r\sin\theta}\right)\left[\cos\left(k\frac{l}{2}\cos\theta\right)-\cos\left(k\frac{l}{2}\right)\right], \tag{6.28a}$$

$$\mathbf{H}=\hat{\phi}\frac{1}{\eta}E_\theta. \tag{6.28b}$$

As in the short dipole cases, the radiation fields are linearly polarized and Poynting's vector is $\hat{r}$-directed. When the length of the antenna is smaller than a wavelength, the radiation pattern is similar to that for a short dipole, except that the beam is narrower and stronger in the direction perpendicular to $\hat{z}$. As l becomes larger than a wavelength, there will be nulls in the radiation pattern in directions other than $\theta=0$. These null directions can be determined from (6.28a) by letting the numerator equal zero. For instance, a 1.5-wavelength antenna has nulls at $\cos\theta=\pm1/3$, and a two-wavelength antenna has a null at $\pi/2$.

The resultant field pattern of an antenna array is a summation of the fields arising from all individual elements. Consider a linear array of N dipoles lined up along the x axis and pointed in the $\hat{z}$ direction:

$$\mathbf{J}=\hat{z}\sum_{n=0}^{N-1}I_n e^{i\alpha_n}\delta(x-d_n)\delta(y)\delta(z). \tag{6.29}$$

Each element n has phase α_n and amplitude I_n and is situated at $x=d_n$. Using (6.13) and (6.14), we obtain the magnitude of the radiated **E** field:

$$E=\left|\sum_{n=0}^{N-1} E_n e^{i\alpha_n - i\psi_n}\right|, \tag{6.30}$$

where E_n is the amplitude due to element n and ψ_n is given by

$$\psi_n = kd_n \sin\theta \cos\phi.$$

This represents the phase shift caused by separation distance d_n as viewed in the radiation field. The radiation pattern of a linear array antenna can be controlled by the amplitude of individual elements, the phase of individual elements, and the distance between adjacent elements.

The simplest array consists of identical elements spaced equally. Let $E_0(\theta,\phi)$ denote the field radiated by each individual element at a constant **r** in the radiation zone; then

$$E=E_0(\theta,\phi)\left|\sum_{n=0}^{N-1} e^{inkd\sin\theta\cos\phi}\right|$$

$$=E_0(\theta,\phi)\left|\frac{\sin(Nkd\sin\theta\cos\phi/2)}{\sin(kd\sin\theta\cos\phi/2)}\right|. \tag{6.31}$$

The first factor, $E_0(\theta,\phi)$, gives the radiation field pattern of a unit element and is called the *unit pattern*; the second factor is termed the *array factor*. The array factor gives rise to a group pattern of the array as if all elements are isotropic.

Consider two dipoles having identical amplitude and phase and spaced a half-wavelength apart, $kd=\pi$. The group pattern is shown in Fig. 6.2*a*, which has rotational symmetry around the x axis. In the x-y plane perpendicular to the dipoles, $\theta=\pi/2$ and the unit pattern is isotropic. The resultant pattern is a product of the unit pattern and the group pattern (see Fig. 6.2*b*). In the plane containing the dipoles and parallel to $\hat{z}$, $\phi=0$. The resultant pattern in this plane is shown in Fig. 6.2*c*. When we focus our attention on the plane $\theta=\pi/2$, the linear array as described by (6.31) has

$$E=E_0(\theta,\phi)\left|\frac{\sin(Nkd\cos\phi/2)}{\sin(kd\cos\phi/2)}\right|. \tag{6.32}$$

The principal maximum of E is equal to NE_0 and occurs at $\phi=\pi/2$. The nulls of the field pattern occur at $Nkd\cos\phi=\pm 2m\pi$, where m is an integer.

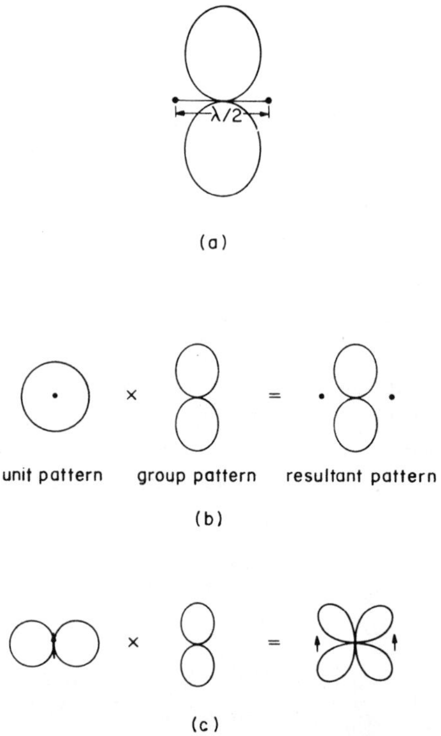

Figure 6.2 Field patterns of a simple array composed of two elements driven in phase with the same amplitude and spaced a half-wavelength apart. (*a*) Group pattern of a two-element array. (*b*) Field pattern in the plane $\theta=\pi/2$. (*c*) Field pattern in the plane containing the two dipoles.

The secondary maxima are side lobes and occur between adjacent nulls. The number of side lobes is also controlled by N; we must have $\cos\phi \leqslant 1$ and thus $2m\pi/Nkd \leqslant 1$. Because the principal direction is perpendicular to the line, this array is called a *broadside array*. Physically, it can be appreciated that the radiation patterns must be broadside because all elements are in phase and they interfere constructively in the broadside direction. Note that, when the spacing d between elements is large, the principal maximum appears again at $\cos\phi = 2m\pi/kd < 1$. Note also that the pattern does not have rotational symmetry with respect to the x axis because the unit factor $E_0(\theta,\phi)$ has a null in the $\pm\hat{z}$ direction.

It should be evident that we can make the radiation pattern *endfire* by imposing on each element a progressive phase shift α_n with respect to the first element, so that $\alpha_n = nkd$. This phase shift makes the wave emitted by element $n-1$ reaching element n in phase with the wave to be emitted by element n. Thus at $\phi=0$ along the axis of the array all elements interfere constructively and give rise to a principal maximum. This is also seen from the array factor in (6.32) with $\cos\phi$ replaced by $\cos\phi - 1$. With electronic tuning of the phase we can make the radiation pattern of an array scan over a predetermined range of angles.

The array factor takes different forms if the amplitude or the spacing between elements is not uniform. For instance, a binomial array has amplitude distribution proportional to the coefficients C_n^N of a binomial expansion. With equal spacing and equal phase (6.30) becomes

$$E = E_0 \left| \sum_{n=0}^{N} C_n^N e^{-inkd\cos\phi} \right|$$

$$= E_0 \left| \cos \frac{kd\cos\phi}{2} \right|^N.$$

When $d = \lambda/2$, the array factor yields a broadside radiation pattern that has no side lobes.

6.1d Čerenkov Radiation

In 1934, Čerenkov discovered experimentally that all liquids and solids bombarded by fast electrons emit visible radiation with the **E** vector polarized parallel to the electron beam. Many unsuccessful attempts were made to explain this discovery with various microscopic approaches. In 1937, Frank and Tamm[37] used the macroscopic theory and established that an electron moving uniformly in a medium characterized by a refractive index larger than unity radiates light if the electron velocity is greater than the velocity of light in the medium. Since this phenomenon, known as *Čerenkov radiation*, marked a significant triumph in the macroscopic electromagnetic theory, we shall devote this section to the subject.

The source of radiation is a particle with charge q moving at velocity $\mathbf{v}$ in an isotropic medium. Let the direction of motion be $\hat{z}$; then we have

$$\mathbf{J}(\mathbf{r},t) = \frac{\hat{z} q v\, \delta(z-vt)\, \delta(\rho)}{2\pi\rho}. \tag{6.33}$$

This source is not time-harmonic. We transform to the frequency domain and obtain

$$\mathbf{J}(\mathbf{r},\omega)=\frac{1}{2\pi}\int \mathbf{J}(\mathbf{r},t)e^{i\omega t}\,dt$$

$$=\hat{z}\,\frac{q}{4\pi^2\rho}\,e^{i\omega z/v}\,\delta(\rho). \tag{6.34}$$

For each spectrum component ω, we solve for the electric field $\mathbf{E}(\mathbf{r},\omega)$. The time-domain values are obtained by the inverse Fourier transform:

$$\mathbf{E}(\mathbf{r},t)=\int \mathbf{E}(\mathbf{r},\omega)e^{-i\omega t}\,d\omega. \tag{6.35}$$

The wave equation 6.3 for the electric field becomes

$$\nabla\times\nabla\times\mathbf{E}(\mathbf{r})-k^2\mathbf{E}(\mathbf{r})=\hat{z}\,\frac{i\omega\mu q}{4\pi^2\rho}\,e^{i\omega z/v}\,\delta(\rho). \tag{6.36}$$

This equation is conveniently solved by defining

$$\mathbf{E}(\mathbf{r})=\left(\bar{\mathbf{I}}+\frac{\nabla\nabla}{k^2}\right)\cdot\mathbf{g}. \tag{6.37}$$

We obtain from (6.36) the wave equation for $\mathbf{g}(\rho)$:

$$(\nabla^2+k^2)\mathbf{g}=-\hat{z}\,\frac{i\omega\mu q}{4\pi^2\rho}\,e^{i\omega z/v}\,\delta(\rho). \tag{6.38}$$

In view of the z dependence on the right-hand side and the azimuthal symmetry of the problem, we write the wave equation in the cylindrical coordinate system. Let

$$\mathbf{g}=\hat{z}g(\rho)\frac{i\omega\mu q}{2\pi}e^{i\omega z/v}. \tag{6.39}$$

Then we obtain

$$\left[\frac{1}{\rho}\frac{d}{d\rho}\left(\rho\frac{d}{d\rho}\right)-\frac{\omega^2}{v^2}+k^2\right]g(\rho)=\frac{-\delta(\rho)}{2\pi\rho}. \tag{6.40}$$

This inhomogeneous Helmholtz wave equation is solved by choosing the outgoing solution:

$$g(\rho)=CH_0^{(1)}(k_\rho\rho),\qquad k_\rho=\sqrt{k^2-\frac{\omega^2}{v^2}}\;, \tag{6.41}$$

for $\rho \neq 0$. The constant C is determined by matching the boundary condition at $\rho \to 0$. Integrating (6.40) over an infinitesimal area, letting $\rho \to 0$, and using the asymptotic formula for $H_0^{(1)}(k_\rho \rho)$, we obtain $C = i/4$. Thus

$$g(\rho) = \frac{i}{4} H_0^{(1)}(k_\rho \rho). \tag{6.42}$$

For two-dimensional problems independent of z, $k_\rho = k$ and (6.42) gives the scalar Green's function.

The solution for the electric field is determined from (6.39) and (6.37):

$$\mathbf{E}(\mathbf{r}) = \frac{-q}{8\pi\omega\epsilon}\left(\hat{z}k^2 + i\frac{\omega}{v}\nabla\right) H_0^{(1)}(k_\rho \rho) e^{i\omega z/v}. \tag{6.43}$$

Since we are interested in radiation from the charge, we use the asymptotic values of $H_0^{(1)}$ to find the far-field solutions. In the radiation zone, $k_\rho \rho \gg 1$. We obtain

$$\mathbf{E}(r) \approx \frac{q}{8\pi\omega\epsilon}\sqrt{\frac{2k_\rho}{\pi\rho}}\left(\hat{\rho}\frac{\omega}{v} - \hat{z}k_\rho\right) e^{-i\pi/4} e^{i(k_\rho \rho + \omega z/v)}. \tag{6.44}$$

This represents a plane wave with wave vector $\mathbf{k} = \hat{\rho}k_\rho + \hat{z}\omega/v$, provided that k_ρ is real. By (6.41), we see that k_ρ is real if $k^2 = \omega^2\mu\epsilon > \omega^2/v^2$, namely,

$$v > \frac{1}{\sqrt{\mu\epsilon}} = \frac{c}{n}. \tag{6.45}$$

Thus plane waves are radiated if the velocity of the charge is larger than the velocity of light in the medium. When the charge velocity v is smaller than the light velocity, k_ρ is imaginary and the wave is evanescent in the $\hat{\rho}$ direction.

The constant phase front of the plane waves forms a cone around the $\hat{z}$ direction. The direction θ that $\mathbf{k}$ makes with $\hat{z}$ is determined from

$$\cos\theta = \frac{k_z}{k} = \frac{1}{n\beta}. \tag{6.46}$$

Note that θ has a real value only if $n\beta > 1$.

Although the analysis is carried out for a charged particle, it applies also to other cases characterized by an inhomogeneous Helmholtz wave equation. The wave front of a supersonic boom, for instance, is a Čerenkov type wave front. We can imagine that the point source emits spherical waves that propagate with speed $u = c/n$ (Fig. 6.3). When the speed of the particle v is larger than u, the various spherical waves interfere constructively with one

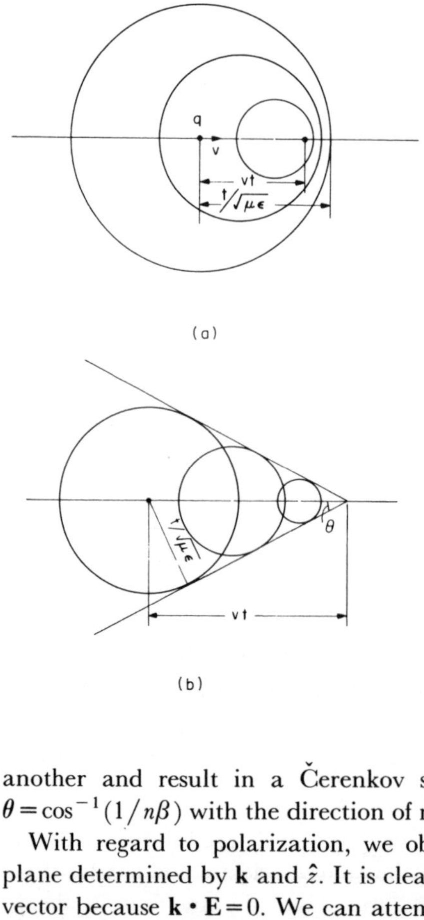

Figure 6.3 Charged particle moving in an isotropic medium. (*a*) Particle velocity *v* is less then the velocity of light in the medium. (*b*) Particle velocity *v* is greater than the velocity of light in the medium.

another and result in a Čerenkov shock wave front, making an angle $\theta = \cos^{-1}(1/n\beta)$ with the direction of motion.

With regard to polarization, we observe from (6.44) that **E** lies in the plane determined by **k** and $\hat{z}$. It is clear that **E** is also perpendicular to the **k** vector because $\mathbf{k} \cdot \mathbf{E} = 0$. We can attempt an explanation by imagining that when a charged particle is at rest the electric fields point radially. When it is moving, the field lines are bent and the degree of bending is proportional to the speed of the particle. As the speed exceeds the velocity of light in the medium, the field lines can actually break away from the charge and constitute a radiated wave. Therefore the electric field vector lies in the plane determined by the direction of motion and the direction of radiation.

To calculate radiated power, we first compute the magnetic field from Faraday's law, which gives

$$\mathbf{H} = \hat{\phi} \frac{q}{8\pi} \sqrt{\frac{2k_\rho}{\pi\rho}} \, e^{-i\pi/4} e^{i(k_\rho \rho + \omega z/v)} \tag{6.47}$$

when terms of order $\rho^{-3/2}$ are neglected.

Consider a cylinder of length l and radius ρ. The total energy radiated through the surface of the cylinder is given by

$$S_\rho = 2\pi\rho l \int_{-\infty}^{\infty} [(\mathbf{E}(\mathbf{r},t)\times\mathbf{H}(\mathbf{r},t))]_\rho dt.$$

To calculate S_ρ, we have to determine the **E** and **H** fields in real space-time. Since only the ρ component of Poynting's vector is required, we need the field component E_z. Using the inverse Fourier transform (6.35), we find

$$E_z = -\frac{q}{4\pi}\sqrt{\frac{2}{\pi\rho}}\int_0^\infty d\omega \frac{\sqrt{k_\rho^3}}{\omega\epsilon}\cos\left(\omega t - k_\rho\rho - \frac{\omega z}{v} + \frac{\pi}{4}\right).$$

By the same token,

$$H_\phi = \frac{q}{4\pi}\sqrt{\frac{2}{\pi\rho}}\int_0^\infty d\omega\sqrt{k_\rho}\cos\left(\omega t - k_\rho\rho - \frac{\omega z}{v} + \frac{\pi}{4}\right).$$

Note that the integration limits are from 0 to ∞. Now we are able to calculate S_ρ, which is a triple integral. First we integrate with respect to t. Note that

$$\int_{-\infty}^{\infty} dt \cos\left[\omega'(t+\alpha)+\frac{\pi}{4}\right]\cos\left[\omega(t+\alpha)+\frac{\pi}{4}\right]$$

$$= \frac{1}{2}\int_{-\infty}^{\infty} dt \cos[(\omega-\omega')(t+\alpha)] = \pi\delta(\omega-\omega').$$

Thus we obtain

$$S_\rho = \frac{q^2 l}{4\pi}\int_0^\infty \frac{d_\rho^2}{\omega\epsilon} d\omega = \frac{\mu q^2 l}{4\pi}\int_0^\infty d\omega\, \omega\left(1 - \frac{1}{n^2\beta^2}\right). \tag{6.48}$$

The energy radiated per unit length of the electron path can be calculated for frequency ranges of interest and for dispersive media.

6.1e Dyadic Green's Functions for Bianisotropic Media

In a bianisotropic medium, we derive the wave equation for **E** from Maxwell's equations:

$$\overline{\nabla} \cdot \mathbf{E} = i\omega(\overline{\mu} \cdot \mathbf{H} + \overline{\zeta} \cdot \mathbf{E}), \tag{6.49a}$$

$$\overline{\nabla} \cdot \mathbf{H} = -i\omega(\overline{\xi} \cdot \mathbf{H} + \overline{\epsilon} \cdot \mathbf{E}) + \mathbf{J}, \tag{6.49b}$$

where $\overline{\nabla}$ is defined by $\overline{\nabla} \cdot \mathbf{A} = \nabla \times \mathbf{A}$ for any vector **A**. Elimination of **H** from these two equations gives

$$\left[(\overline{\nabla} + i\omega\overline{\xi}) \cdot \overline{\mu}^{-1} \cdot (\overline{\nabla} - i\omega\overline{\zeta}) - \omega^2\overline{\epsilon}\right] \cdot \mathbf{E} = i\omega\mathbf{J}(\mathbf{r}). \tag{6.50}$$

Introducing (6.1), we obtain

$$\left[(\overline{\nabla} + i\omega\overline{\xi}) \cdot \overline{\mu}^{-1} \cdot (\overline{\nabla} - i\omega\overline{\zeta}) - \omega^2\overline{\epsilon}\right] \cdot \overline{\mathbf{G}}(\mathbf{r},\mathbf{r}') = i\omega\delta(\mathbf{r}-\mathbf{r}')\overline{\mathbf{I}}. \tag{6.51}$$

This is the wave equation to be satisfied by the dyadic Green's function.

The solution of Green's function $\overline{\mathbf{G}}(\mathbf{r},\mathbf{r}')$ for (6.51) can be expressed in the form of an integral representation by means of Fourier transforms. The Fourier transform pairs for $\overline{\mathbf{G}}(\mathbf{r},\mathbf{r}')$ are

$$\overline{\mathbf{G}}(\mathbf{k},\mathbf{r}') = \int_{-\infty}^{\infty} d^3r\,\overline{\mathbf{G}}(\mathbf{r},\mathbf{r}')e^{-i\mathbf{k}\cdot\mathbf{r}}, \tag{6.52a}$$

$$\overline{\mathbf{G}}(\mathbf{r},\mathbf{r}') = \frac{1}{8\pi^3}\int_{-\infty}^{\infty} d^3k\,\overline{\mathbf{G}}(\mathbf{k},\mathbf{r}')e^{i\mathbf{k}\cdot\mathbf{r}}. \tag{6.52b}$$

In view of

$$\delta(\mathbf{r}-\mathbf{r}') = \frac{1}{8\pi^3}\int d^3k\,e^{i\mathbf{k}\cdot(\mathbf{r}-\mathbf{r}')}, \tag{6.53}$$

(6.51) becomes

$$\left[(\overline{\mathbf{k}} + \omega\overline{\xi}) \cdot \overline{\mu}^{-1} \cdot (\overline{\mathbf{k}} - \omega\overline{\zeta}) + \omega^2\overline{\epsilon}\right] \cdot \overline{\mathbf{G}}(\mathbf{k},\mathbf{r}') = -i\omega e^{-i\mathbf{k}\cdot\mathbf{r}'}\overline{\mathbf{I}}. \tag{6.54}$$

Note that the $\overline{\nabla}$ operator is simply replaced with $i\overline{\mathbf{k}}$. We write

$$\overline{\mathbf{G}}(\mathbf{k},\mathbf{r}') = -i\omega\overline{\mathbf{W}}^{-1}(\mathbf{k})e^{-i\mathbf{k}\cdot\mathbf{r}'}, \tag{6.55}$$

where

$$\overline{\mathbf{W}}(\mathbf{k})=(\bar{\mathbf{k}}+\omega\bar{\xi})\cdot\bar{\mu}^{-1}\cdot(\bar{\mathbf{k}}-\omega\bar{\zeta})+\omega^2\bar{\epsilon} \tag{6.56}$$

and $\overline{\mathbf{W}}^{-1}$ is the inverse of $\overline{\mathbf{W}}$. Taking the inverse Fourier transform of $\overline{\mathbf{G}}(\mathbf{k},\mathbf{r}')$, we then evaluate Green's function from the integral:

$$\overline{\mathbf{G}}(\mathbf{r},\mathbf{r}')=-\frac{i\omega}{8\pi^3}\int d^3k\overline{\mathbf{W}}^{-1}(\mathbf{k})e^{i\mathbf{k}\cdot(\mathbf{r}-\mathbf{r}')}. \tag{6.57}$$

We can put the integration into a simpler form by noting that

$$\overline{\mathbf{W}}^{-1}(\mathbf{k})=\frac{\operatorname{adj}\overline{\mathbf{W}}(\mathbf{k})}{\det\overline{\mathbf{W}}(\mathbf{k})}. \tag{6.58}$$

where adj stands for "adjoint," and det for "determinant." Also note that in Cartesian coordinates each of the nine components in $\operatorname{adj}\overline{\mathbf{W}}(\mathbf{k})$ is a polynomial in k_x, k_y, and k_z. The only $\mathbf{r}$ dependence under the integral is displayed by the exponential term $e^{i\mathbf{k}\cdot(\mathbf{r}-\mathbf{r}')}$. Thus the k_x, k_y, and k_z appearing in the polynomials can be conveniently replaced by the differential operators $-i\partial/\partial x$, $-i\partial/\partial y$, and $-i\partial/\partial z$, respectively. With this observation, we write

$$\overline{\mathbf{G}}(\mathbf{r},\mathbf{r}')=-\frac{i\omega}{8\pi^3}\left[\operatorname{adj}\overline{\mathbf{W}}(-i\nabla)\right]\cdot\int d^3k\frac{e^{i\mathbf{k}\cdot(\mathbf{r}-\mathbf{r}')}}{\det\overline{\mathbf{W}}(\mathbf{k})}. \tag{6.59}$$

The dyadic operator $\operatorname{adj}\overline{\mathbf{W}}(-i\nabla)$ operates on the integral. Formally, we can find $\overline{\mathbf{G}}(\mathbf{r},\mathbf{r}')$ either by solving differential equation 6.54 or evaluating integral 6.59. In practice, however, simplification of the dyad $\overline{\mathbf{W}}$ must be sought before evaluation. For moving isotropic media, dyadic Green's functions have been studied by Tai.[101]

6.2 DIPOLE ANTENNAS AND STRATIFIED MEDIA

6.2a Asymptotic Series for Hankel Functions

To treat radiation and scattering of dipole antennas in the presence of stratified media, we need to use the saddle-point method. In this section we shall introduce this method to study the asymptotic behavior of Hankel functions with large arguments. Although use has been made of the Hankel functions, we have not yet formally defined them. Using wave concepts, we argue first that a cylindrical wave can be represented by a superposition of plane waves emerging from all angles, real and complex. Then we define Hankel functions in terms of a contour integration on a complex plane (Sommerfeld[96]).

The wave equation in cylindrical coordinates takes the form

$$\left[\frac{1}{\rho}\frac{\partial}{\partial\rho}\left(\rho\frac{\partial}{\partial\rho}\right)+\frac{1}{\rho^2}\frac{\partial^2}{\partial\phi^2}+k^2\right]u(\rho,\phi)=0. \tag{6.60}$$

The solution to this equation is

$$u(\rho,\phi)=H_n(k\rho)e^{\pm in\phi},$$

where $H_n(k\rho)$ is the Hankel function. The same wave equation in rectangular coordinates takes the form

$$\left(\frac{\partial^2}{\partial x^2}+\frac{\partial^2}{\partial y^2}+k^2\right)u(x,y)=0. \tag{6.61}$$

The two wave equations are related by a coordinate transformation. The familiar plane wave $e^{ik_x x+ik_y y}$ is a solution to (6.61). The solution to (6.60) represents a cylindrical wave. In cylindrical coordinates, the plane wave becomes

$$e^{ik_x x+ik_y y}=e^{ik\rho\cos(\psi-\phi)},$$

where

$$\mathbf{k}=\hat{x}k\cos\psi+\hat{y}k\sin\psi,$$

$$\boldsymbol{\rho}=\hat{x}\rho\cos\phi+\hat{y}\rho\sin\phi,$$

with the wave vector $\mathbf{k}$ indicating the direction of the plane wave propagation, and the position vector $\boldsymbol{\rho}$ representing the observation point. (See Fig. 6.4.)

We can view a cylindrical wave as a superposition of plane waves, uniform and nonuniform, emerging from all angles ψ, real and complex. Denoting wave amplitudes by $C_n e^{in\psi}$, we write

$$u(\rho,\phi)=\int_\Gamma d\psi\, C_n e^{in\psi}e^{ik\rho\cos(\psi-\phi)}.$$

It is a straightforward matter to show that this integral is indeed a solution to wave equation 6.60 in cylindrical coordinates. The path of integration Γ

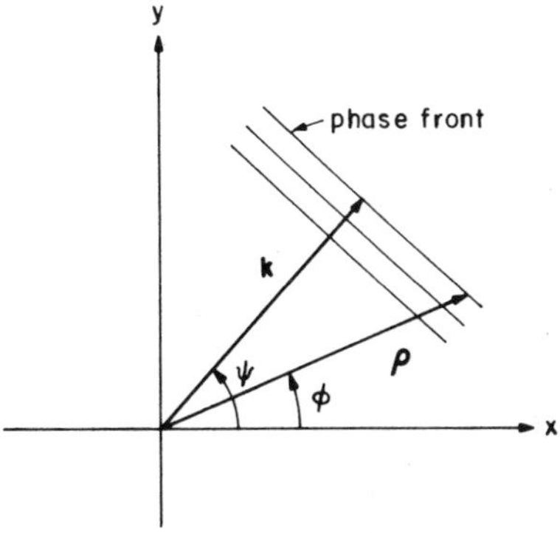

Figure 6.4 Phase front of a plane wave with wave vector k reaching the observation point at ρ.

and the amplitude constant C_n are still to be specified. The path Γ on the complex ψ plane must be chosen to ensure that this integral converges properly. To put the solution into the desired form, we set $\alpha=\psi-\phi$. Thus

$$u(\rho,\phi)=e^{in\phi}\int_{\Gamma}d\alpha\, C_n e^{in\alpha+ik\rho\cos\alpha}.$$

We now investigate the convergence of the integral in the neighborhood of infinity. On the complex plane $\alpha=\alpha'+i\alpha''$,

$$i\cos\alpha=i\cos\alpha'\cosh\alpha''+\sin\alpha'\sinh\alpha''. \tag{6.62}$$

In Fig. 6.5, the unshaded regions are the regions where the second term in (6.62) is negative. The path of integration Γ is restricted asymptotically to these regions. Note that in the subsequent treatment of complex variables, we shall use single prime to denote real part and double prime to denote imaginary part. These should not be confused with the derivatives used elsewhere in the book.

The Hankel function of the first kind is defined as

$$H_n^{(1)}(k\rho)=\frac{1}{\pi}\int_{\Gamma_1}d\alpha\, e^{i(k\rho\cos\alpha+n\alpha-n\pi/2)}. \tag{6.63}$$

The path of integration Γ_1 is shown in Fig. 6.5. The Hankel function of the second kind is defined by the same integral but follows a different path, Γ_2:

$$H_n^{(2)}(k\rho)=\frac{1}{\pi}\int_{\Gamma_2}d\alpha\, e^{i(k\rho\cos\alpha+n\alpha-n\pi/2)}. \tag{6.64}$$

The path of integration Γ_2 is also shown in Fig. 6.5.

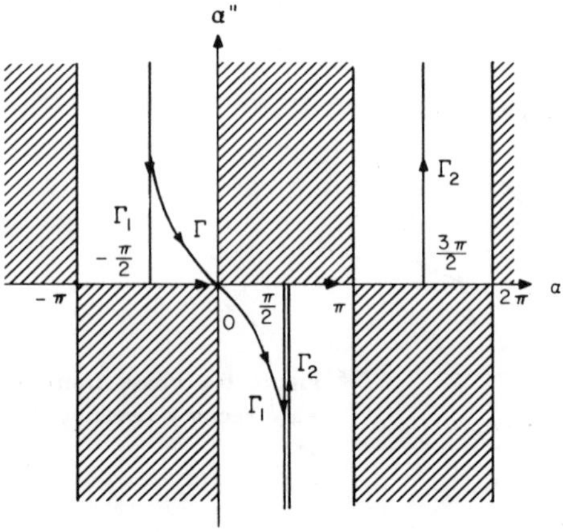

Figure 6.5 Integration paths for Hankel functions: Γ_1 is for the Hankel function of the first kind; Γ_2, for the Hankel function of the second kind.

We now determine asymptotic values of $H_n^{(1)}(k\rho)$ as $k\rho \to \infty$ by using the saddle-point method. In this limit the integrand is dominated by the exponential term $e^{ik\rho\cos\alpha}$. Saddle points are found from

$$\frac{d}{d\alpha}\cos\alpha = 0.$$

Thus, for the Hankel function of the first kind, a saddle point occurs at $\alpha = 0$. We note that away from the saddle point the integrand ascends into the shaded regions and descends into the unshaded regions. Since integration may be viewed as finding the area under a curve, we can find a new path of integration Γ passing the saddle point in such a way that most of the contribution to the integral comes from a small portion of the new path. By Cauchy's theorem, we can deform the path of integration Γ_1 to the new path Γ.

We choose the new path by requiring that on it the imaginary part of $i\cos\alpha$ be a constant equal to its value at the saddle point. In view of (6.62), we have

$$\cos\alpha' \cosh\alpha'' = 1. \tag{6.65}$$

This choice eliminates the oscillatory behavior caused by the imaginary part in the exponential term $e^{ik\rho\cos\alpha}$. The real part of the exponential term has its maximum value at the saddle point and decreases most rapidly along this path. Thus the major contribution to the integral comes from the region near the saddle point. The new path is called the *steepest descent path*. We see from (6.65) that, as $\alpha''\to\pm\infty$, $\alpha'\to\pm\pi/2$, and, as $\alpha\to 0$, $\alpha'=\pm\alpha''$, by expanding the cosine and the hyperbolic cosine functions. Among the possible choices, it is easily determined that the steepest descent path passes the saddle point at $-45°$ with respect to the real α' axis (Fig. 6.5).

To determine the asymptotic values, we let

$$-s^2=i(\cos\alpha-1). \tag{6.66}$$

Integral 6.63 becomes

$$H_n^{(1)}(k\rho)=\frac{1}{\pi}e^{i(k\rho-n\pi/2)}\int_\Gamma ds\frac{d\alpha}{ds}e^{in\alpha}e^{-k\rho s^2}. \tag{6.67}$$

Since contributions to the integral come from near the saddle point, we may expand $\cos\alpha$ to obtain $s^2=i\alpha^2/2$. We then have

$$\alpha=\sqrt{\frac{2}{i}}\;s,$$

with s a real variable. The integration interval is very near the saddle point $\alpha=0$, $e^{in\alpha}\approx 1$, and (6.67) becomes

$$H_n^{(1)}(k\rho)\approx\frac{1}{\pi}e^{i(k\rho-n\pi/2)}\sqrt{\frac{2}{i}}\int ds\,e^{-k\rho s^2}.$$

Since, away from $\alpha=0$, $e^{-k\rho s^2}$ decays very rapidly, we may restore the integration limits to $-\infty$ and $+\infty$. We thus obtain

$$H_n^{(1)}(k\rho)\approx\sqrt{\frac{2}{i\pi k\rho}}\;e^{i(k\rho-n\pi/2)}. \tag{6.68}$$

6.2b Saddle-Point Method

The procedure outlined in Section 6.2a for the saddle-point method can easily be generalized. In essence, the saddle-point method evaluates asymptotic values for an integral when its integrand contains a parameter that is very large. Consider an integral of the form

$$I(r)=\int_{\Gamma_1}F(\alpha)e^{rf(\alpha)}\,d\alpha. \tag{6.69}$$

The parameter r is assumed to be very large. Here $F(\alpha)$ and $f(\alpha)$ are analytical functions of the complex variable α, and Γ_1 is the path of integration on the complex α plane.

The saddle-point method has the following steps:

1. Determine the saddle point α_0 from

$$\frac{df(\alpha)}{d\alpha}\Big|_{\alpha=\alpha_0}=0. \tag{6.70}$$

On the complex plane, a saddle point is neither a maximum nor a minimum, since both the real and imaginary parts of $f(\alpha)$ satisfy the Laplace equation. Around the saddle point, when the imaginary part $f_i(\alpha)$ of the function $f(\alpha)$ is a constant, the real part $f_r(\alpha)$ undergoes the most rapid change, and vice versa.

2. Determine the steepest descent path Γ by

$$f_i(\alpha)=f_i(\alpha_0). \tag{6.71}$$

This requires that the imaginary part $f_i(\alpha)$ be a constant along the path and equal the imaginary part at the saddle point.

3. The contribution to the integral caused by the saddle point α_0 will be shown to be

$$I_D(r)=F(\alpha_0)e^{rf(\alpha_0)}\left[\frac{2\pi}{-rf''(\alpha_0)}\right]^{1/2}$$

$$\times\left\{1+\frac{1}{2r}\left[\frac{f'''}{(f'')^2}\frac{F'}{F}+\frac{1}{4}\frac{f^{\text{iv}}}{(f'')^2}-\frac{5}{12}\frac{(f''')^2}{(f'')^3}-\frac{F''}{Ff''}\right]+\cdots\right\}, \tag{6.72}$$

where primes denote derivatives of a function. We must note that single and double primes are also used to denote the real and imaginary parts of a complex number. In the process of deforming the path of integration to the path of steepest descent, contributions attributable to encountered singularities must be taken into account separately; for example, singularities lying between the new and the old paths, and singularities lying near the saddle point.

To derive the result for I_D, we make the transformation

$$-s^2=f(\alpha)-f(\alpha_0). \tag{6.73}$$

For the new path of integration

$$I_D=e^{rf(\alpha_0)}\int_\Gamma \Phi(s)e^{-rs^2}\,ds,$$

where $\Phi(s)=F[\alpha(s)]d\alpha/ds$. Assume that $\Phi(s)$ is a smooth function of s; then I_D is dominated by the exponential e^{-rs^2}, which possesses maximum value at $s=0$ and decreases rapidly on both sides of s. We can expand $\Phi(s)$ in powers of s and apply the formulas

$$\int_{-\infty}^{\infty} s^{2m+1}e^{-rs^2}ds=0, \tag{6.74a}$$

$$\int_{-\infty}^{\infty} s^{2m}e^{-rs^2}ds=\frac{(2m)!}{m!2^{2m}r^m}\sqrt{\frac{\pi}{r}}\,, \tag{6.74b}$$

to evaluate contributions from the saddle point I_D.

We first expand $\alpha(s)$ around $s=0$ where the integral is to be evaluated:

$$\alpha(s)-\alpha_0=\sum_{n=1}^{\infty} a_n s^n.$$

There is no constant term because $\alpha(s=0)=\alpha_0$. We use (6.73) to determine the coefficients a_n:

$$\begin{aligned}-s^2&=\frac{1}{2!}f''(\alpha_0)(\alpha-\alpha_0)^2+\frac{1}{3!}f'''(\alpha_0)(\alpha-\alpha_0)^3+\frac{1}{4!}f^{\text{iv}}(\alpha_0)(\alpha-\alpha_0)^4\\&\quad+\cdots\\&=\tfrac{1}{2}f''(\alpha_0)\left[a_1^2s^2+2a_1a_2s^3+(a_2^2+2a_1a_3)s^4+\cdots\right]\\&\quad+\tfrac{1}{6}f'''(\alpha_0)(a_1^3s^3+3a_1^2a_2s^4+\cdots)\\&\quad+\tfrac{1}{24}f^{\text{iv}}(\alpha_0)(a_1^4s^4+\cdots)+\cdots.\end{aligned}$$

Comparison of coefficients for s^n yields

$$a_1 = \left[\frac{2}{-f''(\alpha_0)} \right]^{1/2},$$

$$a_2 = -\frac{f'''}{6f''} a_1^2,$$

$$a_3 = \frac{1}{24}\left[\frac{5}{3}\left(\frac{f'''}{f''}\right)^2 - \frac{f^{\text{iv}}}{f''} \right] a_1^3,$$

$$\cdots .$$

We can now expand $\Phi(s)$, but because of (6.74a) we evaluate only coefficients for even powers of s:

$$\Phi(s) = \sum_{k=0}^{\infty} \frac{1}{k!} F^{(k)}(\alpha_0) \left(\sum_{n=1}^{\infty} a_n s^n \right)^k \left(\sum_{m=1}^{\infty} m a_m s^{m-1} \right)$$

$$= F(\alpha_0)(a_1 + 2a_2 s + 3a_3 s^2 + \cdots) + F'(\alpha_0)(a_1 s + a_2 s^2 + \cdots)$$

$$\times (a_1 + 2a_2 s + \cdots) + \tfrac{1}{2} F''(\alpha_0)(a_1^2 s^2 + \cdots)(a_1 + \cdots) + \cdots$$

$$= F(\alpha_0) a_1 + (\text{coef}) s + \left[3a_3 F(\alpha_0) + 3a_1 a_2 F'(\alpha_0) + \tfrac{1}{2} a_1^3 F''(\alpha_0) \right] s^2$$

$$+ \cdots .$$

Making use of formulas 6.74 and substituting the values of coefficients a_n, we obtain result 6.72 for I_D.

Note that the procedure illustrated can be used to determine coefficients a_n up to any order and that I_D can be expanded in higher inverse powers of r. This series, which is an asymptotic one, diverges for any fixed r. In spite of its divergent behavior, an asymptotic series is very useful. The sum of the first few terms of the series approaches the value of the function that the series represents, and then diverges as more terms are added. The error introduced in representing the function by the first n terms is of the order of the $(n+1)^{\text{th}}$ term.When the first few terms of the asymptotic series converge to the actual value of the function, the convergence is often faster than a convergent series expansion of the function. The first term of the asymptotic series may be considered as the leading behavior of the integral. The leading behavior of an integral can be evaluated readily by the saddle-point method. When the integral is expressed along either the real or the imaginary axis and can be evaluated without deforming the path of integration, the method of Laplace and the method of stationary phase are also very convenient.

As an example, we find saddle-point contributions to the integral

$$I(kr) = \int_\Gamma d\alpha\, F(\alpha) e^{ikr\cos(\alpha-\alpha_0)} \tag{6.75}$$

under the condition $kr \to \infty$. The integration path is shown in Fig. 6.6.

We let

$$f(\alpha) = i\cos(\alpha - \alpha_0).$$

The saddle point occurs at $\alpha = \alpha_0$. Using result 6.72, we find

$$I_D(kr) = F(\alpha_0) e^{ikr} \sqrt{\frac{2\pi}{ikr}} \left\{ 1 - \frac{i}{2kr}\left[\frac{1}{4} + \frac{F''(\alpha_0)}{F(\alpha_0)} \right] + \cdots \right\}. \tag{6.76}$$

As before, the subscript D denotes the saddle-point contribution. In this evaluation, bear in mind that the original path of integration Γ has been deformed into the steepest descent path passing the saddle point.

We now determine the path of steepest descent. We assume α_0 to be real:

$$f(\alpha) = i[\cos(\alpha' - \alpha_0)\cosh\alpha'' - i\sin(\alpha' - \alpha_0)\sinh\alpha''].$$

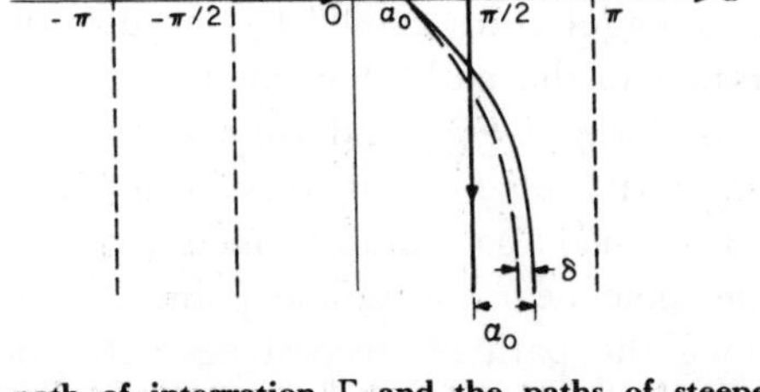

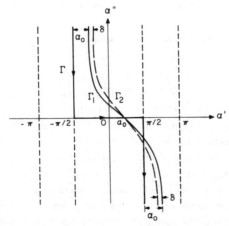

Figure 6.6 Original path of integration Γ and the paths of steepest descent Γ_1 and Γ_2 corresponding to k real and k complex.

In view of the transformation 6.73, we see that $\mathrm{Re}f(\alpha)<0$, which yields the convergence condition

$$\sin(\alpha'-\alpha_0)\sinh\alpha''<0.$$

The imaginary part at the saddle point is i. Thus the steepest descent path is determined by

$$\cos(\alpha'-\alpha_0)\cosh\alpha''=1.$$

Near the saddle point, an expansion gives

$$(\alpha'-\alpha_0)^2-\alpha''^2\approx 0.$$

Because of the convergence condition, we have $\alpha'-\alpha_0=-\alpha''$ and the path is tilted with slope -1. As $\alpha''\to\infty$, we find that $\alpha'\to\alpha_0-\pi/2$, whereas, as $\alpha''\to-\infty$, $\alpha'\to\alpha_0+\pi/2$. The steepest descent path Γ_1 is depicted in Fig. 6.6.

We have assumed k to be real. In case $k=k'+ik''$ has a small imaginary part k'', result 6.76 is still valid as $k'r\to\infty$, but the path of steepest descent will change. We write

$$f(\alpha)=\sin(\alpha'-\alpha_0)\sinh\alpha''-\frac{k''}{k'}\cos(\alpha'-\alpha_0)\cosh\alpha''$$

$$+i\left[\cos(\alpha'-\alpha_0)\cosh\alpha''+\frac{k''}{k'}\sin(\alpha'-\alpha_0)\sinh\alpha''\right].$$

The path of steepest descent Γ_2 for this case is determined by

$$\cos(\alpha'-\alpha_0)\cosh\alpha''+\frac{k''}{k'}\sin(\alpha'-\alpha_0)\sinh\alpha''=1.$$

We find that Γ_2 is at the right of Γ_1 and is separated by a distance δ as $\alpha''\to\infty$, where $\delta=\tan^{-1}(k''/k')$. Near the saddle point, Γ_2 leans more toward the vertical. The steepest descent path Γ_2 is also depicted in Fig. 6.6.

Contributions from singularities caused by the deformation of the original path of integration to the path of steepest descent are determined by the detailed functional form of $F(\alpha)$ and are taken into account separately. Moreover, if a singularity of $F(\alpha)$ is very near the saddle point, the saddle-point method outlined above must be modified. For instance, consider the case of a simple pole near the saddle point, and imagine the cross-sectional profile along the path of steepest descent. Contributions to the integral come not only from the saddle point but also from the nearby pole, which affects the integrand significantly as it gets closer. In this case we can separate the contribution due to the pole from the contribution due to the saddle point by writing

$$F(\alpha)=\left[F(\alpha)-\frac{\text{Res}}{\alpha-\alpha_p}\right]+\frac{\text{Res}}{\alpha-\alpha_p},$$

where Res denotes the residue of the function $F(\alpha)$ at the pole α_p. The bracketed term is now free of pole effects and can be evaluated with the saddle-point method as described above. The last term can be evaluated separately and expressed in terms of the incomplete error function.

We now apply the saddle-point method to find asymptotic forms for the Hankel function of the first kind. A similar process applies to the Hankel function of the second kind, and we shall list only the results. As defined in (6.63), the asymptotic forms depend on the relative magnitudes of $k\rho$ and n. To avoid the impression that n represents integers, we replace n with ν and rewrite (6.63) as

$$H_\nu^{(1)}(k\rho)=\frac{1}{\pi}\int_{\Gamma_1}d\alpha\, e^{i(k\rho\cos\alpha+\nu\alpha-\nu\pi/2)}.$$

We assume both $k\rho$ and ν to be real and large. There are two saddle points within the interval $-\pi/2$ and $3\pi/2$; they are found to occur at

$$\alpha_1=\begin{cases}\sin^{-1}\dfrac{\nu}{k\rho}, & \nu<k\rho,\\[2ex] \dfrac{\pi}{2}+i\cosh^{-1}\dfrac{\nu}{k\rho}, & \nu>k\rho,\end{cases}$$

and

$$\alpha_2=\pi-\alpha_1.$$

The asymptotic forms are determined from (6.72) with the identification of

$$F(\alpha)=\frac{1}{\pi}e^{-i\nu\pi/2},$$

$$rf(\alpha)=ik\rho\cos\alpha+i\nu\alpha.$$

In accordance with the path defined for $H_\nu^{(1)}(k\rho)$, the proper saddle point is $\alpha_1=\sin^{-1}(\nu/k\rho)$ for $\nu<k\rho$ and $\alpha_2=\pi/2-i\cosh^{-1}(\nu/k\rho)$ for $\nu>k\rho$. In deforming the path of integration to the path of steepest descent, no singularities are encountered. We find the leading term in the asymptotic series of $H_\nu^{(1)}$ to be

$$H_\nu^{(1)}(k\rho) \sim \sqrt{\frac{2}{\pi(k^2\rho^2-\nu^2)^{1/2}}}\; e^{i\left(\sqrt{k^2\rho^2-\nu^2}\; -\nu\cos^{-1}(\nu/k\rho)-\pi/4\right)}, \tag{6.77a}$$

for $\nu < k\rho$, and

$$H_\nu^{(1)}(k\rho) \sim -i\sqrt{\frac{2}{\pi(\nu^2-k^2\rho^2)^{1/2}}}\; e^{-\sqrt{\nu^2-k^2\rho^2}\; +\nu\cosh^{-1}(\nu/k\rho)} \tag{6.77b}$$

for $\nu > k\rho$. For the case $\nu \sim k\rho$, the two saddle points α_1 and α_2 begin to merge into one and we need to evaluate the contributions due to both saddle points.

The asymptotic forms for the Hankel functions of the second kind, $H^{(2)}(k\rho)$, can be obtained in a similar manner. The result is

$$H_\nu^{(2)}(k\rho) \sim \sqrt{\frac{2}{\pi(k^2\rho^2-\nu^2)^{1/2}}}\; e^{-i\left(\sqrt{k^2\rho^2-\nu^2}\; -\nu\cos^{-1}(\nu/k\rho)-(\pi/4)\right)} \tag{6.78a}$$

for $\nu < k\rho$, and

$$H_\nu^{(2)}(k\rho) \sim i\sqrt{\frac{2}{\pi(k^2\rho^2-\nu^2)^{1/2}}}\; e^{-\sqrt{\nu^2-k^2\rho^2}\; +\nu\cosh^{-1}(\nu/k\rho)} \tag{6.78b}$$

for $\nu > k\rho$. The difference is in the sign before i.

The most frequently used asymptotic form for the Hankel function occurs when $\nu \ll k\rho$. The result can be obtained from (6.76) with the identification of

$$F(\alpha) = \frac{1}{\pi} e^{i\nu\alpha - i\nu\pi/2}.$$

The saddle point occurs at $\alpha = 0$. Carrying to the second term, from (6.76), we find

$$H_\nu^{(1)}(k\rho) \sim \sqrt{\frac{2}{\pi k\rho}}\; e^{i(k\rho-\nu\pi/2-\pi/4)}\left[1-\frac{i}{2k\rho}\left(\frac{1}{4}-\nu^2\right)+\cdots\right]. \tag{6.79}$$

The first term is identical to (6.68).

6.2c Dipole Antennas and Stratified Media

In this section we shall formulate the problem of radiation of various dipole antennas in the presence of a stratified medium (Fig. 6.7). We employ the cylindrical coordinate system, and let the z axis be perpendicular to the planes of stratification. Four types of dipole configurations are studied: vertical electric dipole (VED), vertical magnetic dipole (VMD), horizontal electric dipole (HED), and horizontal magnetic dipole (HMD). The dipole source is assumed to be located at the origin of the coordinate system. The electromagnetic fields above the stratified medium are determined.

We study first fields arising from both electric and magnetic dipoles in the absence of the stratified medium. We decompose the total field into TE and TM to $\hat{z}$ components and treat TE and TM waves separately. Use is made of the identity (see Problem 6.4)

$$\frac{e^{ikr}}{r}=\frac{i}{2}\int_{-\infty}^{\infty} dk_\rho \frac{k_\rho}{k_z} H_0^{(1)}(k_\rho\rho)e^{\pm ik_z z}. \tag{6.80}$$

For electric dipoles, the electric field vector is found from (6.16a).

1. For a VED with dipole moment Il in the $\hat{z}$ direction,

$$E_z=-\left(\frac{Il}{8\pi\omega\epsilon}\right)\int_{-\infty}^{\infty} dk_\rho \frac{k_\rho^3}{k_z} H_0^{(1)}(k_\rho\rho)e^{\pm ik_z z}, \tag{6.81a}$$

$$H_z=0. \tag{6.81b}$$

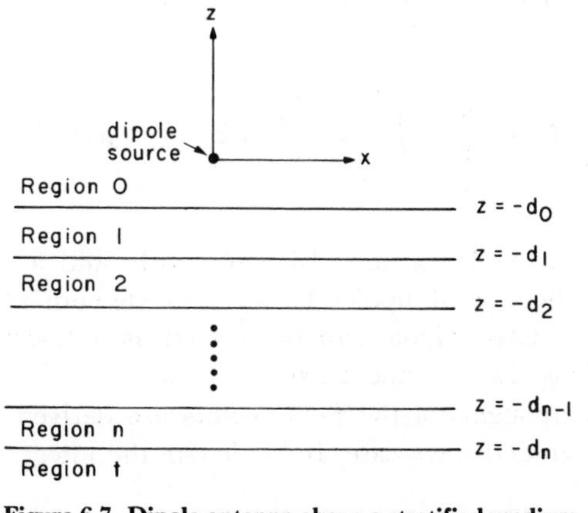

Figure 6.7 Dipole antenna above a stratified medium.

The plus sign denotes $z>0$; the minus sign $z<0$. The field is purely TM to $\hat{z}$ and is derivable from E_z alone.

2. For a HED in the $\hat{x}$ direction, the field has both TM and TE components and is derivable from E_z and H_z:

$$E_z = \pm i\left(\frac{Il}{8\pi\omega\epsilon}\right)\int_{-\infty}^{\infty} dk_\rho k_\rho^2 \cos\phi H_1^{(1)}(k_\rho\rho)e^{\pm ik_z z}, \tag{6.82a}$$

$$H_z = i\left(\frac{Il}{8\pi}\right)\int_{-\infty}^{\infty} dk_\rho \frac{k_\rho^2}{k_z}\sin\phi H_1^{(1)}(k_\rho\rho)e^{\pm ik_z z}. \tag{6.82b}$$

Note that $E_z=0$ at $z=0$. This can be seen by differentiating the left-hand side of (6.80) with respect to ρ and z and setting $z=0$; after the differentiation, the right-hand side yields an integral identical to that in (6.82a). It is also easily appreciated from the field patterns produced by the HED.

3. For a VMD in the $\hat{z}$ direction made of a small current loop with area a and current I (see Problem 6.3),

$$H_z = -i\frac{Ia}{8\pi}\int_{-\infty}^{\infty} dk_\rho \frac{k_\rho^3}{k_z} H_0^{(1)}(k_\rho\rho)e^{\pm ik_z z}. \tag{6.83}$$

4. For a HMD in the $\hat{x}$ direction,

$$H_z = \mp\frac{Ia}{8\pi}\int_{-\infty}^{\infty} dk_\rho k_\rho^2 \cos\phi H_1^{(1)}(k_\rho\rho)e^{\pm ik_z z}, \tag{6.84a}$$

$$E_z = \frac{\omega\mu Ia}{8\pi}\int_{-\infty}^{\infty} dk_\rho \frac{k_\rho^2}{k_z}\sin\phi H_1^{(1)}(k_\rho\rho)e^{\pm ik_z z}. \tag{6.84b}$$

We note that a VED excites TM waves only and a VMD excites TE waves only. The horizontal dipoles, however, excite both TM and TE waves. An arbitrarily oriented dipole can be treated as a linear combination of three dipoles along the x, y, and z axes.

The other components of the field vectors are derived from E_z and H_z, similarly to the guided-wave case. If we denote the integrands of E_z and H_z

by $E_z(k_\rho)$ and $H_z(k_\rho)$, so that

$$E_z = \int_{-\infty}^{\infty} dk_\rho E_z(k_\rho),$$

$$H_z = \int_{-\infty}^{\infty} dk_\rho H_z(k_\rho),$$

we find that similar to (5.30) the transverse components are

$$\mathbf{E}_s(k_\rho) = \frac{1}{k_\rho^2}\left[\nabla_s \frac{\partial E_z(k_\rho)}{\partial z} + i\omega\mu \nabla_s \times \mathbf{H}_z(k_\rho) \right], \tag{6.85a}$$

$$\mathbf{H}_s(k_\rho) = \frac{1}{k_\rho^2}\left[\nabla_s \frac{\partial H_z(k_\rho)}{\partial z} - i\omega\epsilon \nabla_s \times \mathbf{E}_z(k_\rho) \right], \tag{6.85b}$$

where

$$\nabla_s = \hat{\rho}\frac{\partial}{\partial \rho} + \hat{\phi}\frac{1}{\rho}\frac{\partial}{\partial \phi}$$

in cylindrical coordinates. Carrying out a process similar to that in Chapter 4 for the reflection and transmission of a plane wave by a stratified medium, we can match boundary conditions at the interfaces and find reflection coefficients characterizing the stratified medium. When the dipole is on the surface of the stratified medium, $d_0 = 0$ and the reflection coefficients for the TM and TE waves are identical to those discussed in Chapter 4. The longitudinal field components in region 0 are as follows:

1. VED

$$E_z = -\frac{Il}{8\pi\omega\epsilon}\int_{-\infty}^{\infty} dk_\rho \frac{k_\rho^3}{k_z}(1 + R^{\mathrm{TM}})H_0^{(1)}(k_\rho\rho)e^{ik_z z}. \tag{6.86}$$

2. HED

$$E_z = i\frac{Il}{8\pi\omega\epsilon}\int_{-\infty}^{\infty} dk_\rho k_\rho^2 \cos\phi(1 - R^{\mathrm{TM}})H_1^{(1)}(k_\rho\rho)e^{ik_z z}, \tag{6.87a}$$

$$H_z = i\frac{Il}{8\pi}\int_{-\infty}^{\infty} dk_\rho \frac{k_\rho^2}{k_z}\sin\phi(1+R^{\mathrm{TE}})H_1^{(1)}(k_\rho\rho)e^{ik_z z}. \tag{6.87b}$$

3. VMD

$$H_z = -i\frac{Ia}{8\pi}\int_{-\infty}^{\infty} dk_\rho \frac{k_\rho^3}{k_z}(1+R^{\mathrm{TE}})H_0^{(1)}(k_\rho\rho)e^{ik_z z}. \tag{6.88}$$

4. HMD

$$H_z = -\frac{Ia}{8\pi}\int_{-\infty}^{\infty} dk_\rho k_\rho^2 \cos\phi(1-R^{\mathrm{TE}})H_1^{(1)}(k_\rho\rho)e^{ik_z z}, \tag{6.89a}$$

$$E_z = \frac{\omega\mu Ia}{8\pi}\int_{-\infty}^{\infty} dk_\rho \frac{k_\rho^2}{k_z}\sin\phi(1+R^{\mathrm{TM}})H_1^{(1)}(k_\rho\rho)e^{ik_z z}. \tag{6.89b}$$

In these equations, the first term represents the primary field, and the second term the scattered waves caused by the presence of the stratified medium.

It is appropriate to point out at this time that this formulation can be easily generalized to the case of a uniaxial stratified medium with the optic axes parallel to the z axis. In such a case the TE waves derived from H_z are Type I waves, denoted by superscript (m), and the TM waves derived from E_z are Type II waves, denoted by superscript (e). We may consider both the permittivity and the permeability to be uniaxial, so that

$$\bar{\epsilon} = \begin{bmatrix} \epsilon & 0 & 0 \\ 0 & \epsilon & 0 \\ 0 & 0 & \epsilon_z \end{bmatrix},$$

$$\bar{\mu} = \begin{bmatrix} \mu & 0 & 0 \\ 0 & \mu & 0 \\ 0 & 0 & \mu_z \end{bmatrix}.$$

The source-free wave equations for E_z and H_z read as

$$\left[\nabla_s^2 + \frac{\epsilon_z}{\epsilon}\left(\frac{\partial^2}{\partial z^2} + k^2\right)\right] E_z = 0,$$

$$\left[\nabla_s^2 + \frac{\mu_z}{\mu}\left(\frac{\partial^2}{\partial z^2} + k^2\right)\right] H_z = 0.$$

From the dispersion relations, we obtain

$$k_z^{(e)} = \sqrt{k^2 - \frac{\epsilon k_\rho^2}{\epsilon_z}},$$

$$k_z^{(m)} = \sqrt{k^2 - \frac{\mu k_\rho^2}{\mu_z}},$$

where

$$k^2 = \omega^2 \mu \epsilon.$$

Results 6.86–6.89 all hold with k_z replaced by $k_z^{(e)}$ in the TM case and by $k_z^{(m)}$ in the TE case. We also note that both permittivities and permeabilities can assume complex values. In fact, convergence requirements for the integrals often call for an imaginary part. Several special cases of this formulation are discussed in the following sections.

6.2d Dipole on a Half-Space Medium

To illustrate the formulation, we consider the case of a vertical magnetic dipole placed on the surface of a half-space medium. The electromagnetic field above the surface is TE to $\hat{z}$ and is determined by the H_z component:

$$H_z = -i\frac{Ia}{8\pi}\int_{-\infty}^{\infty} dk_\rho \frac{k_\rho^3}{k_z}(1 + R^{\mathrm{TE}}) H_0^{(1)}(k_\rho \rho)\, e^{ik_z z}, \tag{6.90}$$

where

$$1 + R^{\mathrm{TE}} = 1 + \frac{k_z - k_{tz}}{k_z + k_{tz}} = \frac{2k_z}{k_z + k_{tz}} \tag{6.91}$$

under the assumption that $\mu_t = \mu$, and R^{TE} is the Fresnel reflection coefficient for TE waves.

We observe that, on the complex k_ρ plane, branch points occur at $k_\rho = \pm k$ and $\pm k_t$ because of k_z and k_{tz} in $1 + R^{\text{TE}}$, and at $k_\rho = 0$ because of $H_0^{(1)}(k_\rho \rho)$. The path of integration is assumed to be slightly above the negative real k'_ρ axis for $k'_\rho \leqslant 0$, and slightly below the positive real k'_ρ axis for $k'_\rho > 0$. We choose a branch cut for which $\operatorname{Im} k_z = 0$ and manipulate on the Riemann sheet where $\operatorname{Im} k_z > 0$ to ensure that $e^{ik_z z}$ converges. Since $k_z = [k^2 - (k'_\rho + ik''_\rho)^2]^{1/2}$, the equation for the branch cut is $k'_\rho k''_\rho = 0$. For $k'_\rho = 0$, k''_ρ takes any value between 0 and ∞. For $k''_\rho = 0$, $k'_\rho < k$, so that k_z is real. For the branch point at $k_\rho = k_t$, we choose a branch cut for which $\operatorname{Re} k_{tz} = 0$. Similarly, the branch cut is determined by $k''_\rho = 0$ and $k'_\rho > k_t$. Both branch cuts originating from k and k_t are shown in Fig. 6.8*a* for k real and k_t having a small imaginary part. Other singularities, including those from the Hankel function, are not shown because, as we shall see, they will not affect our solutions.

To facilitate the use of the saddle-point method at a distant observation point $r = (\rho^2 + z^2)^{1/2}$ as $kr \to \infty$, we make the transformation

$$k_\rho = k \sin\alpha, \qquad k_z = k\cos\alpha, \tag{6.92a}$$

$$\rho = r\sin\alpha_0, \qquad z = r\cos\alpha_0. \tag{6.92b}$$

In view of the asymptotic form for the Hankel function of the first kind in

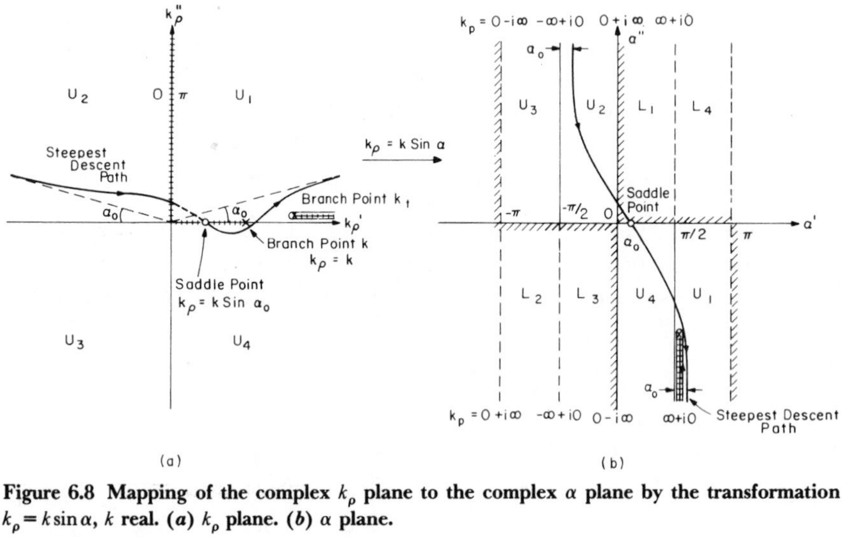

Figure 6.8 **Mapping of the complex k_ρ plane to the complex α plane by the transformation $k_\rho = k\sin\alpha$, k real. (*a*) k_ρ plane. (*b*) α plane.**

(6.68), integral 6.90 for H_z is transformed to

$$H_z = \int d\alpha F(\alpha)\, e^{ikr\cos(\alpha-\alpha_0)}, \tag{6.93a}$$

where

$$F(\alpha) = -i\frac{Ia}{4\pi}\frac{k^3\sin^3\alpha\cos\alpha}{\cos\alpha+\sqrt{k_t^2/k^2-\sin^2\alpha}}\sqrt{\frac{2}{i\pi kr\sin\alpha\sin\alpha_0}}\ . \tag{6.93b}$$

A saddle point occurs at $\alpha=\alpha_0$. The saddle-point contribution to the integral can be calculated by applying 6.76. We obtain, to order of $1/r$,

$$H_z = -\frac{Ia\,e^{ikr}}{2\pi r}\frac{k^2\sin^2\alpha_0\cos\alpha_0}{\cos\alpha_0+\sqrt{k_t^2/k^2-\sin^2\alpha_0}}\ . \tag{6.94}$$

We now examine the location of the steepest descent path and other contributions to the integral.

We consider transformation 6.92a for the case of k real. The transformation from the k_ρ plane to the α plane, as dictated by (6.92a), is explicitly written as

$$k_\rho = k_\rho' + ik_\rho'' = k\sin\alpha'\cosh\alpha'' + ik\cos\alpha'\sinh\alpha''. \tag{6.95}$$

Thus the real axis k_ρ' is mapped onto $\alpha''=0$ and $\alpha'=(2n+1)\pi/2$ for integer n because $k_\rho''=0$. The imaginary axis k_ρ'' is mapped onto $\alpha'=\pm n\pi$ because $k_\rho'=0$. At $\alpha'=0$, $k_\rho''\to\pm\infty$ as $\alpha''\to\pm\infty$. At $\alpha'=\pm\pi$, $k_\rho''\to\pm\infty$ as $\alpha''\to\mp\infty$. At $\alpha'=\pi/2$, $k_\rho'\to\infty$ as $\alpha''\to\pm\infty$. At $\alpha'=-\pi/2$, $k_\rho'\to-\infty$ as $\alpha''\to\pm\infty$. The result of the mapping is shown in Fig. 6.8*b*. On the α plane, the two Riemann sheets associated with the branch cut $\operatorname{Im} k_z=0$ are unfolded. The upper Riemann sheet, where $\operatorname{Im} k_z \geqslant 0$, is indicated by U and the lower Riemann sheet, where $\operatorname{Im} k_z \leqslant 0$, by L.

The path of steepest descent is determined by

$$\cos(\alpha'-\alpha_0)\cosh\alpha''=1. \tag{6.96}$$

We require that $\operatorname{Im}[\cos(\alpha-\alpha_0)]>0$, which gives the convergence condition

$$-\sin(\alpha'-\alpha_0)\sinh\alpha''>0. \tag{6.97}$$

Near the saddle point, (6.96) can be expanded to give $\alpha''\cong\pm(\alpha'-\alpha_0)$. Equation 6.97 dictates the choice of the minus sign; thus the path has a slope of -1 at $\alpha=\alpha_0$. As $\alpha''\to\pm\infty$, $\alpha'\to\alpha_0\pm\pi/2$. On the k_ρ plane,

$k'_\rho = k\sin\alpha' \cosh\alpha''$ and $k''_\rho = k\cos\alpha' \sinh\alpha''$, the steepest descent path goes asymptotically to $k''_\rho / k'_\rho = \pm\tan\alpha_0$. It crosses the imaginary axis at $\alpha' = 0$ and $\alpha'' = \cosh^{-1}(\sec\alpha_0)$, which corresponds to $k''_\rho = k\tan\alpha_0$. It crosses the real k'_ρ axis at the saddle point and the point $\alpha' = \pi/2$ and $\alpha'' = -\cosh^{-1}(\csc\alpha_0)$, which corresponds to $k'_\rho = k\csc\alpha_0$. The steepest descent path is depicted in Fig. 6.8 on both k_ρ and α planes.

In the process of deforming the original path of integration to the path of steepest descent, we observe that the branch point $k_\rho = k_t$ is intercepted. We shall show that the contribution from the branch point is of order $1/r^2$. From $k\sin\alpha_b = k_t$ and assuming k_t real, we obtain

$$\alpha_b = \frac{\pi}{2} + i\cosh^{-1}\left(\frac{k_t}{k}\right). \tag{6.98}$$

The integration paths are shown in Fig. 6.8. To evaluate (6.93a) around the branch point, we deform the path of integration to that parallel to the steepest descent path and collect contributions from the neighborhood of the branch point. The argument is similar to that for the saddle-point method. Since no saddle point is involved, however, the approach may be called a steepest descent method. We let

$$\cos(\alpha - \alpha_0) = \cos(\alpha_b - \alpha_0) + is^2 \tag{6.99a}$$

and change the integration variable to s. Near the branch point, we expand $\cos(\alpha-\alpha_0) \approx \cos(\alpha_b - \alpha_0) - (\alpha - \alpha_b)\sin(\alpha_b - \alpha_0) + \cdots$ and obtain

$$\alpha - \alpha_b = -is^2\csc(\alpha_b - \alpha_0). \tag{6.99b}$$

On the side near the imaginary axis s is positive. On the side near the steepest descent path, s is negative. The radical k_{tz} is expanded into

$$\begin{aligned} k_{tz} &= \left(k_t^2 - k^2\sin^2\alpha\right)^{1/2} \\ &\approx k\left(-(\alpha - \alpha_b)\sin 2\alpha_b\right)^{1/2} \\ &= ks\left(i\sin 2\alpha_b / \sin(\alpha_b - \alpha_0)\right)^{1/2} \end{aligned}$$

so that $\mathrm{Re}\, k_{tz} \gtrless 0$ when $s \gtrless 0$. As α moves along the deformed path, s increases from $-\infty$ to $+\infty$ and integral 6.93a becomes

$$H_z \approx e^{ikr\cos(\alpha_b - \alpha_0)} \int_{-\infty}^{\infty} ds \left[-i\frac{2s}{\sin(\alpha_b - \alpha_0)} \right] \tilde{F}(s) e^{-krs^2},$$

where

$$\tilde{F}(s) \approx -i\frac{Ia}{4\pi}\sqrt{\frac{2}{i\pi kr\sin\alpha_b\sin\alpha_0}}\,\frac{k^3\sin^3\alpha_b\cos\alpha_b}{1-k_t^2/k^2}$$

$$\cdot\left[\cos\alpha_b - s\sqrt{i\sin 2\alpha_b/\sin(\alpha_b-\alpha_0)}\,\right]$$

is transformed from $F(\alpha)$ and expanded to first order in s. The first term in $\tilde{F}(s)$ does not contribute, and the second term is readily evaluated by using (6.74):

$$\int_{-\infty}^{\infty} ds\, s^2 e^{-as^2} = -\frac{d}{da}\int_{-\infty}^{\infty} ds\, e^{-as^2} = \frac{\sqrt{\pi}}{2a^{3/2}}.$$

We obtain

$$H_z \approx -\frac{Ia}{2\pi}\frac{e^{ikr\cos(\alpha_b-\alpha_0)}}{(k_t^2-k^2)r^2}\,\frac{k^3\sin^3\alpha_b\cos^{3/2}\alpha_b}{\sin^{1/2}\alpha_0\sin^{3/2}(\alpha_b-\alpha_0)}. \qquad (6.100)$$

As we stated, the branch-point contribution to H_z is of order $(1/r)^2$.

The branch-point contribution becomes important when the observation point is on the surface of the half-space medium, in which case the solution for H_z as given by (6.94) vanishes and we must resort to second-order terms in $1/r^2$. We see from Fig. 6.8a that, as $\alpha_0=\pi/2$, the path of steepest descent on the k_ρ plane follows a vertical path starting from $k+i\infty$ to $k+i0$, goes around $k+i0$, and returns to $k+i\infty$ (Fig. 6.9). We may thus choose a branch cut from $k_\rho=k$ following the vertical path. Note that now $z=0$ and the convergence condition is simply $\text{Im}\,k_\rho>0$. By the same token, we choose a branch cut from $k_\rho=k_t$ to follow a vertical path and the path around the branch point is also a steepest descent path. Setting $\alpha_0=\pi/2$ in (6.100), we already have the contribution from the branch point at $k_\rho=k_t$. Because of symmetry, the contribution from the branch point at $k_\rho=k$ can be obtained by interchanging k and k_t. Therefore, on the surface of the half-space medium,

$$H_z = -i\frac{Ia}{2\pi(k_t^2-k^2)\rho^2}\left(k_t^3 e^{ik_t\rho} - k^3 e^{ik\rho}\right). \qquad (6.101)$$

The reader should convince himself of the second term by carrying out the saddle-point evaluation to order $1/\rho^2$ for $z=0$. In (6.101), the two terms

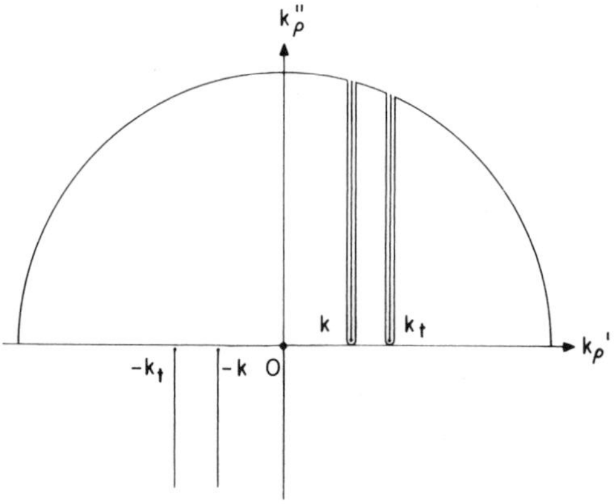

Figure 6.9 Two branch cuts originating from the branch points $k_\rho = k$ and $k_\rho = k_t$; k and k_t are each assumed to have a vanishing imaginary part so that the branch points are located slightly above the real axis. The other branch points, $k_\rho = -k$ and $k_\rho = -k_t$, are not enclosed by the path of integration.

correspond to two waves propagating with different wave numbers k_t and k. As a function of distance along the surface, these two waves interfere and give rise to an interference pattern. The periodicity of the pattern in space is proportional to $2\pi/(k_t - k)$.

6.2e Dipole on a Two-Layer Medium

Consider a dipole antenna located on the surface of a two-layer medium. We evaluate the field H_z arising from a vertical magnetic dipole on the surface. The integral to be evaluated is

$$H_z = -i\frac{Ia}{8\pi}\int_{-\infty}^{\infty} dk_\rho \frac{k_\rho^3}{k_z}(1 + R^{\mathrm{TE}})H_0^{(1)}(k_\rho \rho)e^{ik_z z}. \tag{6.102}$$

The term $1 + R^{\mathrm{TE}}$ for a two-layer medium, in view of (4.55), is given by

$$1 + R^{\mathrm{TE}} = (1 + R_{01})\frac{1 + R_{1t}e^{i2k_{1z}d}}{1 + R_{01}R_{1t}e^{i2k_{1z}d}}. \tag{6.103}$$

We shall solve (6.102) by two approaches; one leads to a geometrical optics interpretation, the other to a guided-wave interpretation.

Since the magnitudes of R_{01} and R_{1t} are less than unity, we may use the fact that $R_{01}=-R_{10}$ and expand the denominator of (6.103) to obtain

$$1+R^{\mathrm{TE}}=(1+R_{01})+(1-R_{01}^2)\sum_{m=1}^{\infty} R_{10}^{m-1}R_{1t}^{m}e^{i2mk_{1z}d}. \tag{6.104}$$

The first term was evaluated in Section 6.2d and yields the half-space solution. The summation term accounts for contributions from the subsurface layer. We observe that each term in the series can be evaluated by the saddle-point method. First, we make the transformation

$$k_\rho=k_1\sin\alpha, \qquad k_{1z}=k_1\cos\alpha, \tag{6.105}$$

$$\rho=R_m\sin\alpha_m, \qquad 2md=R_m\cos\alpha_m, \tag{6.106}$$

where

$$R_m=\sqrt{\rho^2+(2md)^2}\ . \tag{6.107}$$

The integral for the mth term becomes

$$H_z=\int d\alpha\, G_m(\alpha)\, e^{ik_1R_m\cos(\alpha-\alpha_m)}, \tag{6.108}$$

where

$$G_m(\alpha)=-i\frac{Ia}{8\pi}\frac{k_1^3\sin^3\alpha\cos\alpha}{\sqrt{k^2/k_1^2-\sin^2\alpha}}(1-R_{01}^2)R_{10}^{m-1}R_{1t}^{m}\sqrt{\frac{2}{i\pi kR_m\sin\alpha\sin\alpha_m}}\quad e^{ik_zz}$$

after using the asymptotic form for the Hankel function. A saddle point occurs at $\alpha=\alpha_m$. For $|k_1|R_m\gg 1$, we calculate the saddle-point contribution according to (6.76). We retain only the first term, which is of the form $e^{ik_1R_m}/R_m$.

We assume that k_1 has a small imaginary part. Transformation 6.105,

$$k_\rho=(k_1'+ik_1'')(\sin\alpha'\cosh\alpha''+i\cos\alpha'\sinh\alpha''),$$

connects the complex k_ρ plane with the complex α plane. For the real k_ρ' axis, $k_\rho''=0$, we have $k_1'\cos\alpha'\sinh\alpha''+k''\sin\alpha'\cosh\alpha''=0$. As $\alpha''\to+\infty$, $\alpha'\to\cot^{-1}(-k_1''/k_1')=\delta-\pi/2$; as $\alpha''\to-\infty$, $\alpha'\to-\delta+\pi/2$, where $\delta=\tan^{-1}(k_1''/k_1')$. The transformed k_ρ' axis is labeled the original integration path in Fig. 6.10.

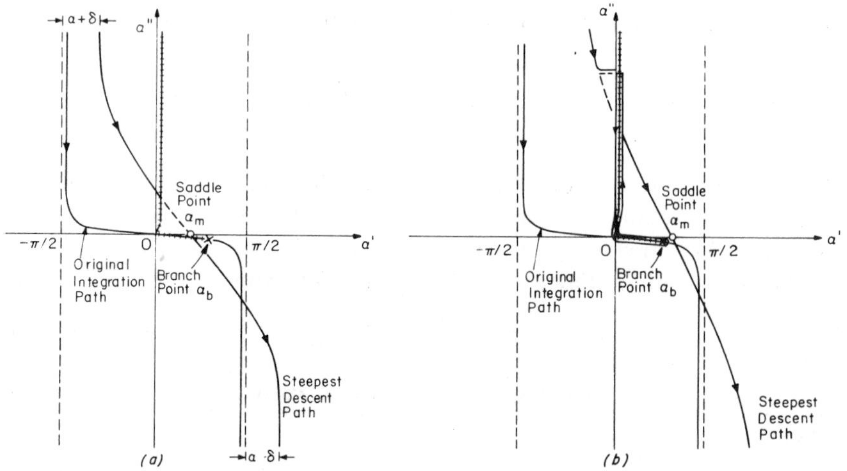

Figure 6.10 **(*a*) The path of steepest descent is at the left of the branch point α_b; the saddle point α_m occurs on the lower Riemann sheet, where $\operatorname{Im} k_z < 0$. (*b*) The path of steepest descent is at the right of the branch point α_b; the saddle point α_m occurs on the upper Riemann sheet, where $\operatorname{Im} k_z > 0$.**

The steepest descent path is determined from $\operatorname{Im}[ik_1 \cos(\alpha - \alpha_m)] = ik_1'$ or

$$\cos(\alpha' - \alpha_m)\cosh\alpha'' + \frac{k_1''}{k_1'}\sin(\alpha' - \alpha_m)\sinh\alpha'' = 1.$$

Near $\alpha = \alpha_m$, we have $\alpha'' \approx -(k_1'' \pm |k_1|)(\alpha' - \alpha_m)/k_1'$. The plus sign must be chosen so that it reduces to the case in which $k_1'' = 0$. Thus the path leans more toward the vertical axis. As $\alpha'' \to \pm\infty$, $\alpha' \to \alpha_m \pm (\delta - \pi/2)$. The path of steepest descent is also shown in Fig. 6.10.

Transformation 6.105 unfolds the two Riemann sheets caused by the branch points at $\pm k_1$. In deforming the original integration path to the steepest descent path, the branch cut originated from the branch point $k_\rho = k$ is crossed. The branch cut is crossed twice when the steepest descent path passing through the saddle point α_m is to the left of branch point α_b with $\alpha_b = \sin^{-1}(k/k_t)$ (Fig. 6.10*a*). In this case, the saddle point is on the lower sheet, corresponding to $\operatorname{Im} k_z < 0$. When the steepest descent path is at the right of α_b, we have to cross the branch cut in such a manner that both ends of the steepest descent path will be on the top sheetof $\operatorname{Im} k_z > 0$ to ensure convergence (Fig. 6.10*b*). The branch-cut contribution needs to be considered, and we can show that it is of order $1/R_m^2$ and negligible. The saddle point is on the top sheet, and we have $\operatorname{Im} k_z > 0$.

Physically, each term in the summation after evaluation by the saddle-point method corresponds to a radiation field caused by a dipole source at a distance $R_m = (\rho^2 + (2md)^2)^{1/2}$. The dipole source can be viewed as an image of the original dipole, situated at a distance $2md$ below the surface. It may also be viewed as being due to the original dipole field having bounced at the second boundary for m times (Fig. 6.11). The critical angle occurs at $\alpha_b = \sin^{-1} k/k_1$. When the saddle point $\alpha_m = \sin^{-1} \rho/R_m$ is at the right of α_b, the angle of reflection is larger than α_b and the wave is essentially guided. The wave decays away from the surface because $\text{Im}\, k_z > 0$. When the saddle point is at the left of α_b, the angle of reflection is smaller than the critical angle. The wave above the surface grows exponentially because $\text{Im}\, k_z < 0$, and it may be termed a *leaky wave*. We note that in this approach, which may be called a *geometrical optics approximation*, the summation series converges faster when the layer is thicker and medium 1 is lossier. As the layer thickness decreases, more terms are needed. We shall now give an alternative approach.

Instead of expanding (6.103), we find all poles from $1 + R^{\text{TE}}$ and express integral 6.104 in terms of a residue series. Poles are determined by setting the denominator in (6.103) equal to zero. We write

$$R_{10} R_{1t} e^{i2k_{1z}d} = e^{i2l\pi}, \tag{6.109}$$

where l is any integer. Equation 6.109 is similar to (5.17) and (5.18) in calculating guidance conditions for a slab waveguide. We make the transformation $k_\rho = k \sin\alpha$ and locate the poles on the complex α plane (Fig. 6.12). Each pole corresponds to a wave mode. For the poles that lie between

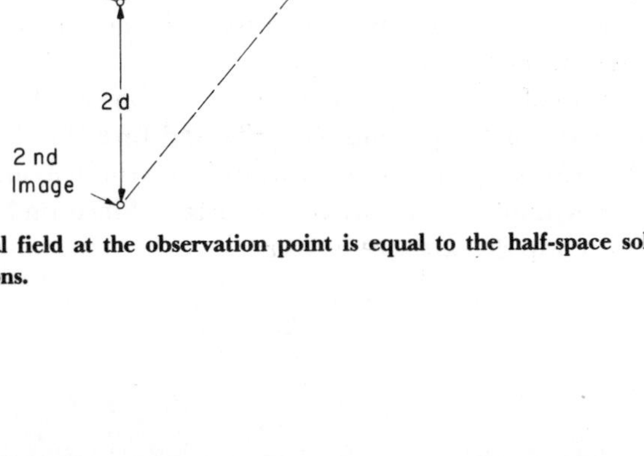

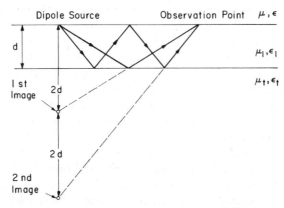

Figure 6.11 **Total field at the observation point is equal to the half-space solution plus the image contributions.**

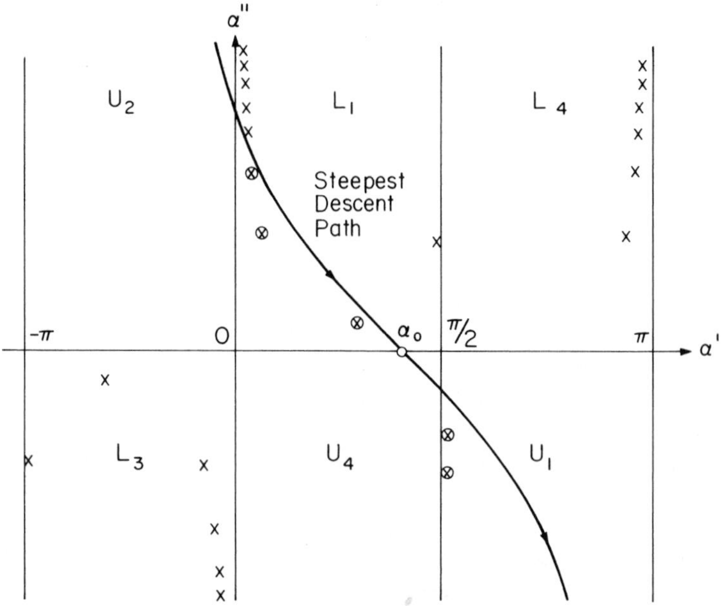

Figure 6.12 Location of poles on the complex α plane.

the original path of integration and the path of steepest descent, the corresponding modes are excited. The saddle point occurs at the observation angle $\alpha = \alpha_0$. For the poles that lie close to the saddle point, their effect on the saddle-point contribution must be taken into account by using the modified saddle-point method. The saddle-point contribution corresponds to a direct wave from the transmitter to the receiver.

In Fig. 6.12, the regions in which $\mathrm{Im}\, k_z > 0$ are labeled U and those in which $\mathrm{Im}\, k_z < 0$ are labeled L. The excited modes in region U_1 are surface-wave modes. They have wave vector components $k < \mathrm{Re}\, k_\rho < k_1$, and $\mathrm{Im}\, k_z > 0$. These surface waves are associated with the guided-wave modes inside the slab region. They decay exponentially away from the surface and propagate along the interface. Because $\mathrm{Im}\, k_\rho \approx 0$, they decay slowly with ρ and are the significant ones when $k_\rho \rho \gg 1$.

The other group of excited modes in region L_1 consists of the leaky waves. They have wave vector components $\mathrm{Re}\, k_\rho < k$ and $\mathrm{Im}\, k_z < 0$. The leaky waves reach the observation point from beneath the surface and increase in magnitude exponentially as they leave the surface. Since $\mathrm{Im}\, k_\rho > 0$, the leaky waves decay very rapidly with ρ.

Other poles do not contribute because they lie outside the region between the original path and the steepest descent path. The branch point caused by k_{1z} at $k_\rho = k_1$ does not contribute to the integral because the integrand is an even function of k_{1z}. The contribution from the branch point at $k_\rho = k_t$ gives rise to a surface wave that decays with $\exp(-2(k_t^2 - k_1^2)^{1/2} d)$ away from the bottom surface. The surface wave is important only when the slab region is sufficiently thin. As the layer becomes thinner, fewer modes are excited and fewer terms are needed in the residue series. As the layer becomes thicker and medium 1 becomes more lossy, the geometrical optics approximation is more useful because the two boundaries are less strongly coupled.

6.3 SCATTERING

6.3a Scattering by Periodic Corrugated Surfaces

Consider a plane wave incident upon a periodic corrugated surface with period $2b$ in the $\hat{z}$ direction, and let the incident wave vector be $\mathbf{k} = -\hat{x}k_x + \hat{z}k_z$. The scattered wave can be expanded in terms of Floquet modes that are characteristic waves for periodic structures. For a TM wave

$$\mathbf{H} = \hat{y}H_0 \left\{ e^{ik_z z - ik_x x} + \sum_{n=-\infty}^{\infty} R_n e^{i[k_z + (n\pi/b)]z + ik_{xn}x} \right\}, \tag{6.110a}$$

where

$$k_{xn} = \sqrt{k^2 - \left(k_z + \frac{n\pi}{b}\right)^2}\,. \tag{6.110b}$$

The first term in (6.110a) denotes the incident wave. The summation term represents a superposition of the Floquet modes. We observe that as $k^2 < [k_z + (n\pi/b)]^2$ the Floquet modes are essentially evanescent waves that decay exponentially in the $\hat{x}$ direction. The $n=0$ mode is the specular reflected wave. At normal incidence, $k_z = 0$. If the wavelength is larger than the period, $k < n\pi/b$, then all Floquet modes are evanescent.

Consider a conducting surface that is corrugated periodically with rectangular grooves (Fig. 6.13) that are infinite in the $\hat{y}$ direction and have

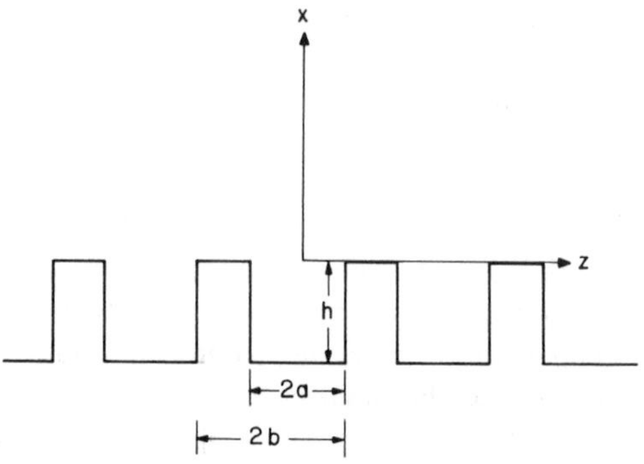

Figure 6.13 Corrugated conducting surface with rectangular grooves.

width $2a$ and depth h. The period of the corrugation is $2b$. This boundary value problem has been solved rigorously by matching mode solution above and inside the grooves.[26] Consider normal incidence for which $k_z=0$. In the groove regions the field can be expanded in terms of waveguide modes:

$$\mathbf{H}=\hat{y}H_0\sum_{m=0}^{\infty}G_m\cos\frac{m\pi(z+a)}{2a}\,\frac{\cos k_{xm}^{(2)}(x+h)}{\cos k_{xm}^{(2)}h},\qquad x<0, \tag{6.111a}$$

where

$$k_{xm}^{(2)}=\sqrt{k^2-\left(\frac{m\pi}{2a}\right)^2}\,. \tag{6.111b}$$

The electric fields are derived from the magnetic field **H** by using Maxwell's equation

$$\mathbf{E}=\frac{i}{\omega\epsilon}\nabla\times\mathbf{H}.$$

It is straightforward to show that the fields in the grooves indeed satisfy the boundary conditions of tangential **E** fields vanishing at the conducting surfaces.

At $x=0$, tangential electric and magnetic fields are continuous except at the conducting surfaces, where tangential electric fields vanish. For $-b<z$

$< b$, we have

$$1+\sum_{n=-\infty}^{\infty} R_n e^{i(n\pi/b)z} = \sum_{m=0}^{\infty} G_m \cos\frac{m\pi(z+a)}{2a}, \qquad -a \leqslant z \leqslant a, \quad (6.112\text{a})$$

$$1-\sum_{n=-\infty}^{\infty} \frac{k_{xn}}{5_x} R_n e^{i(n\pi/b)z} = \begin{cases} -\sum_{m=0}^{\infty} i\frac{k_{xm}^{(2)}}{k_x} G_m \tan k_{xm}^{(2)} h \cos\frac{m\pi(z+a)}{2a}, \\ \qquad -a \leqslant z \leqslant a, \\ 0, \qquad a \leqslant |z| \leqslant b \end{cases}$$

(6.112b)

The task is to determine R_n and G_m. We use the orthogonality properties for cosine functions by multiplying both sides of (6.112a) by $\cos[m\pi(z+a)/2a]$ and integrating over the interval $-a \leqslant z \leqslant a$. We obtain

$$\varepsilon_m a G_m = \int_{-a}^{a} dz \cos\frac{m\pi(z+a)}{2a}\left[1+\sum_{n=-\infty}^{\infty} R_n e^{i(n\pi/b)z}\right], \qquad (6.113)$$

where ε_m, called *Neumann's constant*, equals 2 when $m=0$ and equals 1 when m is any integer other than zero. We multiply (6.112b) by $e^{-in\pi z/b}$ and integrate over the interval $-b \leqslant z \leqslant b$. Because of the orthogonality relations and the fact that the right-hand side is now zero only in the interval $-a \leqslant z \leqslant a$, we obtain

$$2b\left(\delta_{n0}-\frac{k_{xn}}{k_x}R_n\right) = -i\int_{-a}^{a} dz\, e^{-i(n\pi/b)z} \sum_{m=0}^{\infty} \frac{k_{xm}^{(2)}}{k_x} G_m \tan k_{xm}^{(2)} h \cos\frac{m\pi(z+a)}{2a}$$

(6.114)

We define

$$P_{mn} = \int_{-a}^{a} dz\, e^{i(m\pi/b)z} \cos\frac{n\pi(z+a)}{2a}$$

$$= \begin{cases} \left.\begin{array}{ll} \dfrac{2(m\pi/b)}{(m\pi/b)^2-(n\pi/2a)} \sin\dfrac{m\pi a}{b}, & n \text{ even} \\ i\dfrac{2(m\pi/b)}{(m\pi/b)^2-(n\pi/2a)^2} \cos\dfrac{m\pi a}{b}, & n \text{ odd} \end{array}\right\}, & m \neq 0, \\ 2a\delta_{0n}, & m=0. \end{cases} \qquad (6.115)$$

Equations 6.113 and 6.114 can be written as

$$\varepsilon_m a G_m = P_{0m} + \sum_{n=-\infty}^{\infty} P_{nm} R_n, \tag{6.116}$$

$$R_n = \left[\frac{k_x}{k_{xn}} \delta_{n0} + \sum_{m=0}^{\infty} \frac{ik_{xm}^{(2)}}{2bk_x} (\tan k_{xm}^{(2)} h) P_{nm}^* G_m \right]. \tag{6.117}$$

This represents a set of matrix equations to be solved for R_n. Substituting (6.117) in (6.116), we find

$$\sum_{l=0}^{\infty} \left[\varepsilon_l a \delta_{ml} - i (\tan k_{xl}^{(2)} h) Q_{ml} \right] G_l = 2P_{0m}, \tag{6.118}$$

where

$$Q_{ml} = \sum_{n=-\infty}^{\infty} \frac{k_x^{(2)}}{2bk_{xn}} P_{nm} P_{nl}^*$$
$$= \begin{cases} 0, & m+l=\text{odd}, \\ \dfrac{1}{2b} \left[P_{0m} P_{0l}^* + 2 \displaystyle\sum_{n=1}^{\infty} \frac{k_x^{(2)}}{k_{xn}} P_{nm} P_n^* \right], & m+l=\text{even}. \end{cases} \tag{6.119}$$

The mode amplitudes G_l are solved by a straightforward matrix inversion. The number of groove modes needed to calculate the reflection coefficient R_n is determined by the width of the grooves a. For sufficiently narrow grooves with $ka \ll 1$, we can use the lowest mode amplitude G_0 to calculate R_n.

This mode-matching technique can be used to solve rigorously problems involving periodic structures. The use of Floquet modes also greatly facilitates the discussion of scattered waves. As another example, consider a similar structure made of parallel conducting plates $2(b-a)$ thick and separated by distance $2a$. The conducting plane at $x=-h$ in Fig. 6.10 is now removed and the plates extend to $x \to -\infty$. For an incident TM wave with magnetic field **H** in the $\hat{y}$ direction, the TEM waveguide mode in the parallel-plate regions is excited. The reflectivity is always less than unity. For an incident TE wave with electric field **E** in the $\hat{y}$ direction, the excited guided waves in the plate regions are all TE modes. Thus, if the plate separation is such that $2ka < \pi$, all guided-wave modes are evanescent and all the incident power will be scattered.

6.3b Scattering by a Conducting Cylinder

Consider a plane wave incident upon a conducting cylinder (Fig. 6.14). The incident wave is linearly polarized with electric vector **E** parallel to the axis of the cylinder. The incident **k** vector is perpendicular to the axis of the cylinder. In terms of cylindrical coordinates, we have

$$\mathbf{E}=\hat{z}E_0e^{-ikx}=\hat{z}E_0e^{-ik\rho\cos\phi}. \tag{6.120}$$

To match boundary conditions at $\rho=a$, we transform the plane wave solution into a superposition of cylindrical waves satisfying the Helmholtz wave equation in cylindrical coordinates:

$$e^{-ik\rho\cos\phi}=\sum_{n=-\infty}^{\infty}a_nJ_n(k\rho)e^{in\phi}.$$

The constant a_n can be determined by using orthogonality relations for $e^{in\phi}$. We multiply both sides by $e^{-in\phi}$ and integrate over ϕ from 0 to 2π. In view of the integral representation for the Bessel function,

$$J_n(k\rho)=\frac{1}{2\pi}\int_0^{2\pi}d\phi\, e^{-ik\rho\cos\phi-in\phi+in\pi/2}.$$

We obtain $a_n=e^{-in\pi/2}$ and

$$e^{-ik\rho\cos\phi}=\sum_{n=-\infty}^{\infty}J_n(k\rho)e^{in\phi-in\pi/2}. \tag{6.121}$$

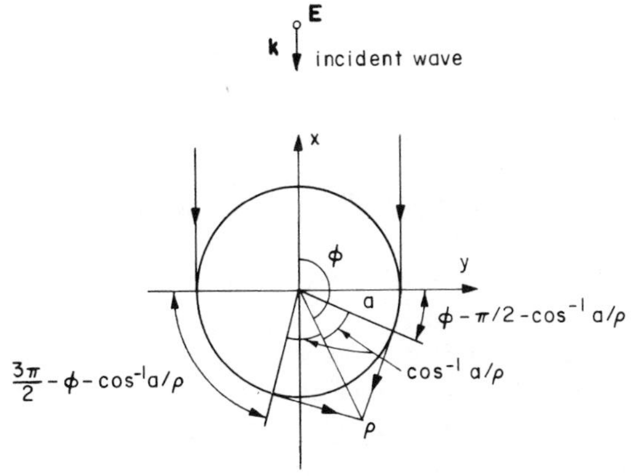

Figure 6.14 Plane wave incident upon a perfectly conducting cylinder. Creeping waves exist in the shadow region.

This expression is referred to as the *wave transformation* that represents a plane wave in terms of cylindrical waves.

The scattered wave can also be expressed as a superposition of the cylindrical functions satisfying the Helmholtz wave equation. Expecting outgoing waves, we write the solution in terms of Hankel functions of the first kind. The sum of the incident wave and the scattered wave satisfies the boundary condition of a vanishing tangential electric field at $\rho = a$. We find the total solution to be

$$\mathbf{E} = \hat{z}E_0 \sum_{n=-\infty}^{\infty} \left[J_n(k\rho) - \frac{J_n(ka)}{H_n^{(1)}(ka)} H_n^{(1)}(k\rho) \right] e^{in\phi - in\pi/2}. \qquad (6.122)$$

The first summation term represents the incident wave; the second summation term, the scattered wave. At $\rho = a$, (6.122) gives $\mathbf{E}(\rho = a) = 0$.

In the far-field zone, $k\rho \gg 1$, we make use of the asymptotic formula for $H_n^{(1)}(k\rho)$ and find that the scattered wave takes the form

$$\mathbf{E}_s \xrightarrow{k\rho\gg 1} \hat{z}E_0 \sum_{n=-\infty}^{\infty} -\sqrt{\frac{2}{\pi k\rho}} \frac{J_n(ka)}{H_n^{(1)}(ka)} e^{ik\rho + in(\phi - \pi) - i\pi/4}.$$

We can expand this result with respect to ka:

$$\mathbf{E}_s = \hat{z}iE_0 \sqrt{\frac{\pi}{2k\rho}} \left[\frac{1}{\ln ka} + (ka)^2 \cos\phi - \frac{(ka)^4}{8} \cos 2\phi + \cdots \right] e^{ik\rho - i\pi/4}.$$

$$(6.123)$$

Observe that this series converges rapidly when the radius of the cylinder is small compared with the wavelength, $ka \ll 1$. The first term is angle-independent and signifies that the scattered wave caused by a thin wire is isotropic. We can show, however, that for an incident wave with magnetic field $\mathbf{H}$ parallel to the axis of the cylinder the scattered wave will be angle-dependent (Problem 6.22).

The series in (6.123) converges very slowly when the radius of the cylinder is not small compared to the wavelength. In this case, we use the Watson transformation to convert the solution into a fast convergent series. The Watson transformation relates a residue series to a contour integration. Let the contour C in the complex ν plane be as depicted in Fig. 6.15. We find

$$\sum_{n=-\infty}^{\infty} e^{in\phi}B_n = \frac{i}{2}\oint_C d\nu \frac{e^{i\nu(\phi-\pi)}}{\sin\nu\pi} B_\nu \tag{6.124}$$

under the assumption that B_ν has no singularities on the real axis. The singularities from $\sin\nu\pi$ are all first-order poles located on the real axis at $\nu=0,\ \pm1, \pm2, \ldots$. Note that because of the direction of the contour the contour integral is equal to $-2\pi i$ times the residues of the function $e^{i\nu(\phi-\pi)}B_\nu/\sin\nu\pi$, which are $e^{in\phi}B_n/\pi$ for all integer values of n.

To make use of the Watson transformation, we identify B_n according to (6.122):

$$\begin{aligned} B_n &= \frac{E_0}{H_n^{(1)}(ka)}\left[J_n(k\rho)H_n^{(1)}(ka) - J_n(ka)H_n^{(1)}(k\rho)\right]e^{-in\pi/2} \\ &= \frac{E_0}{2H_n^{(1)}(ka)}\left[H_n^{(2)}(k\rho)H_n^{(1)}(ka) - H_n^{(2)}(ka)H_n^{(1)}(k\rho)\right]e^{-in\pi/2}. \end{aligned} \tag{6.125}$$

The singularities of B_n are caused by the zeros of $H_n^{(1)}(ka)$, which are illustrated in Fig. 6.16.

The contour integration on the right-hand side of (6.124) can be carried out by converting into an integral from $-\infty$ to $+\infty$ and then closing the contour on the upper half-plane (Fig. 6.16). It can be shown that integration along the large semicircle does not contribute. The entire contribution comes from the singularities of B_ν. Using the relations

$$H_{-\nu}^{(1)}(x) = e^{i\nu\pi}H_\nu^{(1)}(x),$$

$$H_{-\nu}^{(2)}(x) = e^{-i\nu\pi}H_\nu^{(2)}(x),$$

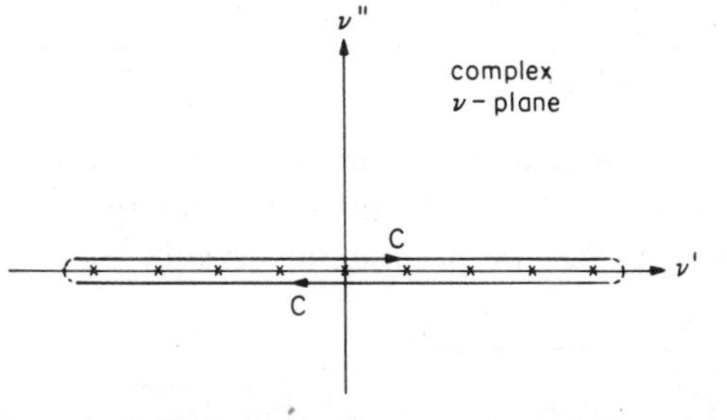

Figure 6.15 Complex integration on the complex ν plane.

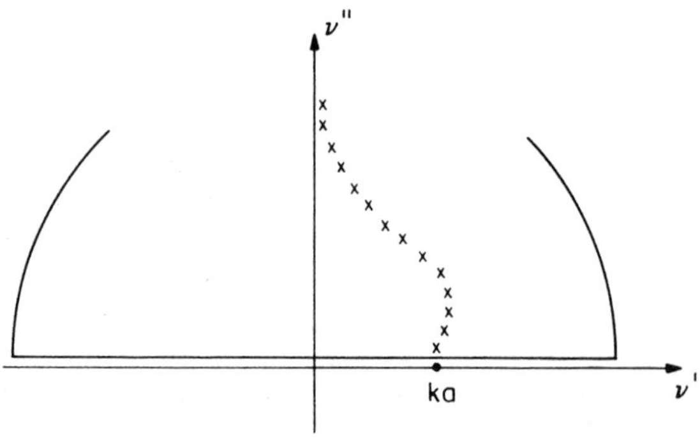

Figure 6.16 Crosses denote zeros of $H_\nu^{(1)}(ka)$.

we find $B_{-\nu}=B_\nu$. Equation 6.124 yields

$$\mathbf{E}=\hat{z}i\int_{-\infty}^{\infty}d\nu\frac{\cos\nu(\phi-\pi)}{\sin\nu\pi}B_\nu$$

$$=\hat{z}\pi E_0\sum_{n=1}^{\infty}\frac{H_{\nu_n}^{(2)}(ka)}{[(\partial/\partial\nu)H_\nu^{(1)}(ka)]_{\nu=\nu_n}}\frac{\cos\nu_n(\phi-\pi)e^{-i\nu_n\pi/2}}{\sin\nu_n\pi}H_{\nu_n}^{(1)}(k\rho),\tag{6.126}$$

where ν_n denotes zeros of $H_\nu^{(1)}(ka)$. Note that the first term of B_n in (6.125) does not contribute because $H_n^{(1)}(ka)$ in the numerator and the denominator cancel each other.

The series in (6.126) converges rapidly when $\pi/2<\phi<3\pi/2$. This is due to the fact that

$$\frac{\cos\nu_n(\phi-\pi)e^{-i\nu_n\pi/2}}{\sin\nu_n\pi}=\frac{-i\left[e^{i\nu_n(\phi-\pi/2)}+e^{i\nu_n(3\pi/2-\phi)}\right]}{1-e^{i\nu_n2\pi}}\tag{6.127a}$$

and ν_n takes positive imaginary values. The convergent range is in the shadow region of the cylinder. An interesting interpretation can be given to the terms involved. We use asymptotic formula 6.77a for $H_{\nu_n}^{(1)}(k\rho)$ when $k\rho>\nu_n$:

$$H_{\nu_n}^{(1)}(k\rho)\sim\sqrt{\frac{2}{\pi\left(k^2\rho^2-\nu_n^2\right)^{1/2}}}\exp\left[i\left(\sqrt{k^2\rho^2-\nu_n^2}-\nu_n\cos^{-1}\frac{\nu_n}{k\rho}-\frac{\pi}{4}\right)\right]\tag{6.127b}$$

Note that the imaginary part of ν_n is positive and increases for increasing n. For the first few dominant terms we may approximate $\nu_n \approx ka$

$$\sqrt{k^2\rho^2 - \nu_n^2} \approx k\sqrt{\rho^2 - a^2} \qquad \text{and} \qquad \cos^{-1}\frac{\nu_n}{k\rho} \approx \cos^{-1}\frac{a}{\rho}.$$

In view of (1.127), the exponential dependence of (6.126) takes the form

$$e^{ik(\rho^2-a^2)^{1/2}}\left\{e^{ika\left[\phi-\pi/2-\cos^{-1}(a/\rho)\right]}+e^{ika\left[3\pi/2-\phi-\cos^{-1}(a/\rho)\right]}\right\}. \qquad (6.128)$$

This provides an interesting interpretation of how the waves reach an observation point at (ρ,ϕ) in the shadow region (Fig. 6.14). The terms $(\phi-\pi/2-\cos^{-1}a/\rho)$ and $(3\pi/2-\phi-\cos^{-1}a/\rho)$ correspond to the paths of two rays traveling along the surface of the cylinder. The attenuation of the two rays is determined by the imaginary part of ν_n. The term $k(\rho^2-a^2)^{1/2}$ corresponds to the path along which the rays travel after leaving the cylinder surface. The two rays recombine at ρ in the shadow region. Because of this vivid picture, they are called *creeping waves*.

In the illuminated region, $-\pi/2<\phi<\pi/2$, a useful formula can be obtained by noting that

$$e^{i\nu(\phi-\pi)}+e^{-i\nu(\phi-\pi)} = -i2e^{i\nu\phi}\sin\nu\pi + 2e^{i\nu\pi}\cos\nu\phi.$$

When this relation is used, the contour integral in (6.124) gives

$$\mathbf{E} = \hat{z}\left(\int_{-\infty}^{\infty} d\nu\, e^{i\nu\phi}B_\nu + i\int_{-\infty}^{\infty} d\nu\,\frac{\cos\nu\phi}{\sin\nu\pi}\,e^{i\nu\pi}B_\nu\right)$$

instead of (6.126). Upon closing the contour in the upper half-plane, the second integral is evaluated in terms of residues from $H_n^{(1)}(ka)$. Because the zeros of $H_n^{(1)}(ka)$ have positive imaginary parts, we can show that this contribution will not be as significant as the first integral in the bracket. The evaluation of the first integral is left as an exercise to the reader (Problem 6.24). The result is

$$\mathbf{E} \sim \hat{z}E_0\left[e^{-ik\rho\cos\phi} - \sqrt{\frac{a\cos(\phi/2)}{2\rho}}\,e^{ik(\rho-2a\cos(\phi/2))}\right]. \qquad (6.129)$$

This can be interpreted in terms of geometrical optics. The first term is the incident wave, and the second term corresponds to the ray reflected at the surface of the cylinder. Upon striking the surface at $\rho=a$, the incident ray has a phase factor $e^{-ika\cos(\phi/2)}$. Upon reflection from the cylinder the ray reaches the observation point and gains another phase factor, $e^{ik\rho-ika\cos(\phi/2)}$.

6.3c Scattering by a Moving Column

In this section, we illustrate the formal solution to the problem of a plane wave scattered by a uniaxial dielectric of cylindrical shape and moving along its axial direction $\hat{z}$. The optic axis is also in the $\hat{z}$ direction. A plane wave linearly polarized with electric field parallel to the z axis is normally incident upon the cylinder. In view of (6.120) and (6.121), the incident wave is written as

$$\mathbf{E}=\hat{z}\,e^{-ikz}=\hat{z}\sum_{m=-\infty}^{\infty} E_0 J_m(k\rho)\,e^{im\phi-im\pi/2}.$$

For the space outside the cylinder, the $\hat{z}$-directed fields are

$$E_z=E_0\sum_{m=-\infty}^{\infty}\left[J_m(k\rho)+A_m H_m^{(1)}(k\rho)\right]e^{im\phi-im\pi/2}, \tag{6.130a}$$

$$H_z=E_0\sum_{m=-\infty}^{\infty} B_m H_m^{(1)}(k\rho)\,e^{im\phi-im\pi/2}. \tag{6.130b}$$

The first term in (6.130a) is the incident field; the second term, the scattered field. Making use of the formulation in Section 5.2a, we find the transvere fields from (5.30).

For the region of the moving medium, we use the formulation in Section 5.2e and let $\epsilon_g'=0$. Note that $k_z=0$ in all regions. Equations 5.72 and 5.71 reduce to

$$\left(\nabla_s^2+\frac{\epsilon_z'}{\epsilon}p\frac{\omega^2}{c^2}\right)E_{1z}=0, \tag{6.131a}$$

$$\left(\nabla_s^2+p\frac{\omega^2}{c^2}\right)H_{1z}=0; \tag{6.131b}$$

$$\mathbf{E}_s=i\frac{c}{\omega p}\left(l\nabla_s E_z+c\mu'\nabla_s\times\mathbf{H}_z\right), \tag{6.132a}$$

$$\mathbf{H}_s=i\frac{c}{\omega p}\left(l\nabla_s H_z-c\epsilon'\nabla_s\times\mathbf{E}_z\right), \tag{6.132b}$$

where $p=(n^2-\beta^2)/(1-\beta^2)$ and $l=\beta(n^2-1)/(1-\beta^2)$. The $\hat{z}$-directed field components take the form

$$E_z = E_0 \sum_{m=-\infty}^{\infty} C_m J_m\left(k_1^{(e)}\rho\right) e^{im\rho - im\pi/2}, \tag{6.133a}$$

$$H_z = E_0 \sum_{m=-\infty}^{\infty} D_m J_m\left(k_1^{(m)}\rho\right) e^{im\phi - im\pi/2}. \tag{6.133b}$$

The wave numbers $k_1^{(e)}$ and $k_1^{(m)}$ satisfy the dispersion relations for the type I and type II waves. According to (6.131),

$$k_1^{(e)} = p\frac{\epsilon_z'}{\epsilon'}\frac{\omega^2}{c^2}, \tag{6.134a}$$

$$k_1^{(m)} = p\frac{\omega^2}{c^2}. \tag{6.134b}$$

The transverse field components follow from (6.132).

The relative field amplitudes A_m, B_m, C_m, and D_m are determined by the boundary condition. At the boundary $\rho=a$, tangential components of both the electric and magnetic fields are continuous. After some manipulations, we obtain

$$C_m = A_m\frac{H_m^{(1)}(ka)}{J_m\left(k_1^{(e)}a\right)} + \frac{J_m(ka)}{J_m\left(k_1^{(e)}a\right)}, \tag{6.135a}$$

$$D_m = B_m\frac{H_m^{(1)}(ka)}{J_m\left(k_1^{(m)}a\right)}, \tag{6.135b}$$

$$\begin{bmatrix} -i\dfrac{ml}{a} & \dfrac{pk}{c\epsilon}\dfrac{H_m^{(1)\prime}(ka)}{H_m^{(1)}(ka)} - c\mu' k_1\dfrac{J_m'\left(k_1^{(m)}a\right)}{J_m\left(k_1^{(m)}a\right)} \\ \dfrac{pk}{c\mu}\dfrac{H_m^{(1)\prime}(ka)}{H_m^{(1)}(ka)} - c\epsilon' k_1^{(e)}\dfrac{J_m'\left(k_1^{(e)}a\right)}{J_m\left(k_1^{(e)}a\right)} & i\dfrac{ml}{a} \end{bmatrix} \cdot \begin{bmatrix} A_m \\ B_m \end{bmatrix}$$

$$= \begin{bmatrix} i\dfrac{ml}{a}\dfrac{J_m(ka)}{H_m^{(1)}(ka)} \\ c\epsilon' k_1^{(e)}\dfrac{J_m'\left(k_1^{(e)}a\right)}{J_m\left(k_1^{(e)}a\right)}\dfrac{J_m(ka)}{H_m^{(1)}(ka)} - \dfrac{pk}{c\mu}\dfrac{J_m'(ka)}{H_m^{(1)}(ka)} \end{bmatrix}. \tag{6.135c}$$

We observe that, for the incident TM to $\hat{z}$ wave with only an E_z component, both TM and TE to $\hat{z}$ waves inside and outside of the cylinder are excited. However, when the moving column in its rest frame is a uniaxial plasma, $l=0$ and we find $D_m = B_m = 0$. The excited waves have the same linear polarization as the incident wave. These observations also hold when the incident wave is a TE wave. In view of this formal treatment, we must also appreciate that scattering by a cylindrically stratified medium can be solved in closed form with similar procedures and with much more complicated manipulations.

6.3d Scattering by Spheres

The problem of a plane wave scattered by a sphere can also be solved rigorously by matching boundary conditions. To facilitate the solutions, we decompose spherical waves into TM to $\hat{r}$ and TE to $\hat{r}$ components by introducing the Debye potentials π_e and π_m so that

$$\mathbf{H} = \nabla \times (\hat{r}\pi_e) \tag{6.136a}$$

for TM to $\hat{r}$ waves, and

$$\mathbf{E} = \nabla \times (\hat{r}\pi_m) \tag{6.136b}$$

for TE to $\hat{r}$ waves.

The equation to be satisfied by π_e and π_m is the Helmholtz wave equation:

$$(\nabla^2 + k^2)\begin{Bmatrix} \pi_e \\ \pi_m \end{Bmatrix} = 0. \tag{6.137}$$

The solution to this wave equation is composed of superpositions of spherical Bessel functions, associated Legendre polynomials, and harmonic functions. Using Maxwell's equations and (6.136), we find that the field components in spherical coordinates take the forms

$$E_r = \frac{i}{\omega\epsilon}\left(\frac{\partial^2}{\partial r^2} r\pi_e + k^2 r\pi_e\right), \tag{6.138a}$$

$$E_\theta = \frac{i}{\omega\epsilon}\frac{1}{r}\frac{\partial^2}{\partial r\,\partial\theta} r\pi_e + \frac{1}{\sin\theta}\frac{\partial}{\partial\phi}\pi_m, \tag{6.138b}$$

$$E_\phi = \frac{i}{\omega\epsilon}\frac{1}{r\sin\theta}\frac{\partial^2}{\partial r\,\partial\theta} r\pi_e - \frac{\partial}{\partial\theta}\pi_m; \tag{6.138c}$$

$$H_r = -\frac{i}{\omega\mu}\left(\frac{\partial^2}{\partial r^2} r\pi_m + k^2 r\pi_m\right), \tag{6.139a}$$

$$H_\theta = -\frac{i}{\omega\mu}\frac{1}{r}\frac{\partial^2}{\partial r\,\partial\theta} r\pi_m + \frac{1}{\sin\theta}\frac{\partial}{\partial\phi}\pi_e, \tag{6.139b}$$

$$H_\phi = -\frac{i}{\omega\mu}\frac{1}{r\sin\theta}\frac{\partial^2}{\partial r\,\partial\phi} r\pi_m - \frac{\partial}{\partial\theta}\pi_e. \tag{6.139c}$$

The total electromagnetic fields are now decomposed into TE and TM components and expressed in terms of the Debye potentials π_e and π_m.

Consider a sphere with radius a located at the origin of a coordinate system (Fig. 6.17). The sphere has permittivity ϵ_t and permeability μ_t. A plane wave,

$$\mathbf{E} = \hat{x}E_0 e^{ikz} = \hat{x}E_0 e^{ikr\cos\theta},$$

$$\mathbf{H} = \hat{y}\sqrt{\frac{\epsilon}{\mu}}\,E_0 e^{ikr\cos\theta},$$

is incident upon the sphere. Note that the direction of the plane wave is along $\hat{z}$. The coordinate system is different from that used for Rayleigh scattering in Section 6.1a, where the z axis is in the direction of the linearly polarized electric field.

To match boundary conditions at the sphere surface, we expand the incident wave in terms of spherical harmonics by using the wave transformation (Problem 6.23):

$$e^{ikr\cos\theta} = \sum_{n=0}^{\infty} i^n(2n+1)j_n(kr)P_n(\cos\theta). \tag{6.140}$$

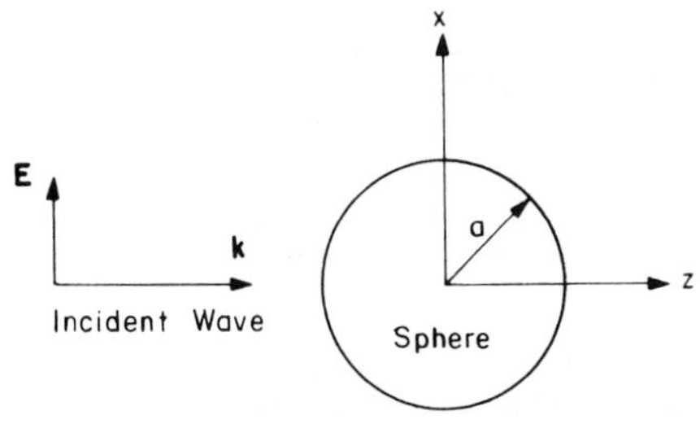

Figure 6.17 Scattering of a plane wave by a sphere.

To determine the Debye potentials for the incident wave, we note that

$$E_r = E_0 \sin\theta \cos\phi\, e^{ikr\cos\theta}$$

$$= -\frac{iE_0\cos\phi}{(kr)^2}\sum_{n=1}^{\infty} i^n(2n+1)\hat{J}_n(kr)P_n^1(\cos\theta),$$

where $\hat{J}_n(kr) = krj_n(kr)$.

The summation now starts with $n=1$ because $P_0^1(\cos\theta)=0$. The potential π_e that satisfies (6.138a) is shown to be

$$\pi_e = -\frac{E_0\cos\phi}{\omega\mu}\sum_{n=1}^{\infty}\frac{i^n(2n+1)}{n(n+1)}\hat{J}_n(kr)P_n^1(\cos\theta). \tag{6.141a}$$

By duality, the potential π_m is found to be

$$\pi_m = \frac{E_0\sin\phi}{k}\sum_{n=1}^{\infty}\frac{i^n(2n+1)}{n(n+1)}\hat{J}_n(kr)P_n^1(\cos\theta). \tag{6.141b}$$

The scattered field can be characterized by Debye potentials:

$$\pi_e^s = -\frac{E_0\cos\phi}{\omega\mu}\sum_{n=1}^{\infty} a_n\hat{H}_n^{(1)}(kr)P_n^1(\cos\theta), \tag{6.142a}$$

$$\pi_m^s = \frac{E_0\sin\phi}{k}\sum_{n=1}^{\infty} b_n\hat{H}_n^{(1)}(kr)P_n^1(\cos\theta). \tag{6.142b}$$

The total field outside the sphere is equal to the sum of the incident and scattered fields.

The field inside the sphere can also be expressed in terms of the Debye potentials:

$$\pi_e^i = -\frac{E_0\cos\phi}{\omega\mu_t}\sum_{n=1}^{\infty} c_n\hat{J}_n(kr)P_n^1(\cos\theta), \tag{6.143a}$$

$$\pi_m^i = \frac{E_0\sin\phi}{k_t}\sum_{n=1}^{\infty} d_n\hat{J}_n(kr)P_n^1(\cos\theta). \tag{6.143b}$$

The boundary conditions at $r=a$ require that E_θ, E_ϕ, H_θ, and H_ϕ be continuous, with the result that four equations are solvable for the unknown coefficients a_n, b_n, c_n, and d_n. In view of (6.138) and (6.139), the coefficients are determined as

$$a_n = \frac{i^n(2n+1)}{n(n+1)} \frac{-\sqrt{\epsilon_t\mu}\,\hat{J}_n'(ka)\hat{J}_n(k_t a) + \sqrt{\epsilon\mu_t}\,\hat{J}_n(ka)\hat{J}_n'(k_t a)}{\sqrt{\epsilon_t\mu}\,\hat{H}_n^{(1)\prime}(ka)\hat{J}_n(k_t a) - \sqrt{\epsilon\mu_t}\,\hat{H}_n^{(1)}(ka)\hat{J}_n'(k_t a)}, \tag{6.144a}$$

$$b_n = \frac{i^n(2n+1)}{n(n+1)} \frac{-\sqrt{\epsilon_t\mu}\,\hat{J}_n(ka)\hat{J}_n'(k_t a) + \sqrt{\epsilon\mu_t}\,\hat{J}_n'(ka)\hat{J}_n(k_t a)}{\sqrt{\epsilon_t\mu}\,\hat{H}_n^{(1)}(ka)\hat{J}_n'(k_t a) - \sqrt{\epsilon\mu_t}\,\hat{H}_n^{(1)\prime}(ka)\hat{J}_n(k_t a)}, \tag{6.144b}$$

$$c_n = \frac{i^n(2n+1)}{n(n+1)} \frac{i\sqrt{\epsilon_t\mu}}{\sqrt{\epsilon_t\mu}\,\hat{H}_n^{(1)\prime}(ka)\hat{J}_n(k_t a) - \sqrt{\epsilon\mu_t}\,\hat{H}_n^{(1)}(ka)\hat{J}_n'(k_t a)}, \tag{6.144c}$$

$$d_n = \frac{i^n(2n+1)}{n(n+1)} \frac{-i\sqrt{\epsilon\mu_t}}{\sqrt{\epsilon_t\mu}\,\hat{H}_n^{(1)}(ka)\hat{J}_n'(k_t a) - \sqrt{\epsilon\mu_t}\,\hat{H}_n^{(1)\prime}(ka)\hat{J}_n(ka)}. \tag{6.144d}$$

In the case of small spheres, $ka \ll 1$, only the $n=1$ terms dominate, and

$$a_n \to -(ka)^3 \frac{\epsilon_t - \epsilon}{\epsilon_t + 2\epsilon}, \qquad b_n \to -(ka)^3 \frac{\mu_t - \mu}{\mu_t + 2\mu}.$$

The results reduce to those of Rayleigh scattering. For the scattering of electromagnetic waves by spheres of finite radius, where the Rayleigh limit $ka \ll 1$ is not met, the phenomenon is known as *Mie scattering*.

PROBLEMS

6.1. In view of Gauss' magnetic field law, a vector potential $\mathbf{A}$ can be introduced such that

$$\mathbf{B} = \nabla \times \mathbf{A}.$$

This equation does not uniquely specify $\mathbf{A}$ because if we define a new $\mathbf{A}$, related to $\mathbf{A}$ by $\mathbf{A}' = \mathbf{A} + \nabla\chi$ with χ representing any scalar function, then both $\mathbf{A}$ and $\mathbf{A}'$ yield the same $\mathbf{B}$. The transformation of $\mathbf{A}$ to $\mathbf{A}'$ is called a *gauge transformation*. To uniquely define $\mathbf{A}$, the divergence of $\mathbf{A}$ must also be specified. The equation that specifies the divergence of $\mathbf{A}$ is called the *gauge condition*.

Introduce a scalar potential ϕ and show that

$$\mathbf{E} = -\frac{\partial \mathbf{A}}{\partial t} - \nabla\phi.$$

For isotropic media, the *Lorentz gauge* is given by

$$\nabla \cdot \mathbf{A} + \mu\epsilon \frac{\partial \phi}{\partial t} = 0.$$

Show that with this gauge condition the vector and the scalar potentials satisfy the inhomogeneous wave equations

$$\left(\nabla^2 - \mu\epsilon \frac{\partial^2}{\partial t^2}\right)\phi = \frac{-\rho}{\epsilon},$$

$$\left(\nabla^2 - \mu\epsilon \frac{\partial^2}{\partial t^2}\right)\mathbf{A} = -\mu\mathbf{J}.$$

In view of the arbitrariness introduced by gauge transformation, the potential functions appear to have no direct physical meaning such as the field vectors do. The problem also becomes involved when a choice of potential functions is attempted in the case of anisotropic and bianisotropic media. However, for isotropic media, the use of potentials simplifies the mathematics considerably. The rectangular components of $\mathbf{A}$ are in the same direction as the rectangular components of the source current $\mathbf{J}$. Whereas $\mathbf{E}$ and $\mathbf{H}$ must both be divergence-free outside the source, there is no such restriction on $\mathbf{A}$. Ascertain these statements by calculating the vector potential $\mathbf{A}$ with a given source distribution $\mathbf{J}$. Show that the result for the electric field is identical to (6.11).

6.2. For a Hertzian dipole in $\hat{z}$ direction, it is reasonable to expect that in the cylindrical coordinate system the electromagnetic fields have no ϕ variation. Show that the differential equation for H_ϕ is

$$\left(\frac{\partial^2}{\partial \rho^2} + \frac{1}{\rho}\frac{\partial}{\partial \rho} - \frac{1}{\rho^2} + \frac{\partial^2}{\partial z^2}\right)H_\phi + k^2 H_\phi = 0.$$

The boundary condition requires

$$\lim_{\rho \to 0} \int_0^{2\pi} H_\phi \rho \, d\phi = Il\,\delta(z) = \frac{Il}{2\pi}\int_{-\infty}^{\infty} dk_z\, e^{ik_z z}.$$

Show that the solution for H_ϕ that satisfies both the differential equation and the boundary condition is

$$H_\phi = -i\frac{Il}{8\pi}\int_{-\infty}^{\infty} k_\rho H_1^{(1)}(k_\rho \rho)\, e^{ik_z z}\, dk_z.$$

Using the relation $k_\rho^2 + k_z^2 = k^2$, write the identity in (6.80) in the form of integration over k_z. Show that

$$H_\phi = -\frac{\partial}{\partial \rho}\left(\frac{Il}{4\pi}\frac{e^{ikr}}{r}\right)$$

and that this solution is identical to (6.16b).

6.3. Consider radiation occurring with a small loop antenna with radius R and carrying current I:

$$\mathbf{J} = \hat{\phi} I \delta(\rho - R)\delta(z).$$

Apply the Green's function formalism (6.11) derived in Section 6.1, and find the electric field vector $\mathbf{E}$.

As another approach similar to Problem 6.2, show that the differential equation for E_ϕ is

$$\left(\frac{\partial^2}{\partial \rho^2} + \frac{1}{\rho}\frac{\partial}{\partial \rho} - \frac{1}{\rho^2} + \frac{\partial^2}{\partial z^2}\right)E_\phi + k^2 E_\phi = 0.$$

Write the solution for E_ϕ as

$$E_\phi = \begin{cases} \displaystyle\int_{-\infty}^{\infty} A J_1(k_\rho \rho)\, e^{ik_z z}\, dk_z, & \rho \leqslant R, \\ \displaystyle\int_{-\infty}^{\infty} B H_1^{(1)}(k_\rho \rho)\, e^{ik_z z}\, dk_z, & \rho \geqslant R. \end{cases}$$

Using Faraday's law, calculate $\mathbf{H} = (1/i\omega\mu)\nabla \times \mathbf{E}$. Match the boundary condition at $\rho = R$ by requiring that E_ϕ be continuous and that discontinuity in H_z gives rise to the current in the loop, show that, for $\rho > a$,

$$\mathbf{E} = \hat{\phi} - \frac{\omega\mu I \pi R^2}{8\pi}\int_{-\infty}^{\infty} k_\rho H_1^{(1)}(k_\rho \rho)\, e^{ik_z z}\, dk_z.$$

Compare this solution with H_ϕ obtained in Problem 6.2. Observe that $\pi R^2 = a$ is equal to the area of the loop and that the two solutions are duals of each other with the replacement $Il \to i\omega\mu Ia$ and $H_\phi \to -E_\phi$. Because of this dual relation, the small current loop is also called a *magnetic dipole*.

6.4. Prove[95] the identity in (6.80).

6.5. A turnstile antenna contains two Hertzian dipoles drive 90° out of phase and placed perpendicular to each other at the origin, $\mathbf{J}=(\hat{x}+i\hat{y})Il\delta(\mathbf{r})$. Show that the electric field produced by this antenna is

$$\mathbf{E}=\eta\frac{kIle^{ikr}}{4\pi r}e^{i\phi}\left[\hat{r}\left(\frac{1}{kr}+\frac{i}{5^2r^2}\right)2\sin\theta+\hat{\theta}_i\left(1+\frac{i}{kr}-\frac{1}{k^2r^2}\right)\cos\theta\right.$$
$$\left.-\hat{\phi}\left(1+\frac{i}{kr}-\frac{1}{k^2r^2}\right)\right].$$

Find the radiation field patterns. Show that the power density in the $\hat{z}$ direction is twice the maximum power density radiated in the $x-y$ plane.

6.6. Show that for a linear antenna with current distribution given by (6.26) the radiation resistance is

$$R_r=\frac{\eta}{2\pi}\int_0^\pi d\theta\frac{\left[\cos\left(kl\cos\theta/2\right)-\cos\left(kl/2\right)\right]^2}{\sin\theta}.$$

Plot the radiation resistance as a function of k, and show that R_r for a half-wavelength dipole is about 73Ω. Calculate the limiting case as $kl \to 0$, and compare the radiation resistance of a Hertzian dipole. Explain the difference.

6.7. Consider two Hertzian dipoles with equal current amplitudes placed a distance d apart and pointed in the same direction. The phase difference of the two dipoles is ψ. Sketch the radiation patterns in the planes parallel and perpendicular to the two dipoles. The values of d and ψ are as follows: (*a*) $d=\lambda/4$, $\psi=\pi/2$; (*b*) $d=\lambda/2$, $\psi=\pi/3$; (*c*) $d=\lambda$, $\psi=0$.

6.8. Water has an index of refraction of 1.337 in the visible range. Determine how many photons in the frequency range 500–750 THz are emitted per meter by an electron of energy 5 Bev. Compute the angle of emission of the Čerenkov radiation.

6.9. Calculate the Čerenkov radiation condition caused by a charged particle moving along the optic axis in a uniaxial medium.

6.10. Calculate the Čerenkov radiation condition caused by a charged particle moving along the dc magnetic field in an anisotropic plasma. Is it possible to have Čerenkov radiation in an isotropic plasma?

6.11. Assume that a plane wave at 10-GHz frequency is incident upon a raindrop with 1-mm radius and that the raindrop has dielectric constant $\epsilon = 80\epsilon_0$ and conductivity $\sigma = \omega\epsilon$. Determine the Rayleigh scattering cross section. Find the absorption cross section, defined by

$$\sigma_A = \frac{\int \frac{1}{2}\sigma |E_i|^2 \, dV}{(1/2\eta)|E_0|^2},$$

where E_i is the electric field inside the water droplet and E_0 is the field of the incident wave. Which mechanism, Rayleigh scattering or absorption, is dominant?

6.12. Making use of the radiation vector and (6.14), show that

$$\lim_{r \to \infty} r(\nabla \times \mathbf{E} - ik\hat{r} \times \mathbf{E}) = 0,$$

$$\lim_{r \to \infty} r(\nabla \times \mathbf{H} - ik\hat{r} \times \mathbf{H}) = 0.$$

This is referred to as the radiation condition for electromagnetic fields.

6.13. Derive the dyadic Green's function for a moving isotropic medium. Find the radiation fields of a Hertzian dipole in the moving medium.[101]

6.14. Solve Green's function for the time-dependent wave equation

$$\left(\nabla^2 - \frac{1}{c^2}\frac{\partial^2}{\partial t^2}\right) G(\mathbf{r}, t; \mathbf{r}', t') = -\delta(\mathbf{r} - \mathbf{r}')\delta(t - t'),$$

subject to the constraint that, for $t < t'$, $G = 0$ and, for $t > t'$, G represents an outgoing wave. First consider the Fourier transform of G:

$$G(\mathbf{r}, t; \mathbf{r}', t') = \int d^3k \int d\omega \, g(\mathbf{k}, \omega) \, e^{i\mathbf{k}\cdot(\mathbf{r} - \mathbf{r}') - i\omega(t - t')}$$

and find the function $g(\mathbf{k}, \omega)$. Determine the singularities of $g(\mathbf{k}, \omega)$ and evaluate the integral. Show that

$$G(\mathbf{r},t;\mathbf{r}',t')=\frac{\delta\left(t'+\dfrac{|\mathbf{r}-\mathbf{r}'|}{c}-t\right)}{4\pi|\mathbf{r}-\mathbf{r}'|},$$

which is known as the retarded Green's function.

6.15. Derive Stirling's formula:

$$n!\sim(2\pi)^{1/2}n^{n+1/2}e^{-n}$$

for large n from the defining integral,

$$n!=\int_0^\infty x^n e^{-x}\,dx,$$

by using the Laplace method.

6.16. Using the identity in (6.80), show that the exact solution for H_z in (6.90) on the surface is

$$H_z=-\frac{Ia}{2\pi(k_t^2-k^2)}\left\{\frac{1}{\rho}\frac{\partial}{\partial\rho}\left(\frac{k_t^2}{\rho}e^{ik_t\rho}-\frac{k^2}{\rho}e^{ik\rho}\right)\right.$$
$$\left.+\frac{3}{\rho}\left(\frac{\partial^2}{\partial\rho^2}-\frac{1}{\rho}\frac{\partial}{\partial\rho}\right)\left(\frac{1}{\rho^2}e^{ik_t\rho}-\frac{1}{\rho^2}e^{ik\rho}\right)\right\}.$$

Keeping terms of the order of $1/\rho^2$, we obtain (6.101). The proof is due to Van der Pol.[5]

6.17. Find the field caused by a vertical electric dipole placed on the surface of a half-space medium. Assuming that the medium is nonconductive, plot the radiation patterns caused by the dipole. Assuming that the medium is very conductive, calculate fields near the surface.

6.18. Calculate fields arising from a vertical magnetic dipole above an isotropic plasma with collisions.

6.19. Determine the field solutions caused by a horizontal electric dipole lying on the surface of a half-space dielectric medium. Plot the radiation patterns inside the dielectric in directions parallel and perpendicular to the dipole axis. Show that[105] the direction for maximum gain is equal to the critical angle in the plane perpendicular to the dipole and is equal to $\sin^{-1}[2/(1+\epsilon_t/\epsilon)]^{1/2}$ in the plane containing the dipole.

6.20. Determine the radiation fields caused by a horizontal electric dipole lying on the surface of a two-layer dielectric medium.

6.21. Solve the problem of scattering of a TE incident wave by the corrugated structure in Fig. 6.13.

6.22. Find the scattered wave arising from a wave incident upon a conducting cylinder. The magnetic field **H** of the incident wave is parallel to the axis of the cylinder. Show that, when the radius of the cylinder is infinitesimally small, the scattered wave is not isotropic, as opposed to the result shown by (6.123).

6.23. Derive the wave transformation in (6.140).

6.24. Derive (6.129) in the illuminated region for a wave scattered by a conducting cylinder. The electric field of the incident wave is parallel to the axis of the cylinder.

6.25. Solve the problem of scattering of a plane wave by a dielectric cylinder. Assume normal incidence and that the electric field is parallel to the axis of the cylinder.

6.26. Consider an infinitely long current filament adjacent to a perfectly conducting cylinder of radius a. The incident field caused by the current filament can be written as

$$E_z^i = -\frac{kI}{4\omega\epsilon} H_0^{(1)}(k|\boldsymbol{\rho}-\boldsymbol{\rho}'|),$$

where $\boldsymbol{\rho}'$ is the position of the current filament. Prove the addition theorem for Hankel functions:

$$H_0^{(1)}(k|\boldsymbol{\rho}-\boldsymbol{\rho}'|) = \begin{cases} \sum_{n=-\infty}^{\infty} H_n^{(1)}(k\rho')J_n(k\rho)\, e^{in(\phi'-\phi)}, & \rho<\rho', \\ \sum_{n=-\infty}^{\infty} J_n(k\rho')H_n^{(1)}(k\rho)\, e^{in(\phi'-\phi)}, & \rho>\rho'. \end{cases}$$

Using the addition theorem, find the scattered wave and express it in terms of a summation of cylindrical functions.[35]

7

Theorems of Waves and Media

In this chapter we shall discuss the classical and quantum aspects of electromagnetic theory. The equivalence principle, supported by the uniqueness theorem, enables us to devise many important theorems in practical problem solving. To study reciprocity, we introduce the reaction concept, which is also used to derive several stationary formulas based on the variational technique. Treating field vectors as operators, we postulate, in addition to Maxwell's equations, commutation relations among the field operators in order to quantize electromagnetic waves. As the final topic, we formulate macroscopic electrodynamics in terms of a scalar Lagrangian. Maxwell's equations are derived from Hamilton's principle, and the energy momentum conservation theorems from Noether's theorem. To facilitate the derivation, four-dimensional notation is introduced.

7.1 EQUIVALENCE PRINCIPLE

When we are interested in a limited region of space, we can replace all uninteresting regions outside this space by equivalent sources. We can place equivalent sources in the uninteresting regions, or we can place equivalent current sheets on the boundaries of the region of interest. The equivalent sources are by no means unique, and there are many different ways of constructing them. We need to make sure that all boundary conditions are satisfied and that the original fields and sources in the region of interest are preserved. When two different specifications of sources give the same solution in the region of interest (they certainly will give different solutions outside the region of interest), the two problems are called *equivalent*.

7.1a Equivalent Source Concept

Electric and Magnetic Dipole Sources. A small current loop (Fig. 7.1*a*) can be viewed as a magnetic dipole (Fig. 7.1*b*) if the loop is enclosed by a small volume and we are not interested in the inside of the volume. The current loop and the magnetic dipole yield identical fields; only when one penetrates the interior of the sources can a current loop be distinguished from a magnetic dipole. In the interior the magnetic fields of a loop and a magnetic dipole point in opposite directions. Just as electric dipoles constitute the building blocks of electric current sources, the magnetic dipoles constitute the building blocks of magnetic current sources.

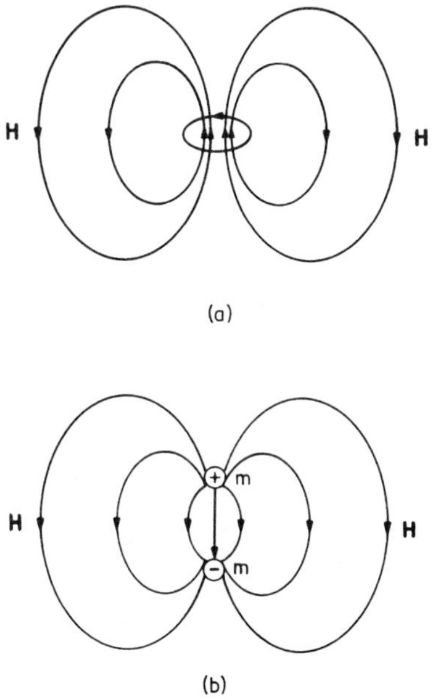

Figure 7.1 **(*a*) Small current loop (*b*) Magnetic dipole.**

Electric and Magnetic Current Sheets. When surface boundaries are replaced by equivalent sources, both electric and magnetic surface current sheets are required. The electric current sheets $\mathbf{J}_s$ are produced by discontinuities in tangential magnetic field components across the boundary

$$\mathbf{J}_s = \hat{n} \times \delta\mathbf{H},$$

where $\hat{n}$ is the surface normal, and $\delta\mathbf{H}$ is the difference between magnetic field components across the boundary. The magnetic surface current sheets are produced by discontinuities in tangential electric field components across the boundary

$$\mathbf{M}_s = -\hat{n} \times \delta\mathbf{E}.$$

Note that, from the definition for $\mathbf{M}_s$, the circulation of electric fields around $\mathbf{M}_s$ follows the left-hand rule, while the circulation of magnetic fields around $\mathbf{J}_s$ follows the right-hand rule.

Consider a surface electric current sheet with surface current $\mathbf{J}_s$ at $z=0$ flowing in the $-\hat{x}$ direction (Fig. 7.2a). This current sheet generates plane

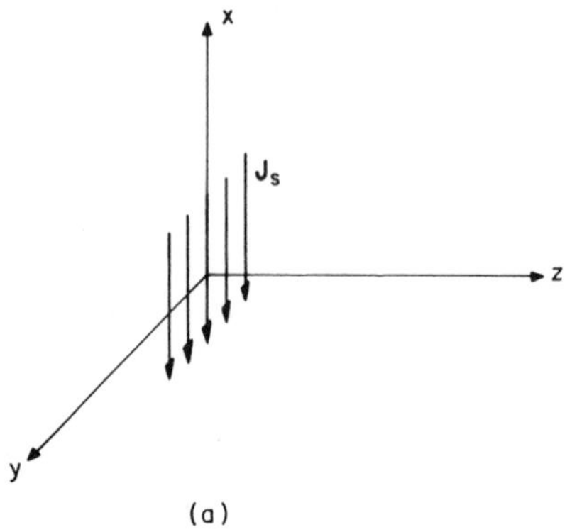

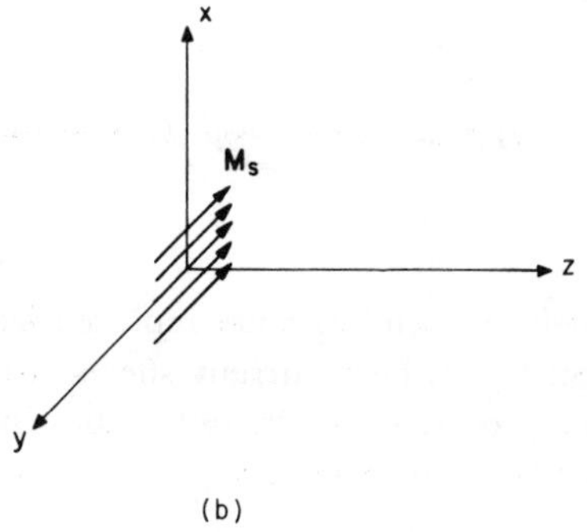

Figure 7.2 (*a*) Electric current sheet. (*b*) Magnetic current sheet.

waves in both positive and negative $\hat{z}$ directions:

$$E_x = \frac{\eta J_s}{2} e^{ikz}, \qquad H_y = \frac{1}{\eta} E_x, \qquad z > 0,$$

$$E_x = \frac{\eta J_s}{2} e^{-ikz}, \qquad H_y = -\frac{1}{\eta} E_x, \qquad z < 0.$$

All of the boundary conditions, which require that tangential electric fields be continuous at $z=0$ and that the discontinuity in tangential magnetic fields on both sides be equal to the strength of the current sheet, are satisfied.

As a dual situation, consider a magnetic surface current sheet M_s at $z=0$ flowing in the $-\hat{y}$ direction (Fig. 7.2*b*). The boundary condition requires that the tangential magnetic fields be continuous across the boundary and that the discontinuity in tangential electric fields be equal to the strength of the current sheet. The solution is as follows:

$$H_y = \frac{M_s}{2\eta} e^{ikz}, \qquad E_x = \eta H_y, \qquad z>0,$$

$$H_y = \frac{M_s}{2\eta} e^{-ikz}, \qquad E_x = -\eta H_y \qquad z<0.$$

Plane waves are radiated in both positive and negative $\hat{z}$ directions. Note that, by properly choosing the phase of the current sheets, we can generate plane waves in any direction. For instance, let $\mathbf{J}_s = \hat{x} J_s e^{ik_y y}$; then plane waves with wave vectors $\mathbf{k} = \hat{y}k_y + \hat{z}k_z$ and $\mathbf{k} = \hat{y}k_y - \hat{z}k_z$ are generated. If k_y is larger than k, the waves are evanescent in the $\hat{z}$ directions. Similar arguments apply to magnetic current sheets.

Using these equivalent current sheet concepts, we can illustrate several equivalent situations for a plane wave propagating in the $+\hat{z}$ direction (Fig. 7.3). Let the electric field be $\hat{x}$-directed:

$$\mathbf{E} = \hat{x} E_0 e^{ikz},$$

$$\mathbf{H} = \hat{y} \frac{1}{\eta} E_0 e^{ikz},$$

and let the region of interest be $z>0$.

Equivalent Problem 1: Put an electric current sheet with $\mathbf{J}_s = -\hat{x}E_0/\eta$ and a magnetic current sheet with $\mathbf{M}_s = -\hat{y}E_0$; then the same field is preserved for $z>0$. In the region $z<0$, there is no field.

Equivalent Problem 2: Put an electric current sheet with $\mathbf{J}_s = -\hat{x}2E_0/\eta$; then the same field is preserved for $z>0$. In the negative $\hat{z}$ direction, a plane wave propagates with $\mathbf{E} = \hat{x}E_0 e^{-ikz}$.

Equivalent Problem 3: Put a magnetic current sheet with $\mathbf{M}_s = -\hat{y}2E_0$; then the same field is preserved for $z>0$. In the negative $\hat{z}$ direction, a plane wave propagates with $\mathbf{H} = \hat{y}(E_0/\eta)e^{-ikz}$.

Equivalent Problem 4: Replace the region $z<0$ with a perfect conductor. Place in front of the conductor an electric current sheet with $\mathbf{J}_s = -\hat{x}E_0/\eta$ and a magnetic current sheet with $\mathbf{M}_s = -\hat{y}E_0$. The electric current sheet does not generate any field, because an equal and opposite electric current sheet is induced on the surface of the electric conductor and this sheet cancels the impressed $\mathbf{J}_s$.

Equivalent Problem 5: As a dual situation, we introduce the concept of a magnetic conductor on the surface of which tangential magnetic fields vanish. The same electric and magnetic current sheets as in Equivalent

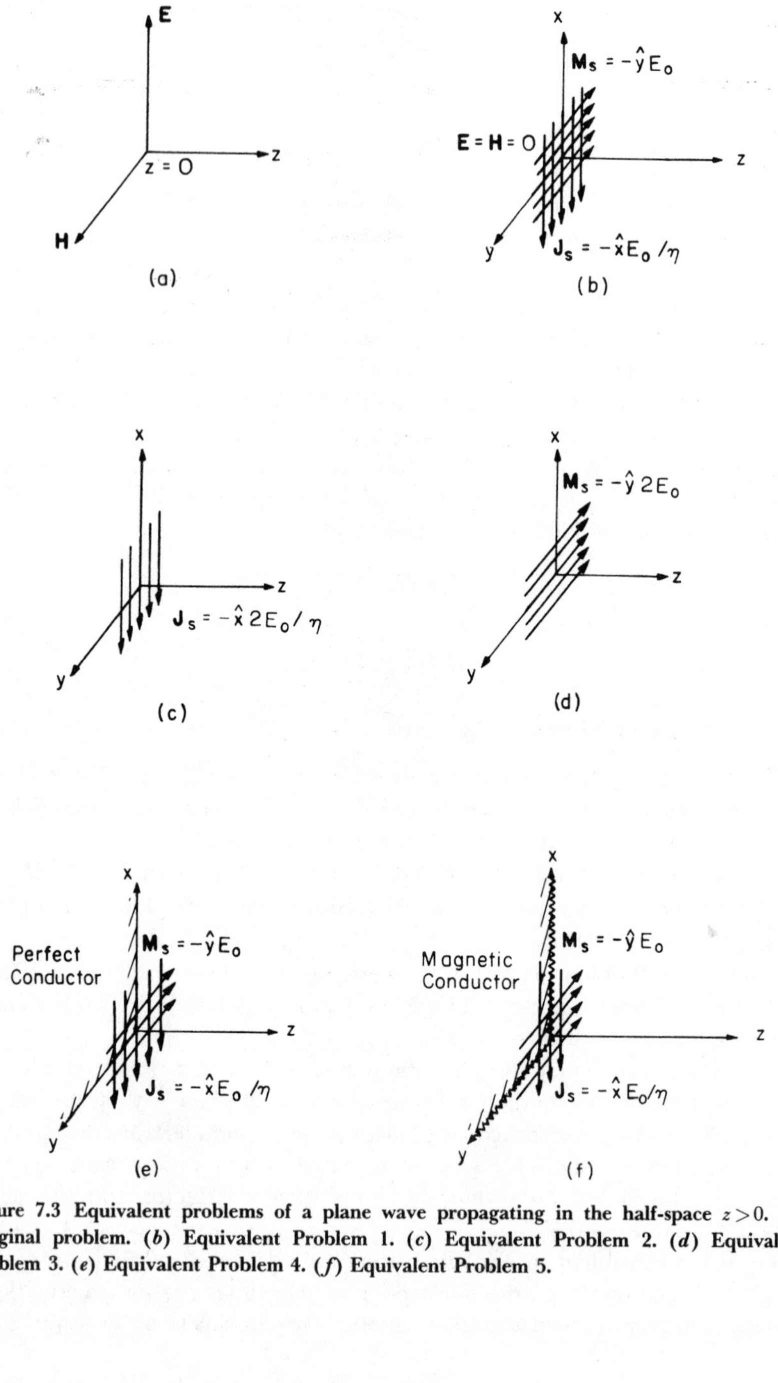

Figure 7.3 Equivalent problems of a plane wave propagating in the half-space $z > 0$. (*a*) Original problem. (*b*) Equivalent Problem 1. (*c*) Equivalent Problem 2. (*d*) Equivalent Problem 3. (*e*) Equivalent Problem 4. (*f*) Equivalent Problem 5.

Problem 4 are placed on the surface of a magnetic conductor occupying the region $z<0$. By a similar argument, the magnetic current sheet does not generate any field. The electric current sheet generates the plane wave for $z>0$.

Impressed and Induced Current Sheets. In these discussions we have used the terms *impressed* and *induced* current sheets. It is important to distinguish these two concepts. On the surface of a conducting body an induced current sheet is physically carried by charged particles attached to the surface of the body, whereas an impressed current sheet is carried by outside agents. When a layer of charge or current is impressed along the surface of a body, induced surface charge and current sheets are generated at the surface of the body so that the boundary conditions are satisfied.

Consider a plane wave normally incident upon the surface of a perfectly conducting half-space (Fig. 7.4). Let the electric field be $\mathbf{E}=\hat{x}E_0e^{ikz}$. An electric current sheet with $\mathbf{J}_s=\hat{x}2E_0/\eta$ is then induced on the surface of the conductor. The conductor can be replaced by the induced current sheet, which radiates into both $z>0$ and $z<0$ half-spaces. This induced current generates a reflected wave with $\mathbf{E}=-\hat{x}E_0e^{-ikz}$, so that at the boundary surface the total electric field is zero. This induced current sheet also generates a plane wave $\mathbf{E}=-\hat{x}E_0e^{ikz}$ in the region $z>0$, which combines with the incident wave to produce a zero field inside the conductor.

We must appreciate that magnetic sources are useful concepts, although in reality they may not exist. It is instructive to add an equivalent magnetic source to Faraday's law:

$$\nabla\times\mathbf{H}=-i\omega\mathbf{D}+\mathbf{J}, \tag{7.1a}$$

$$-\nabla\times\mathbf{E}=-i\omega\mathbf{B}+\mathbf{M}. \tag{7.1b}$$

The added magnetic source is denoted as $\mathbf{M}$. Similarly to the boundary condition for tangential magnetic fields, $\hat{n}\times\delta\mathbf{H}=\mathbf{J}_s$, the boundary condition for tangential electric fields becomes $-\hat{n}\times\delta\mathbf{E}=\mathbf{M}_s$.

The equivalence principle is useful in turning a problem into one with all uninteresting regions replaced by equivalent sources. Note that the equivalent sources are unknown until the problem is completely solved. In practice, equivalent sources are assumed in order to obtain approximate solutions. With the equivalent sources given, solutions can be readily calculated by employing Huygens' principle, In the next section we shall derive mathematical expressions for Huygens' principle and study the problem of diffraction by using the equivalence principle.

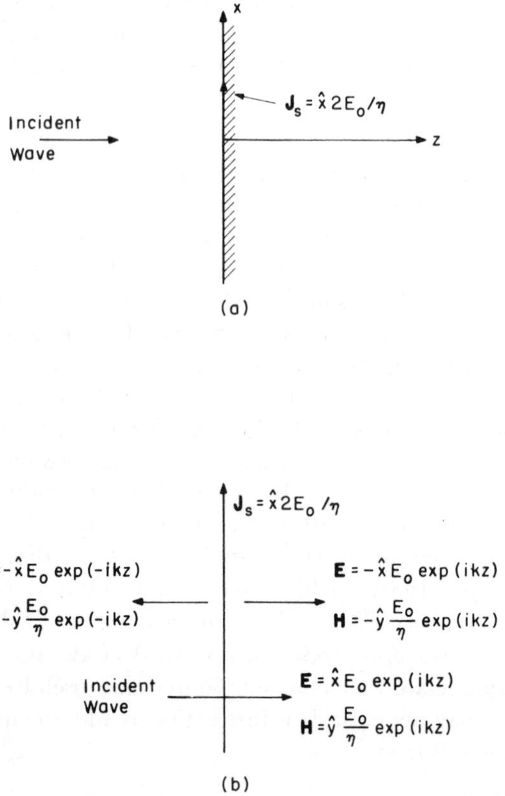

Figure 7.4 (*a*) A plane wave incident upon a perfect conductor induces surface current J_s. (*b*) The induced surface current generates a reflected wave in the $z<0$ half-space and a wave in the $z>0$ half-space which, combined with the incident wave, yield zero total field.

7.1b Huygens' Principle

Formulated in mathematical terms, Huygens' principle expresses fields at an observation point in terms of fields at the boundary surface. Consider a surface S' enclosing a radiation source. Under the assumption that the tangential electric and magnetic fields are known on the surface, the electric and magnetic fields outside this surface S' will be shown to take the following forms:

$$\mathbf{E}(\mathbf{r}) = \oint\!\!\!\oint_{S'} \left\{ i\omega\mu\overline{\mathbf{G}}(\mathbf{r},\mathbf{r}')\cdot[\hat{n}\times\mathbf{H}(\mathbf{r}')] + \nabla\times\overline{\mathbf{G}}(\mathbf{r},\mathbf{r}')\cdot[\hat{n}\times\mathbf{E}(\mathbf{r}')]\right\} dS, \tag{7.2a}$$

$$\mathbf{H}(\mathbf{r}) = \oiint_{S'} \left\{ -i\omega\epsilon\overline{\mathbf{G}}(\mathbf{r},\mathbf{r}') \cdot [\hat{n}\times\mathbf{E}(\mathbf{r}')] \right.$$

$$\left. + \nabla\times\overline{\mathbf{G}}(\mathbf{r},\mathbf{r}') \cdot [\hat{n}\times\mathbf{H}(\mathbf{r}')] \right\} dS, \tag{7.2b}$$

where $\hat{n}$ is the outward normal to the surface S'. Observe that $\hat{n}\times\mathbf{H}=\mathbf{J}_s$ and $-\hat{n}\times\mathbf{E}=\mathbf{M}_s$ are identifiable as the surface current sheets. Equation 7.2a is the dual of (7.2b). The dyadic Green's function $\overline{\mathbf{G}}(\mathbf{r},\mathbf{r}')$ is given by

$$\overline{\mathbf{G}} = \left(\overline{\mathbf{I}} + \frac{1}{5^2}\nabla\nabla\right) g(\mathbf{r},\mathbf{r}'). \tag{7.3}$$

For three-dimensional problems, the scalar Green's function $g(\mathbf{r},\mathbf{r}')$ for isotropic media written in spherical coordinates takes the form [see (6.10)]

$$g(\mathbf{r},\mathbf{r}') = \frac{e^{ik|\mathbf{r}-\mathbf{r}'|}}{4\pi|\mathbf{r}-\mathbf{r}'|}. \tag{7.4}$$

For two-dimensional problems, the scalar Green's function for isotropic media written in cylindrical coordinates takes the form [see (6.42)]

$$g(\boldsymbol{\rho},\boldsymbol{\rho}') = \frac{i}{4} H_0^{(1)}(k|\boldsymbol{\rho}-\boldsymbol{\rho}'|). \tag{7.5}$$

The dyadic Green's function $\overline{\mathbf{G}}$ satisfies the wave equation

$$\nabla\times\nabla\times\overline{\mathbf{G}}(\mathbf{r},\mathbf{r}') - k^2\overline{\mathbf{G}}(\mathbf{r},\mathbf{r}') = \overline{\mathbf{I}}\delta(\mathbf{r}-\mathbf{r}'), \tag{7.6}$$

which is identical to (6.4) except for a factor $i\omega\mu$ on the right-hand side.

To derive (7.2) (see Tai[101]), we integrate the identity

$$\mathbf{P}\cdot\nabla\times\nabla\times\mathbf{Q} - \mathbf{Q}\cdot\nabla\times\nabla\times\mathbf{P} = \nabla\cdot[\mathbf{Q}\times\nabla\times\mathbf{P} - \mathbf{P}\times\nabla\times\mathbf{Q}] \tag{7.7}$$

over the volume V bounded by the surface S' and the surface at infinity. Letting $\mathbf{P}=\mathbf{E}$ and $\mathbf{Q}=\overline{\mathbf{G}}\cdot\mathbf{a}$, where $\mathbf{a}$ is an arbitrary constant vector, we obtain

$$\iiint_V dV \left\{ \mathbf{E}(\mathbf{r})\cdot\nabla\times\nabla\times\overline{\mathbf{G}}(\mathbf{r},\mathbf{r}')\cdot\mathbf{a} - [\nabla\times\nabla\times\mathbf{E}(\mathbf{r})]\cdot\overline{\mathbf{G}}(\mathbf{r},\mathbf{r}')\cdot\mathbf{a} \right\}$$

$$= -\oiint_S dS\,\hat{s}\cdot \left\{ [\nabla\times\mathbf{E}(\mathbf{r})]\times\overline{\mathbf{G}}(\mathbf{r},\mathbf{r}')\cdot\mathbf{a} + \mathbf{E}(\mathbf{r})\times\nabla\times\overline{\mathbf{G}}(\mathbf{r},\mathbf{r}')\cdot\mathbf{a} \right\}.$$

On the right-hand side, the surface at infinity does not contribute, and on the surface S' the surface normal $\hat{s}$ is equal to $-\hat{n}$. On the left-hand side we use (7.6) and the fact that $\nabla\times\nabla\times\mathbf{E}(\mathbf{r})=k^2\mathbf{E}(\mathbf{r})$ because $\mathbf{J}(\mathbf{r})=0$ in V. We then have

$$\iiint_V dV\delta(\mathbf{r}-\mathbf{r}')\mathbf{E}(\mathbf{r})\cdot\mathbf{a}=\iint_{S'} dS\{[\hat{n}\times\nabla\times\mathbf{E}(\mathbf{r})]\cdot\overline{\mathbf{G}}(\mathbf{r},\mathbf{r}')\cdot\mathbf{a}$$

$$+[\hat{n}\times\mathbf{E}(\mathbf{r})]\cdot\nabla\times\overline{\mathbf{G}}(\mathbf{r},\mathbf{r}')\cdot\mathbf{a}\}.$$

Cancelling **a** from both sides of the equation, replacing $\nabla\times\mathbf{E}$ with $i\omega\mu\mathbf{H}$, and interchanging the primed and unprimed variables, we obtain (7.2a). Equation 7.2b for $\mathbf{H}(\mathbf{r})$ may be written by invoking duality.

To illustrate the use of (7.2), we consider diffraction of a plane wave by a slit (Fig. 7.5). Let the slit be infinite in the $\hat{x}$ direction and have width w with $kw>1$. Also let the plane wave be linearly polarized in the $\hat{x}$ direction and normally incident upon the opening from the region $z<0$. In the half-space $z>0$, the wave is diffracted. To solve for the diffracted field, we use the equivalence principle. We assume that the field at the slit is equal to the original field, which is a chopped plane wave. Similar to Equivalent Problem 2 for a plane wave as discussed in Section 7.1a, we assume at the aperture an electric current sheet $\mathbf{J}_s=2\hat{z}\times\mathbf{H}$ and no magnetic current sheet:

$$\mathbf{H}(\mathbf{r}')=\hat{y}\,\frac{E_0}{\eta}\,U\Big(\frac{w}{2}-|y'|\Big),\tag{7.8}$$

where E_0 is the amplitude of the incident wave, and $U[(w/2)-|y'|]$ is a unit step function that is unity for $|y'|\leqslant w/2$ and zero for $|y'|\geqslant w/2$. Equation 7.2a becomes simply

$$\mathbf{E}(\mathbf{r})=2\iint_{S'} dS i\omega\mu\overline{\mathbf{G}}(\mathbf{r},\mathbf{r}')\cdot\hat{z}\times\mathbf{H}(\mathbf{r}').$$

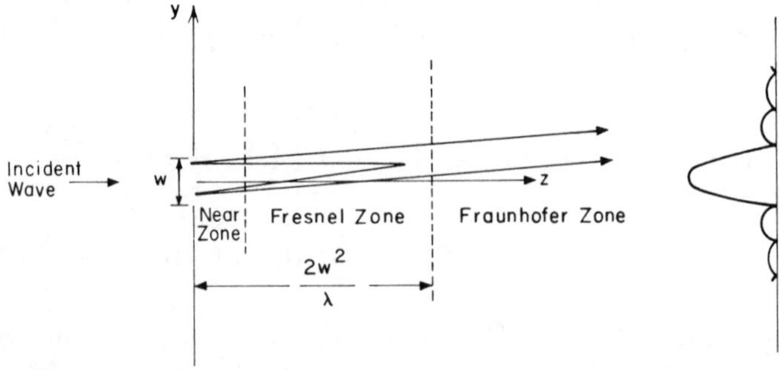

Figure 7.5 Diffraction by a slit.

The problem is independent of the coordinate x, $\partial/\partial x = 0$. We use the two-dimensional scalar Green's function 7.5 and find

$$\mathbf{E}(y,z) = \hat{x}\,\frac{\omega\mu}{2\eta}\int_{-\infty}^{\infty} dy'\,E_0 U\left(\frac{w}{2} - |y'|\right) H_0^{(1)}\left[k\sqrt{(y-y')^2 + z^2}\,\right]. \quad (7.9)$$

In the far-field zone, we use the asymptotic form 6.68 for the Hankel function.

Assume that the observation points are far away from the aperture but very near the z axis. Then $|z| \gg |y - y'|$ and we can expand:

$$k\sqrt{(y-y')^2 + z^2} \approx kz\left[1 + \frac{1}{2}\left(\frac{y-y'}{z}\right)^2 + \cdots\right]. \quad (7.10)$$

Equation 7.9 becomes

$$\mathbf{E}(y,z) = \hat{x}\,\frac{k}{2}\sqrt{\frac{2}{i\pi kz}}\;e^{ikz}\int_{-\infty}^{\infty} dy'\,E_0 U\left(\frac{w}{2} - |y'|\right) e^{ik(y-y')^2/2z}, \quad (7.11)$$

where we use the first two terms of expansion 7.10 in the exponent and only the first term in the denominator of the Hankel function expansion.

Expansion 7.10, which gives rise to (7.11), is referred to as the *Fresnel approximation*. Equation 7.11 is conveniently evaluated in terms of tabulated functions known as Fresnel integrals. The maximum of y' is equal to half the aperture dimension. If we further neglect $ky'^2/2z$ in (7.10), we have the *Fraunhofer approximation*:

$$k\sqrt{(y-y')^2 + z^2} \approx kz + \frac{ky^2}{2z} - \frac{kyy'}{z} + \cdots.$$

Equation 7.11 becomes

$$\mathbf{E}(y,z) = \hat{x}\,\frac{k}{2}\sqrt{\frac{2}{i\pi kz}}\;e^{ikz + iky^2/2z}\int_{-\infty}^{\infty} dy'\,E_0 U\left(\frac{w}{2} - |y'|\right) e^{-i(ky/z)y'}. \quad (7.12)$$

The integral is a Fourier transform of the step function, and the result is

$$\mathbf{E}(y,z) = \hat{x}\,\frac{E_0 kw}{2}\sqrt{\frac{2}{i\pi kz}}\;\frac{\sin(kyw/2z)}{kyw/2z}\,e^{ikz + iky^2/2z}. \quad (7.13)$$

Note that in (7.12) the source function $E_0U[(w/2)-|y'|]$ can be replaced by any general aperture field distribution. The observed field is proportional to the Fourier transform of the aperture field.

According to the distance from the aperture, the diffracted field can be divided into the near zone, the Fresnel zone, and the Fraunhofer zone (Fig. 7.5). The separation between the Fresnel zone and the Fraunhofer zone may be taken at $z_F=2w^2/\lambda$. For an aperture of dimension 1 cm and a wavelength of $0.63\,\mu$, the Fraunhofer zone begins at $z_F\approx 320$ m, a rather stringent constraint for Fourier optics experiments. The use of a convergent lens may eliminate this problem. Under a paraxial approximation, the effect of a lens amounts to giving a wave transmitting through the lens a phase factor proportional to $e^{-iky^2/2f}$, where f is the focal length. If a lens is placed immediately in front of the aperture, we find that (7.11) reduces to (7.12) at $z=f$. Thus, when a screen is placed a focal length from the lens, the Fourier transform of the aperture field is observed.

As another example, consider the diffraction of a plane wave normally incident on a circular aperture. We shall not restrict observation points to be near the z axis. At the aperture, we use Equivalent Problem 3 for a plane wave as discussed in Section 7.1a to assume a magnetic current sheet with $\mathbf{M}_s-2\hat{n}\times\mathbf{E}(\mathbf{r}')$ and no electric current sheet. Equation 7.2a becomes, in view of Green's function 7.4,

$$\mathbf{E}(\mathbf{r})=2\int\int_{S'}\nabla\times\mathbf{G}(\mathbf{r},\mathbf{r}')\cdot[\hat{n}\times\mathbf{E}(\mathbf{r}')]dS'$$

$$=2\nabla\times\int\int_{S'}\hat{n}\times\mathbf{E}(\mathbf{r}')\frac{e^{ik|\mathbf{r}-\mathbf{r}'|}}{4\pi|\mathbf{r}-\mathbf{r}'|}dS' \tag{7.14}$$

where S' denotes the area of the circular aperture.

In the radiation zone, the observation point is so remote from the aperture that all wave vectors originating from the aperture are essentially parallel and we have

$$|\mathbf{r}-\mathbf{r}'|\approx r-\mathbf{r}'\cdot\hat{r},$$

which is identical to (6.12). The integral in (7.14) becomes

$$\mathbf{E}(\mathbf{r})\approx\frac{ie^{ikr}}{2\pi r}\mathbf{k}\times\int\int_{S'}\hat{n}\times\mathbf{E}(\mathbf{r}')e^{-i\mathbf{k}\cdot\mathbf{r}'}dS'.$$

Let the plane wave be polarized in the $\hat{x}$ direction. For a wave vector

$\mathbf{k}=\hat{x}k\sin\theta\cos\phi+\hat{y}k\sin\theta\sin\phi+\hat{z}k\cos\theta$, we have

$$\mathbf{E}(\mathbf{r})=\frac{iE_0e^{ikr}}{2\pi r}(\mathbf{k}\times\hat{y})\int_0^R d\rho'\int_0^{2\pi}\rho'\,d\phi'\,e^{-ik\rho'\sin\theta\cos(\phi-\phi')}$$

$$=(\hat{z}\sin\theta\cos\phi-\hat{x}\cos\theta)\frac{ikE_0e^{ikr}}{2\pi r}\int_0^R\rho'\,d\rho'J_0(k\rho'\sin\theta)$$

$$=(\hat{z}\sin\theta\cos\phi-\hat{x}\cos\theta)\frac{J_1(kR\sin\theta)}{kR\sin\theta}\frac{ikR^2E_0e^{ikr}}{r}$$

where E_0 is the amplitude of the incident wave and R is the radius of the aperture.

7.1c Duality and Complementarity

The duality displayed by (7.1) in terms of the electric and magnetic fields and sources is self-evident. With the replacements

$$\mathbf{E}\to\mathbf{H},\qquad \mathbf{H}\to-\mathbf{E},$$

$$\mathbf{D}\to\mathbf{B},\qquad \mathbf{B}\to-\mathbf{D},$$

$$\mathbf{J}\to\mathbf{M},\qquad \mathbf{M}\to-\mathbf{J},$$

(7.1a) becomes (7.1b), and (7.1b) becomes (7.1a). In actual problem solving, when we obtain the solution for a particular problem the solution to its dual can be immediately written, as we have seen on many occasions in preceding chapters.

Consider the problem of diffraction by electric conductors. Anticipating later developments, we assume that the conductors cover parts of a plane. Suppose that the problem is solved and the field solutions are denoted as $\mathbf{E}_e$ and $\mathbf{H}_e$. The dual to this problem is replacement of the electric conductors with magnetic conductors and sources with their duals. Denote the solution to the dual problem by $\mathbf{E}_m$ and $\mathbf{H}_m$. If the dual replacements are made numerically, then we have numerically $\mathbf{E}_m=\mathbf{H}_e$ and $\mathbf{H}_m=-\mathbf{E}_e$.

Consider a plane screen made of electric conductors (Fig. 7.6*a*). Let the electromagnetic fields in the absence of the screen be $\mathbf{E}$ and $\mathbf{H}$. We use Il to denote electric sources and Kl to denote magnetic sources. In the presence of the screen the fields are denoted by $\mathbf{E}_a$ and $\mathbf{H}_a$, and the boundary condition

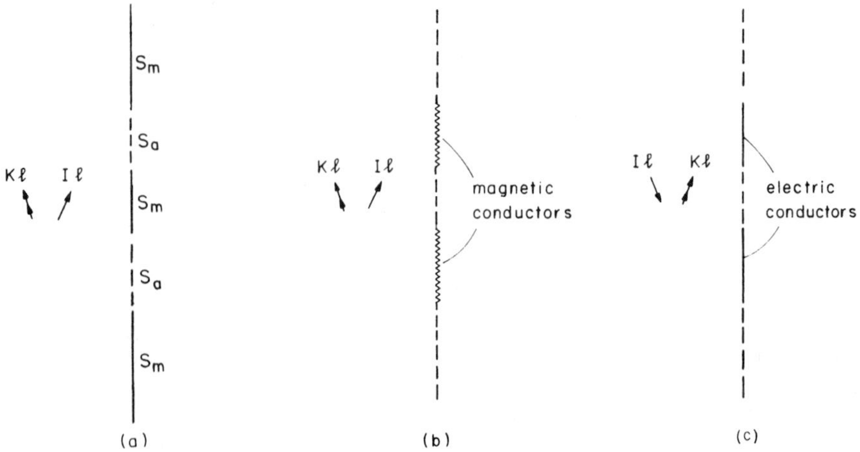

Figure 7.6 Babinet's principle. (*a*) Electric screen. (*b*) Complementary problem of (a). (*c*) Dual problem of (b).

states that

$$\hat{n}\times\mathbf{E}_a=0 \qquad (\text{over } S_m),$$

$$\hat{n}\times\mathbf{H}_a=\hat{n}\times\mathbf{H} \qquad (\text{over } S_a),$$

where S_m denotes the conductor area, and S_a the aperture area. The first condition warrants that the tangential **E** field vanishes over the conducting surface. The second condition is due to the fact that induced current elements on the conducting plane do not give rise to any tangential magnetic component at the aperture area. The tangential magnetic field $\mathbf{H}_a$ is thus equal to the original tangential field.

Consider another plane screen made of magnetic conductors (Fig. 7.6*b*). The two screens are *complementary* to each other if on superposition the two cover the whole plane with no overlap. For the fields $\mathbf{E}_m$ and $\mathbf{H}_m$ in the presence of the magnetically conducting screen, the boundary condition states that

$$n\times\mathbf{E}_m=n\times\mathbf{E} \qquad (\text{over } S_m),$$

$$n\times\mathbf{H}_m=0 \qquad (\text{over } S_a).$$

This is a complementary case of the electrically conducting screen.

When the two complementary screens are superimposed, the boundary conditions are also superimposed, and we have

$$\hat{n}\times(\mathbf{E}_a+\mathbf{E}_m)=\hat{n}\times\mathbf{E} \qquad (\text{over } S_m),$$

$$\hat{n}\times(\mathbf{H}_a+\mathbf{H}_m)=\hat{n}\times\mathbf{H} \qquad (\text{over } S_a).$$

We see that the magnetic current sheets corresponding to the original tangential **E** field are specified over S_m, and the electric current sheets corresponding to the original tangential **H** field are specified over S_a. According to the uniqueness theorem, which will be proved later, this specification of magnetic current on parts of the plane and electric current on the rest of the plane is sufficient to determine the field at the right of its plane uniquely. Thus we have

$$\mathbf{E}_a+\mathbf{E}_m=\mathbf{E}, \tag{7.15a}$$

$$\mathbf{H}_a+\mathbf{H}_m=\mathbf{H}. \tag{7.15b}$$

This is referred to as *Babinet's principle*. We see that, if the diffraction problem involving either an electric or a magnetic screen is solved, the solution to the complementary problem is immediately obtained.

Babinet's principle can be expressed in a different manner by combining the dual and the complementary problems. The dual problem of Fig. 7.6*b* is illustrated in Fig. 7.6*c*. Denoting the solution by $\mathbf{E}_e$ and $\mathbf{H}_e$, we have $\mathbf{E}_m=\mathbf{H}_e$ and $\mathbf{H}_m=-\mathbf{E}_e$. Babinet's principle then states that $\mathbf{E}_a+\mathbf{H}_e=\mathbf{E}$ and $\mathbf{H}_a-\mathbf{E}_e=\mathbf{H}$.

7.1d Image Theorem

When solving problems for electromagnetic sources in front of a perfectly conducting plane, the method of images is useful. The conventional image method consists of the following steps: (1) finding the image sources and removing the conducting plane; (2) assuming the same media over the entire space; and (3) solving for field quantities in the presence of both original and image sources, but in the absence of the plane boundary. The solution thus obtained is valid only in front of the conducting plane.

To find image sources, we must make sure that the boundary conditions are satisfied at the conducting plane. In Fig. 7.7 we illustrate the images to the elementary sources so that the combination of the original and its image makes the tangential electric field vanish at an electrically conducting boundary, and the tangential magnetic field vanish at a magnetically conducting boundary. We use single-headed arrows to represent electric dipoles and double-headed arrows to represent magnetic dipoles. When

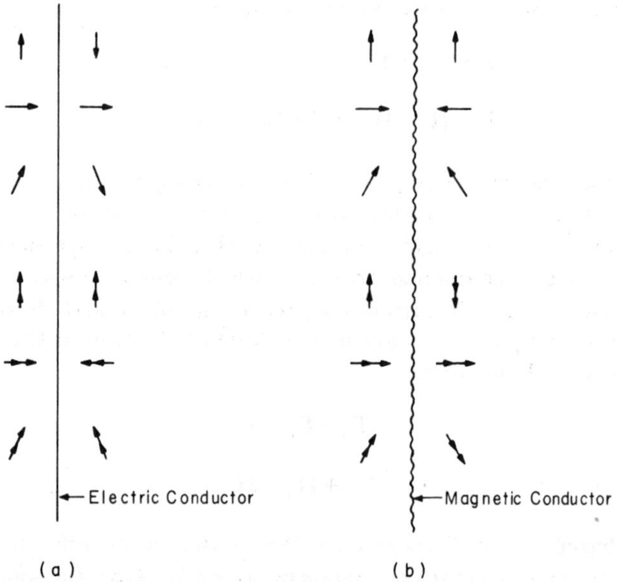

Figure 7.7 **(*a*) Sources in front of a perfect conductor and their images. (*b*) Sources in front of a magnetic conductor and their images.**

more than one plane boundary is involved, multiple images may be constructed. Figure 7.8 illustrates a dipole antenna situated between two plane conductors. The solution obtained by the image method is valid in the region between the two conductors and invalid outside this region.

The construction of the image source in front of a plane electric conductor involves a simple spatial inversion. Let the z axis be perpendicular to the

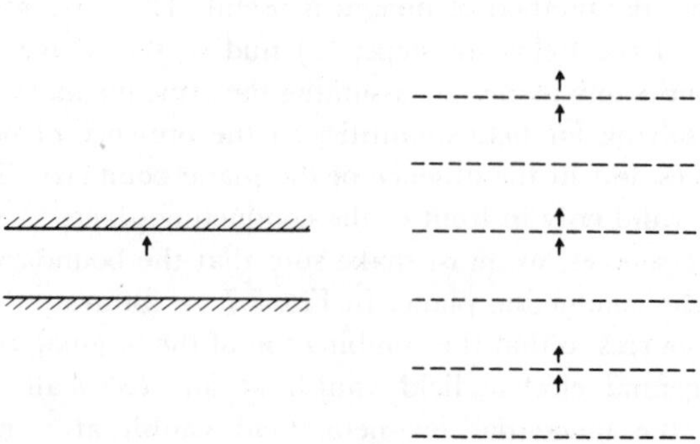

Figure 7.8 **(*a*) An electric dipole in the region between two parallel-plate conductors. (*b*) Multiple images of the dipole.**

plane boundary, and define an inversion operator P:

$$P=\begin{bmatrix} 1 & & \\ & 1 & \\ & & -1 \end{bmatrix}. \tag{7.16}$$

We see that

$$\mathbf{J}'(\mathbf{r}')=-P\mathbf{J}(\mathbf{r}), \tag{7.17a}$$

$$\mathbf{M}'(\mathbf{r}')=P\mathbf{M}(\mathbf{r}), \tag{7.17b}$$

where primes denote quantities in the image region. With the conducting plane removed, Maxwell's equations in the image region are as follows:

$$\nabla'\times\mathbf{E}'=-\frac{\partial\mathbf{B}'}{\partial t'}-\mathbf{M}', \tag{7.18a}$$

$$\nabla'\times\mathbf{H}'=\frac{\partial\mathbf{D}'}{\partial t'}+\mathbf{J}', \tag{7.18b}$$

while in the region in front of the conducting plane we have

$$\nabla\times\mathbf{E}=-\frac{\partial\mathbf{B}}{\partial t}-\mathbf{M}, \tag{7.19a}$$

$$\nabla\times\mathbf{H}=\frac{\partial\mathbf{D}}{\partial t}+\mathbf{J}. \tag{7.19b}$$

Under the inversion transformation, we can show that, for any vector $\mathbf{A}$,

$$\begin{aligned}\nabla'\times\mathbf{A}&=(P\nabla)\times\mathbf{A}\\&=-P[\nabla\times(P\mathbf{A})].\end{aligned} \tag{7.20}$$

In view of this identity and the transformation for $\mathbf{J}$, we find that under P

$$\mathbf{D}'(\mathbf{r}')=-P\mathbf{D}(\mathbf{r}), \tag{7.21a}$$

$$\mathbf{E}'(\mathbf{r}')=-P\mathbf{E}(\mathbf{r}), \tag{7.21b}$$

$$\mathbf{H}'(\mathbf{r}')=P\mathbf{H}(\mathbf{r}), \tag{7.21c}$$

$$\mathbf{B}'(\mathbf{r}')=P\mathbf{B}(\mathbf{r}). \tag{7.21d}$$

At the conducting plane $z=0$, boundary conditions require that both tangential electric field and normal magnetic field vanish, and this is seen to be satisfied with these fields.

In the image method, step 2 is valid only for simple media. We now examine the validity of step 2 for bianisotropic media. Maxwell's equations are form-invariant in the region $z>0$ and in the region $z<0$ after the conducting plane is removed and the half-space is replaced by an image medium. Let the constitutive relations for the image medium be as follows:

$$\begin{bmatrix} c\mathbf{D}' \\ \mathbf{H}' \end{bmatrix} = \begin{bmatrix} \overline{\mathbf{P}}' & \overline{\mathbf{L}}' \\ \overline{\mathbf{M}}' & \overline{\mathbf{Q}}' \end{bmatrix} \cdot \begin{bmatrix} \mathbf{E}' \\ c\mathbf{B}' \end{bmatrix}. \tag{7.22}$$

Requiring that the constitutive relations be form-invariant under P, in view of the transformation formulas for the field vectors in (7.21), we obtain

$$\overline{\mathbf{P}}' = P\overline{\mathbf{P}}P, \tag{7.23a}$$

$$\overline{\mathbf{Q}}' = P\overline{\mathbf{Q}}P, \tag{7.23b}$$

$$\overline{\mathbf{L}}' = -P\overline{\mathbf{L}}P, \tag{7.23c}$$

$$\overline{\mathbf{M}}' = -P\overline{\mathbf{M}}P. \tag{7.23d}$$

In general, the primed constitutive matrices are not equal to the unprimed ones. Only when they are equal will step 2 be valid. When they are not equal, the image method turns the original problem into a problem involving two sets of sources in two different media separated by a plane boundary. The problem may become more difficult than the original one, and then the image method is not useful.

The relations between the image medium and the original medium are derived in (7.23). Since the conventional image method is useful only when the two media are identical, it is important to investigate under what conditions this will be so. Examination of (7.23) reveals that the image medium is identical to the original medium if the latter is described by constitutive relations of the following forms:

$$\overline{\mathbf{P}} = \begin{bmatrix} p_{11} & p_{12} & 0 \\ p_{21} & p_{22} & 0 \\ 0 & 0 & p_{33} \end{bmatrix}, \tag{7.24a}$$

$$\overline{\mathbf{Q}}=\begin{bmatrix} q_{11} & q_{12} & 0 \\ q_{21} & q_{22} & 0 \\ 0 & 0 & q_{33} \end{bmatrix}, \tag{7.24b}$$

$$\overline{\mathbf{L}}=\begin{bmatrix} 0 & 0 & l_{13} \\ 0 & 0 & l_{23} \\ l_{31} & l_{32} & 0 \end{bmatrix}, \tag{7.24c}$$

$$\overline{\mathbf{M}}=\begin{bmatrix} 0 & 0 & m_{13} \\ 0 & 0 & m_{23} \\ m_{31} & m_{32} & 0 \end{bmatrix}. \tag{7.24d}$$

We observe that the conventional image method is applicable if the medium in front of the conducting plane is isotropic. For an anisotropic crystal, one of its three principal axes must be perpendicular to the boundary. For a gyrotropic medium, the magnetostatic field must be perpendicular to the boundary. For biisotropic media or bianisotropic media with the Dzyaloshinskii constitutive matrices, the conventional image method is not applicable. However, when the bianisotropic medium is an anisotropic medium with constitutive matrices $\overline{\mathbf{P}}$ and $\overline{\mathbf{Q}}$, as described in (7.24a) and (7.24b), and is moving uniformly parallel to the perfectly conducting plane, the original medium and the image medium are identical.

7.1e Uniqueness Theorem

The whole foundation of the equivalence principle rests on the validity of the uniqueness theorem, which guarantees that there is one and only one solution to a properly specified problem. Thus we are entitled to make use of all expedient means to find the solution. We shall examine the uniqueness theorem for bianisotropic media.

Let the region of interest be denoted as V and bounded by a closed surface S. Assume that a set of sources inside V produces two different solutions for the field quantities, and let $\mathbf{E}_1$ and $\mathbf{H}_1$ denote one set of solutions and $\mathbf{E}_2$ and $\mathbf{H}_2$ the other set. Both sets of solutions satisfy Maxwell's equation. We form the differences of the two sets of solutions and denote them by $\delta\mathbf{E}=\mathbf{E}_1-\mathbf{E}_2$ and $\delta\mathbf{H}=\mathbf{H}_1-\mathbf{H}_2$. The uniqueness theorem requires that $\delta\mathbf{E}=\delta\mathbf{H}=0$.

Obviously, $\delta\mathbf{E}$ and $\delta\mathbf{H}$ satisfy the source-free Maxwell's equations in V:

$$\nabla\times\delta\mathbf{H}=-i\omega(\bar{\epsilon}\cdot\delta\mathbf{E}+\bar{\xi}\cdot\delta\mathbf{H}), \tag{7.25a}$$

$$-\nabla\times\delta\mathbf{E}=-i\omega(\bar{\zeta}\cdot\delta\mathbf{E}+\bar{\mu}\cdot\delta\mathbf{H}). \tag{7.25b}$$

Using (7.25), we can derive the following equations:

$$\nabla\cdot(\delta\mathbf{E}\times\delta\mathbf{H}^*)=-i\omega(\delta\mathbf{E}\cdot\bar{\epsilon}^*\cdot\delta\mathbf{E}^*+\delta\mathbf{E}\cdot\bar{\xi}^*\cdot\delta\mathbf{H}^* \\ -\delta\mathbf{H}^*\cdot\bar{\zeta}\cdot\delta\mathbf{E}-\delta\mathbf{H}^*\cdot\bar{\mu}\cdot\delta\mathbf{H}), \tag{7.26}$$

$$\nabla\cdot(\delta\mathbf{E}^*\times\delta\mathbf{H})=i\omega(\delta\mathbf{E}^*\cdot\bar{\epsilon}\cdot\delta\mathbf{E}+\delta\mathbf{E}^*\cdot\bar{\xi}\cdot\delta\mathbf{H} \\ -\delta\mathbf{H}\cdot\bar{\zeta}^*\cdot\delta\mathbf{E}^*-\delta\mathbf{H}\cdot\bar{\mu}^*\cdot\delta\mathbf{H}^*). \tag{7.27}$$

When we add the two equations and integrate over the volume, the divergence terms give rise to a surface integral over S:

$$\oint\!\!\oint_S dS\,\hat{n}\cdot(\delta\mathbf{E}\times\delta\mathbf{H}^*+\delta\mathbf{E}^*\times\delta\mathbf{H}) \\ =\int\!\!\int\!\!\int_V dV\,[\delta\mathbf{E}^*\ \delta\mathbf{H}^*]\cdot\left[i\omega\delta\bar{\mathbf{C}}_{\mathbf{EH}}\right]\cdot\begin{bmatrix}\delta\mathbf{E}\\ \delta\mathbf{H}\end{bmatrix}, \tag{7.28}$$

where

$$\delta\bar{\mathbf{C}}_{\mathbf{EH}}=\begin{bmatrix}\bar{\epsilon}-\bar{\epsilon}^+ & \bar{\xi}-\bar{\zeta}^+\\ \bar{\zeta}-\bar{\xi}^+ & \bar{\mu}-\bar{\mu}^+\end{bmatrix},$$

and $\hat{n}$ is the surface normal.

We let $i\omega\delta\bar{\mathbf{C}}_{\mathbf{EH}}$ be negative (positive) definite, which means that the medium is passive (active). If the medium is not passive (active) we treat it as the limiting case of a passive (active) medium. For instance, for a passive isotropic medium, $\epsilon-\epsilon^*=2i\epsilon''$ and $\mu-\mu^*=2i\mu''$. The integrand in the right-hand side of (7.28) becomes $-2\epsilon''|\delta\mathbf{E}|^2-2\mu''|\delta\mathbf{H}|^2$ and is everywhere negative definite. Therefore if the left-hand side of (7.28) is zero, we shall have$\delta\mathbf{E}=\delta\mathbf{H}=0$ and the solution is unique. The same arguement is true for media that are not isotropic.

The left-hand side of (7.28) will be zero if one of the following conditions are met: (*a*) $\hat{n}\times\mathbf{E}$ is specified everywhere over the whole boundary S; (*b*) $\hat{n}\times\mathbf{H}$ is specified everywhere over the whole boundary S; or (*c*) $\hat{n}\times\mathbf{E}$ is specified over part of S and $\hat{n}\times\mathbf{H}$ is specified over the rest of S. To summarize, we state that the solution to a problem in a region V bounded by surface S will be unique if $\hat{n}\times\mathbf{E}$ or $\hat{n}\times\mathbf{H}$ is specified over the entire bounding surface. It is not necessary to specify both $\hat{n}\times\mathbf{E}$ and $\hat{n}\times\mathbf{H}$ at any part of S.

7.2 REACTION AND RECIPROCITY

7.2a Reaction Concept

Consider a time-harmonic source a, denoted by $\mathbf{J}_a$ and $\mathbf{M}_a$, situated in a field $\mathbf{E}_b$ and $\mathbf{H}_b$ produced by another source b, denoted by $\mathbf{J}_b$ and $\mathbf{M}_b$. The reaction of source a with field b is defined as

$$\langle a,b\rangle \equiv \int\int\int_V dV(\mathbf{J}_a\cdot\mathbf{E}_b-\mathbf{M}_a\cdot\mathbf{H}_b). \tag{7.29}$$

Note that in the representation $\langle a,b\rangle$ the first entry, a, associates with the source and the second entry, b with the field; whenever the source is zero, the reaction is zero. The integration extends over the region containing source a, which is composed of volume current densities, as well as surface current densities. In the case of surface current densities, the volume integrals become surface integrals.

The reaction is a complex number. It is different from complex power in two respects; first, in the definition for power the current density is complex-conjugated; second, the reaction is defined for a source with respect to the field produced by another source. When the source is reacting to the field produced by itself, we have the self-reaction $\langle a,a\rangle$.

To understand what reaction means physically, consider the case in which source a is a dipole, $\mathbf{J}_a=\mathbf{I}l\delta(\mathbf{r}-\mathbf{r}_0)$. Then the reaction

$$\langle a,b\rangle=\mathbf{I}l\cdot\mathbf{E}_b(\mathbf{r}_0) \tag{7.30}$$

is proportional to the electric field in the direction of $\mathbf{I}$ as measured by the dipole. It is equal to the field strength produced by source b at the dipole when the dipole moment $\mathbf{I}l$ is unity.

As another example, consider the reaction of a current source and a voltage source in circuits (Fig. 7.9). Let V_b and I_b be caused by an unspecified source b. For the current source (Fig. 7.9a), we have

$$\langle a,b\rangle = \int\int\int \mathbf{J}_a \cdot \mathbf{E}_b \, dV = I_a \int \mathbf{E}_b \cdot d\mathbf{l}$$

$$= -I_a V_b. \tag{7.31a}$$

For the voltage source (Fig. 7.9b), we have

$$\langle a,b\rangle = -\int\int\int \mathbf{M}_a \cdot \mathbf{H}_b \, dV = -V_a \int \mathbf{H}_b \cdot d\mathbf{l}$$

$$= -V_a I_b. \tag{7.31b}$$

Thus, if we use a unit current source, the reaction $\langle a,b\rangle$ is equal to the voltage at a due to source b. If we use a unit voltage source, the reaction $\langle a,b\rangle$ is equal to the current at a due to source b.

7.2b Reciprocity

We define a system as *reciprocal* if, with respect to two sets of sources a and b,

$$\langle a,b\rangle = \langle b,a\rangle. \tag{7.32}$$

Let the system be an isotropic medium; we shall show that isotropic media are reciprocal. We write Maxwell's equations with source a as

$$\nabla \times \mathbf{H}_a = -i\omega\epsilon \mathbf{E}_a + \mathbf{J}_a, \tag{7.33a}$$

$$-\nabla \times \mathbf{E}_a = -i\omega\mu \mathbf{H}_a + \mathbf{M}_a, \tag{7.33b}$$

and Maxwell's equations with source b as

$$\nabla \times \mathbf{H}_b = -i\omega\epsilon \mathbf{E}_b + \mathbf{J}_b, \tag{7.34a}$$

$$-\nabla \times \mathbf{E}_b = -i\omega\mu \mathbf{H}_b + \mathbf{M}_b. \tag{7.34b}$$

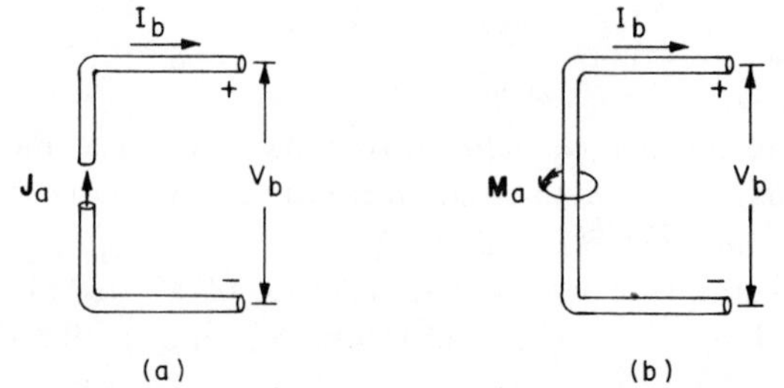

Figure 7.9 (*a*) Reaction of a current source. (*b*) Reaction of a voltage source.

Forming the sum of (7.33a) dot-multiplied by $\mathbf{E}_b$ and (7.34b) dot-multiplied by $\mathbf{H}_a$, and the sum of (7.33b) dot-multiplied by $\mathbf{H}_b$ and (7.34a) dot-multiplied by $\mathbf{E}_a$, we arrive at

$$-\nabla\cdot(\mathbf{E}_b\times\mathbf{H}_a)=-i\omega\epsilon\mathbf{E}_a\cdot\mathbf{E}_b+\mathbf{J}_a\cdot\mathbf{E}_b-i\omega\mu\mathbf{H}_a\cdot\mathbf{H}_b+\mathbf{M}_b\cdot\mathbf{H}_a, \quad (7.35a)$$

$$-\nabla\cdot(\mathbf{E}_a\times\mathbf{H}_b)=-i\omega\epsilon\mathbf{E}_a\cdot\mathbf{E}_b+\mathbf{J}_b\cdot\mathbf{E}_a-i\omega\mu\mathbf{H}_a\cdot\mathbf{H}_b+\mathbf{M}_a\cdot\mathbf{H}_b. \quad (7.35b)$$

Subtracting (7.35b) from (7.35a) and integrating, we obtain

$$\langle a,b\rangle-\langle b,a\rangle=\oint\!\!\oint_S dS\,\hat{s}\cdot(\mathbf{E}_a\times\mathbf{H}_b-\mathbf{E}_b\times\mathbf{H}_a). \quad (7.36)$$

By definition, isotropic media are reciprocal provided that

$$\oint\!\!\oint_S dS\,\hat{s}\cdot(\mathbf{E}_a\times\mathbf{H}_b-\mathbf{E}_b\times\mathbf{H}_a)=0. \quad (7.37)$$

This statement is referred to as the *Lorentz reciprocity theorem*. The theorem is valid when the region is unbounded but all sources and matter are of finite extent. At infinite distances away from the source, the radiation fields are $E_\theta=\eta H_\phi$ and $E_\phi=-\eta H_\theta$. As a consequence, $\mathbf{E}_a\times\mathbf{H}_b-\mathbf{E}_b\times\mathbf{H}_a=0$ and we find that the surface integral in (7.37) vanishes. The theorem is also valid when the region is bounded by perfect conductors.

7.2c Applications of the Reciprocity Theorem

The reciprocity theorem that we have just proved is useful in many situations. We shall apply it in subsequent sections to derive stationary formulas in variational problems based on reciprocity. We now use it to prove a few simple assertions. First, consider an electric current sheet impressed along the surface of a perfect conductor. If the surface of the conductor is a plane, the image theorem assures us that no field is produced by the current. But what if the surface is not a plane? Does it seem reasonable to extend the result and claim that all electric current sheets impressed along the surface of a conductor of any shape do not produce a field? The reciprocity theorem assures us that this is true by a simple argument. Let the impressed source be denoted as source a (Fig. 7.10). Let there be a source b that measures the field produced by source a; for instance, source b can be a dipole antenna. Source b produces no tangential electric field along the surface of the conductor because of the boundary conditions. The reaction of a and b is $\langle a,b\rangle=0$. By the reciprocity theorem,

$$\langle b,a\rangle=\langle a,b\rangle=0.$$

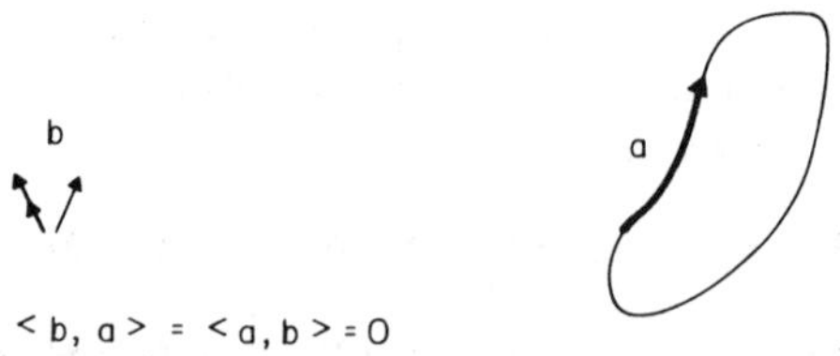

Figure 7.10 Impressed electric current sheets along perfect conductors produce no field.

But $\langle b,a\rangle$ is the field arising from the impressed source a as measured by source b, and b can be any arbitrary source. Therefore source b measures no field, and this proves that impressed electric current sheets along the surface of a perfect conductor produce no field.

As another example, we shall prove that the receiving pattern of an antenna is identical to its radiation pattern. Let antenna a be the antenna under consideration, and antenna b be a testing antenna, constructed so that it is omnidirectional. For both receiving and radiation patterns we are concerned with far fields. By the reciprocity theorem, we see that, in a direction where a as a transmitter radiates a weaker plane wave to b, a as a receiver also receives a weaker plane wave from b. Thus the radiation pattern and the receiving pattern of antenna a are identical. The gain $G_a(\theta,\phi)$ characterizes the radiation pattern of antenna a when it is acting as a transmitter. When a is acting as a receiver, we can define an effective area $A(\theta,\phi)$ to characterize its receiving pattern. From the proof we know that, for any antenna a, the effective area is related to its gain $G(\theta,\phi)$ by a constant. We define

$$A(\theta,\phi)=\pi\left(\frac{1}{k}\right)^2 G(\theta,\phi). \tag{7.38}$$

Note that the proportionality constant π/k^2 is universal for all antennas; it depends only on the frequency and has the dimension of area. In the case of a dipole, $G(\theta,\phi)=\frac{3}{2}\sin^2\theta$, and

$$A(\theta,\phi)=\frac{3}{2}\pi\left(\frac{1}{k}\right)^2 \sin^2\theta.$$

It is important to observe that, for an incident wave with electric field amplitude E and polarized in the plane determined by the dipole axis and the incident $\mathbf{k}$ vector, the incident power per unit area is $S_{\text{inc}}=E^2/2\eta$. The power received by the antenna with a properly matched load $Z_L = R_r - iX_L$ is

$$P = \frac{(\mathbf{E}\cdot \boldsymbol{l})^2}{2(Z_L + Z_L^*)^2} R_r = \frac{E^2 l^2 \sin^2\theta}{8R_r}$$

$$= A(\theta,\phi) S_{\text{inc}}$$

after making use of (6.21) for the dipole radiation resistance R_r. Thus the power received by the antenna is equal to the power per unit area of the incident wave times the effective area $A(\theta,\phi)$.

7.2d Stationary Formulas

Consider a cavity at resonance. We want to calculate the resonant frequency, but we do not know the precise field distribution inside the cavity. Nevertheless, we can find a formula that expresses resonant frequencies in terms of field distributions. We can then assume a field distribution and calculate the resonant frequency in terms of the assumed field. If the formula is stationary, we can come closer to the true resonant frequency than is possible by using a nonstationary formula. What do we mean by a stationary formula? Consider the formula $y = f(x)$. We want to calculate y at $x = x_0$, but we do not know the precise value of x_0. We assume that $x = x_0 + p$, where p is a parameter characterizing the deviation from x_0. The formula $y = f(x) = f(x_0 + p)$ can be expanded around x_0 in a Taylor series. We call the formula stationary at $p = 0$ if

$$\left.\frac{\partial f}{\partial p}\right]_{p=0} = 0. \tag{7.39}$$

When this condition is satisfied, $f(x) = f(x_0) + \frac{1}{2}p^2 f^{(2)}(x_0) + \cdots$ and the deviation of $f(x)$ from $f(x_0)$ is of order p^2. Clearly, the stationary formula has an extremum at $p = 0$. When p is complex, we have a saddle point at $p = 0$.

Stationary Formulas for Resonant Wave Numbers. We shall derive a stationary formula for the resonant frequency of a cavity with an assumed electric field. It is appropriate to mention that a stationary formula involving an assumed magnetic field or a mixture of electric and magnetic fields can be similarly derived. In terms of the assumed electric field (with subscript a), Maxwell's equations give

$$\mathbf{J}_a = i\omega\epsilon\mathbf{E}_a + \nabla\times\mathbf{H}_a$$

$$= \frac{1}{i\omega\mu}\left[-k^2\mathbf{E}_a + \nabla\times(\nabla\times\mathbf{E}_a)\right] \tag{7.40}$$

inside the cavity. On the cavity wall, we have

$$\mathbf{M}_s = \hat{n}\times\mathbf{E}_a, \tag{7.41}$$

where $\hat{n}$ is a unit vector normal to the cavity wall and pointing toward the outside of the cavity. This magnetic surface current sheet will be zero if the **E** field is the exact electric field, because of the boundary conditions. Since the electric field will be assumed, there is no guarantee that it satisfies the boundary conditions.

We form a reaction for the cavity:

$$\langle a,a\rangle = \iiint_V dV\mathbf{J}_a\cdot\mathbf{E}_a - \oiint_S dS\mathbf{M}_s\cdot\mathbf{H}_a$$

$$= \frac{1}{i\omega\mu}\left\{-k^2\iiint_V dV\mathbf{E}_a^2 + \iiint_V dV(\nabla\times\mathbf{E}_a)^2\right.$$

$$\left. + 2\oiint_S dS\hat{n}\cdot[(\nabla\times\mathbf{E}_a)\times\mathbf{E}_a]\right\}. \tag{7.42}$$

In the derivation, we made use of the identities

$$\hat{n}\times\mathbf{E}_a\cdot\nabla\times\mathbf{E}_a = -\hat{n}\cdot[(\nabla\times\mathbf{E}_a)\times\mathbf{E}_a]$$

and

$$\mathbf{E}_a\cdot\nabla\times(\nabla\times\mathbf{E}_a) = (\nabla\times\mathbf{E}_a)^2 + \nabla\cdot[(\nabla\times\mathbf{E}_a)\times\mathbf{E}_a].$$

We require that reaction 7.42 be equal to the correct reaction of the cavity $\langle c,c\rangle$, with c standing for "correct." What is the correct reaction of the cavity? Inside the cavity, where the field is nonzero, the source is zero. On the cavity walls, where the source is nonzero, the field is zero. Thus $\langle c,c\rangle = 0$. When $\langle a,a\rangle$ is set equal to $\langle c,c\rangle$ equal to zero, we obtain from (7.42) a formula for the resonant wave number squared k^2:

$$k^2 = \frac{1}{\iiint_V dV\,\mathbf{E}_a\cdot\mathbf{E}_a}\left[\iiint_V dV(\nabla\times\mathbf{E}_a)^2 + 2\oiint_S dS\hat{n}\cdot(\nabla\times\mathbf{E}_a)\times\mathbf{E}_a\right]. \tag{7.43}$$

Note that this formula is exact if $\mathbf{E}_a$ is the exact field. We would like to find out whether this formula is stationary. Let $\mathbf{E}_a = \mathbf{E} + p\mathbf{e}$ and denote (7.43) by $k^2 = N(p)/D(p)$. We wish to examine

$$\left.\frac{\partial k^2}{\partial p}\right]_{p=0} = \frac{D(0)N'(0) - N(0)D'(0)}{D^2(0)}$$

$$= \frac{N'(0) - k^2 D'(0)}{D(0)},$$

since $N(0) = k^2 D(0)$. Differentiating the numerator of (7.43) and setting $p = 0$, we obtain

$$N'(0) = 2\iiint_V dV(\nabla\times\mathbf{E})\cdot(\nabla\times\mathbf{e}) + 2\oiint_S dS\,\hat{n}\cdot(\nabla\times\mathbf{E})\times\mathbf{e}$$

$$= 2k^2\iiint_V dV\mathbf{e}\cdot\mathbf{E} = k^2 D'(0)$$

Here we used the identity

$$(\nabla\times\mathbf{E})\cdot(\nabla\times\mathbf{e}) = \nabla\cdot(\mathbf{e}\times\nabla\times\mathbf{E}) + \mathbf{e}\cdot\nabla\times\nabla\times\mathbf{E}$$

and the wave equation $\nabla\times\nabla\times\mathbf{E} = k^2\mathbf{E}$. From these results we have proved that $\partial k^2/\partial p|_{p=0} = 0$. Therefore, by requiring that the reaction caused by an assumed field be the same as the reaction attributable to the true field, we obtained a stationary formula for the resonant frequency of a cavity.

Consider a circular cavity as shown in Fig. 5.12. The exact field is known to be, for the fundamental mode, $\mathbf{E} = \hat{z}E_0 J_0(k\rho)$. The exact resonant wave number multiplied by the radius of the cavity a, which is the first root of $J_0(ka)$,

$$ka = 2.405 \qquad \text{or} \qquad k^2a^2 = 5.784, \tag{7.44}$$

is also known. We now use (7.43) to estimate ka. Let us assume a trial field

$$\mathbf{E}_a = \hat{z}\cos\frac{\pi\rho}{2a}. \tag{7.45}$$

This trial field satisfies the boundary condition at $\rho = a$, and it is not a solution to the wave equation. The curl of $\mathbf{E}_a$ can be calculated:

$$\nabla\times\mathbf{E}_a = \hat{\phi}\frac{\pi}{2a}\sin\frac{\pi\rho}{2a}.$$

Stationary formula 7.43 yields

$$k^2a^2 = \frac{\int_0^d dz \int_0^{2\pi} d\phi \int_0^a \rho\, d\rho\, (\pi/2a)^2 \sin^2(\pi\rho/2a)}{\int_0^d dz \int_0^{2\pi} d\phi \int_0^a \rho\, d\rho \cos^2(\pi\rho/2a)} = 5.830.$$

Next, we assume the trial field

$$\mathbf{E}_a = \hat{z}\left(1 + A\frac{\rho}{a}\right), \tag{7.46}$$

where A is a constant. We shall determine the constant A by using the Ritz procedure as illustrated below. Applying (7.43), we obtain

$$k^2a^2 = \frac{\int_0^d dz \int_0^{2\pi} d\phi \int_0^a \rho\, d\rho\, (A/a)^2 - \int_0^d dz \int_0^{2\pi} d\phi\, 2A(1+A)}{\int_0^d dz \int_0^{2\pi} d\phi \int_0^a \rho\, d\rho \left[1 + A(\rho/a)\right]^2}$$

$$= \frac{-2A - \frac{3}{2}A^2}{\frac{1}{2} + \frac{2}{3}A + \frac{1}{4}A^2}. \tag{7.47}$$

We require k^2a^2 to be stationary with respect to A. Setting $\partial(k^2a^2)/\partial A = 0$, we find $A = -1$ or -2. Inserting $A = -1$ into (7.47) yields $k^2a^2 = 6$. The value of $A = -1$ also enables the trial field to satisfy the boundary condition at $\rho = a$. The other value of $A = -2$ gives rise to a negative k^2a^2 and is discarded.

The Ritz procedure can be extended to n parameters A_l, $l = 1, 2, 3, \ldots, n$, which are then determined from the n equations $\partial k^2/\partial A_l = 0, l = 1, 2, \ldots, n$. As an example, we assume a trial field characterized by parameters A_1 and A_2 and write

$$\mathbf{E} = \hat{z}\left(1 + \frac{A_1\rho}{a} + \frac{A_2\rho^2}{a^2}\right). \tag{7.48}$$

Then

$$\nabla \times \mathbf{E} = -\hat{\phi}\frac{(A_1 + 2A_2\rho/a)}{a}.$$

Inserting in the stationary formula and performing the integration, we obtain

$$k^2a^2 = -\frac{10[18A_2^2+(28A_1+24)A_2+9A_1^2+12A_1]}{10A_2^2+(24A_1+30)A_2+15A_1^2+40A_1+30}.$$

The parameters A_1 and A_2 are determined by requiring that

$$\frac{\partial(k^2a^2)}{\partial A_1}=0, \qquad \frac{\partial(k^2a^2)}{\partial A_2}=0;$$

these yield two equations:

$$38A_2^3+(90A_1+84)A_2^2+(51A_1^2+45A_1-60)A_2-45A_1^2-135A_1-90=0,$$

$$(76A_1+150)A_2^2+(180A_1^2+600A_1+540)A_2+102A_1^3+461A_1^2+720A_1+360=0.$$

Solving these equations we obtain $A_1=-0.7817$ and $A_2=-0.1834$ and

$$k^2a^2=5.934.$$

This is closer to the exact solution than the value we obtained by using trial field 7.46. With these values of A_1 and A_2, we see that the trial field satisfies neither the wave equation nor the boundary condition. Solutions from other values of A_1 and A_2 are not calculated and may correspond to higher resonant wave numbers. As a final remark, we note that in using the Ritz procedure it is often advisable to choose the trial field components from an orthogonal complete set of functions.

Stationary Formulas for Antenna Impedance. We now derive a stationary formula for antenna self-impedance. Consider an antenna made of perfect conductor excited by a current source I. The self-reaction of this antenna is due entirely to the current at the terminal because on the conducting surface, where surface current is not zero, the field is zero. The self-reaction is equal to $-VI$ at the terminal. To calculate the self-impedance, we maintain the same terminal current I and assume a current distribution on the surface of the antenna. The self-reaction $\langle a,a\rangle$ is then calculated. We require that this reaction be approximately equal to the correct reaction $\langle c,c\rangle$:

$$\langle a,a\rangle \approx \langle c,c\rangle = -VI. \tag{7.49}$$

The input impedance is approximated by

$$Z_{\text{in}} = -\frac{\langle c,c\rangle}{I^2} \approx -\frac{\langle a,a\rangle}{I^2}. \tag{7.50}$$

Let $a=c+pe$. We see that

$$\frac{\partial Z_{\text{in}}}{\partial p} = -\frac{1}{I^2}(\langle e,c\rangle + \langle c,e\rangle)$$

$$= -\frac{2}{I^2}\langle e,c\rangle = 0.$$

The second equality follows from reciprocity. The last equality is due to the fact that at the terminal, where the field is not zero, the error source e is zero; everywhere else the correct field c is zero. Thus the formula for the input impedance is proved to be stationary.

We apply the stationary formula for input impedance to a radial parallel-plate waveguide excited by a probe of radius a (Fig. 7.11). The source terminal of the probe is outside the waveguide. For the TEM mode, we assume that the trial current is uniform along the probe, $\mathbf{J}_a = \hat{z}I/\pi a$. The electric field generated by this current is

$$\mathbf{E}_a = \hat{z}\left[-\frac{k^2 I}{4\omega\epsilon} H_0^{(1)}(k\rho)\right].$$

Let d denote the distance between the two plates. The input impedance is calculated to be

$$Z_{\text{in}} = \frac{\eta}{4} k\, d H_0^{(1)}\left(\frac{ka}{2}\right)$$

by using stationary formula 7.50.

Stationary Formulas for Scattering. We shall now derive a stationary formula for scattering problems. Consider a transmitter, a receiver, and a conducting scatterer (Fig. 7.12). The receiver receives a wave composed of two components: one directly from the transmitter in the absence of the scatterer, and one originating from currents on the scatterer induced by the transmitter alone. Note that the receiver also induces currents on the scatterer.

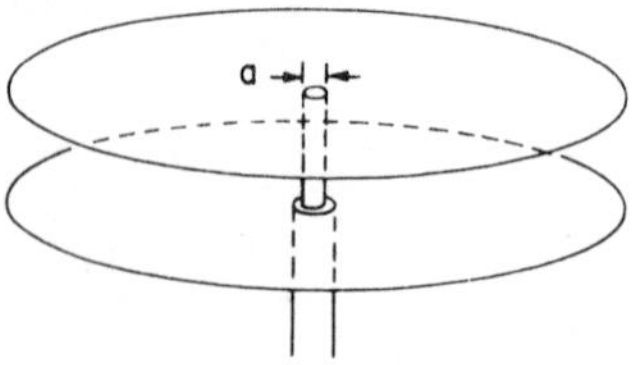

Figure 7.11 Radial parallel-plate waveguide excited by a probe.

We want to find a stationary formula for the scattered field V_s as received by the receiver, $V_s = -\langle i, c_t \rangle$, where i denotes the receiver source current, and c_t the field produced by the transmitter-induced currents on the scatterer. We anticipate that the stationary formula will be composed of surface integrals over the scatterer surface. By reciprocity, $V_s = -\langle i, c_t \rangle = -\langle c_t, i \rangle$, where, in $\langle c_t, i \rangle, c_t$ denotes the current induced by the transmitter on the scatterer, and i the field produced by the receiver in the absence of the scatterer. In the presence of the scatterer, the field on the scatterer produced by the receiver-induced current is equal and opposite to i because the scatterer is a conducting body. Let this field be denoted as c_r. We then have

$$V_s = -\langle c_t, i \rangle = \langle c_t, c_r \rangle. \tag{7.51}$$

Thus far, we have established that the scattered field as received by the receiver is equal to the reaction between the current induced on the scatterer by the transmitter and the field generated on the scatterer by the receiver-induced current. The total field at the receiver is a superposition of the scattered field and the field produced by the transmitter in the absence of the scatterer.

To calculate V_s, we approximate $\langle c_t, c_r \rangle$ and write

$$V_s = \langle c_t, c_r \rangle \approx \langle a_t, a_r \rangle, \tag{7.52}$$

where a stands for "assumed." The formula is a stationary formula provided that the following constraints are met:

$$\langle a_t, a_r \rangle = \langle c_t, a_r \rangle = \langle a_t, c_r \rangle, \tag{7.53}$$

where c_t is the correct current induced on the scatterer by the transmitter, and c_r is the correct field generated on the scatterer by the receiver-induced current.

Instead of applying the stationary formulas to specific examples, which should now seem straightforward, we address ourselves to the more urgent question of how we know that constraints 7.53 can lead us to a stationary

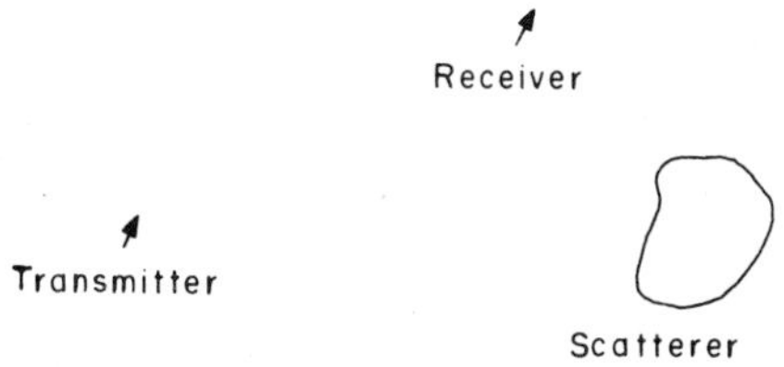

Figure 7.12 Scattering by a conducting body.

formula. We answer this question with a general proposition: a reaction $\langle a,b\rangle$ is stationary if it satisfies the constraints

$$\langle a,b\rangle=\langle c_a,b\rangle=\langle a,c_b\rangle. \tag{7.54}$$

To prove this theorem, let

$$a=c_a+p_ae_a, \qquad b=c_b+p_be_b,$$

where c stands for "correct," and e for "error." By definition, $\langle a,b\rangle$ is stationary if

$$\left.\frac{\partial\langle a,b\rangle}{\partial p_a}\right]_{p_a=p_b=0}=\left.\frac{\partial\langle a,b\rangle}{\partial p_b}\right]_{p_a=p_b=0}=0.$$

Note that the constraint $\langle a,b\rangle=\langle c_a,b\rangle$ implies that $\langle e_a,b\rangle=0$, which gives $\langle e_a,c_b\rangle=-p_b\langle e_a,e_b\rangle$. Then we can prove that

$$\frac{\partial\langle a,b\rangle}{\partial p_a}=\langle e_a,c_b\rangle=-p_b\langle e_a,e_b\rangle.$$

Setting $p_a=p_b=0$, we have $(\partial\langle a,b\rangle/\partial p_a)_{p_a=p_b=0}=0$. A similar proof applies with respect to the parameter p_b.

It is important to note that constraints 7.54 are sufficient conditions for a formula to be stationary; they are not necessary conditions. Recall that we did not use these constraints in establishing stationary formulas for the resonant wave number of a cavity or for the input impedance of an antenna. In fact, the constraints may be contradicted. For instance, consider the input impedance of an antenna. Here the constraint $\langle a,a\rangle=\langle a,c\rangle$ is violated because $\langle a,c\rangle$ is equal to zero but $\langle a,a\rangle$ is not.

7.2e Modified Reciprocity Theorem

In the preceding sections we proved that isotropic media are reciprocal, and we applied the reciprocity theorem to various situations. We shall now examine the validity of the reciprocity theorem for a bianisotropic medium.

Using the Lorentz reciprocity theorem and the same procedure as illustrated in (7.33)–(7.37), we find

$$\langle a,b\rangle-\langle b,a\rangle=\iiint_V i\omega(\mathbf{E}_b\cdot\mathbf{D}_a-\mathbf{E}_a\cdot\mathbf{D}_b+\mathbf{H}_a\cdot\mathbf{B}_b-\mathbf{H}_b\cdot\mathbf{B}_a)\,dV. \tag{7.55}$$

If the right-hand side is zero, the medium is reciprocal. The constitutive relations for bianisotropic media are

$$\mathbf{D}=\bar{\epsilon}\cdot\mathbf{E}+\bar{\xi}\cdot\mathbf{H},$$

$$\mathbf{B}=\bar{\mu}\cdot\mathbf{H}+\bar{\zeta}\cdot\mathbf{E}.$$

Inserting in the right-hand side of (7.55), we obtain

$$\langle a,b\rangle-\langle b,a\rangle=\iiint_V i\omega\Big[\mathbf{E}_b\cdot(\bar{\epsilon}-\bar{\epsilon}^t)\cdot\mathbf{E}_a+\mathbf{H}_a\cdot(\bar{\mu}-\bar{\mu}^t)\cdot\mathbf{H}_b$$

$$+\mathbf{E}_b\cdot(\bar{\xi}+\bar{\zeta}^t)\cdot\mathbf{H}_b+\mathbf{H}_a\cdot(\bar{\zeta}+\bar{\xi}^t)\cdot\mathbf{E}_b\Big]dV.$$

Thus $\langle a,b\rangle=\langle b,a\rangle$ if

$$\bar{\epsilon}=\bar{\epsilon}^t, \tag{7.56a}$$

$$\bar{\mu}=\bar{\mu}^t, \tag{7.56b}$$

$$\bar{\xi}=-\bar{\zeta}^t. \tag{7.56c}$$

These are the conditions for a medium to be reciprocal. Consequently, isotropic media are reciprocal, and anisotropic media with symmetrical permittivity and permeability tensors are reciprocal. Bianisotropic media that satisfy (7.56a), (7.56b), and the lossless condition $\bar{\xi}=\bar{\zeta}^+$ are reciprocal if $\bar{\xi}$ and $\bar{\zeta}$ are purely imaginary matrices.

An anisotropic plasma is an example of a nonreciprocal medium. It possesses a permittivity tensor with $\bar{\epsilon}=\bar{\epsilon}^+$, which contradicts (7.56a). Consider a slab region filled with the plasma with magnetic field $\mathbf{B}_0$ perpendicular to the slab (Fig. 7.13). Assume that, when a linearly polarized plane wave is transmitted through the slab, the Faraday rotation causes the field vector to rotate 45° in the increasing ϕ direction. Let a current sheet with $\mathbf{J}_a$ at the left-hand side of the slab produce a plane wave polarized in the direction $\phi=0$, and a current sheet with $\mathbf{J}_b$ at the right-hand side of the slab produce a plane wave polarized in the direction $\phi=45°$. Let $\mathbf{J}_a$ be source a and $\mathbf{J}_b$ be source b. The reaction of $\langle a,b\rangle$ is seen to be zero because the plane wave as produced by $\mathbf{J}_b$ is polarized with $\mathbf{E}_b$ perpendicular to $\mathbf{J}_a$ after transmitting through the slab, while the reaction of $\langle b,a\rangle$ is nonzero because $\mathbf{E}_a$ and $\mathbf{J}_b$ are in the same direction. Thus $\langle a,b\rangle\neq\langle b,a\rangle$, and the Faraday rotation produces nonreciprocal results.

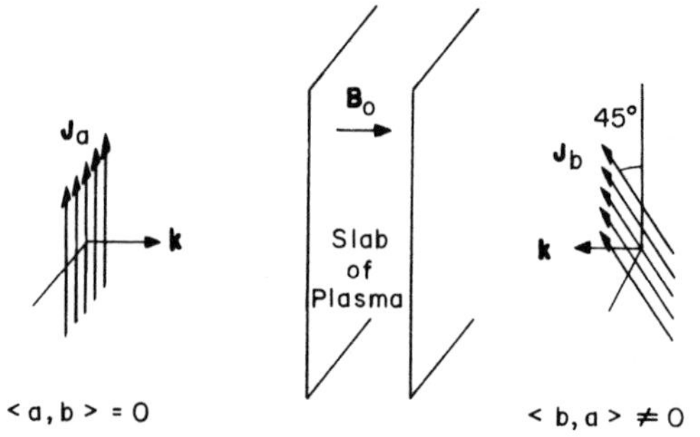

Figure 7.13 An anisotropic plasma is nonreciprocal.

Natural optical activity also rotates field vectors, but the effect is reciprocal. For instance, a quartz crystal exhibits optical rotatory power and can be described as a bianisotropic medium whose constitutive relations satisfy (7.56). Let the slab region in Fig. 7.13 be filled with an optically active medium such as quartz. The electric field vector will be rotated 45° in increasing ϕ when transmitted from left to right and rotated 45° in decreasing ϕ when transmitted from right to left. Thus we have the reactions $\langle a,b\rangle=\langle b,a\rangle$.

The reciprocity theorem can be extended as follows. With respect to source a, we write

$$-\nabla\times\mathbf{E}_a=-i\omega\left(\bar{\boldsymbol{\mu}}\cdot\mathbf{H}_a+\bar{\boldsymbol{\zeta}}\cdot\mathbf{E}_a\right)+\mathbf{M}_a, \tag{7.57a}$$

$$\nabla\times\mathbf{H}_a=-i\omega\left(\bar{\boldsymbol{\epsilon}}\cdot\mathbf{E}_a+\bar{\boldsymbol{\xi}}\cdot\mathbf{H}_a\right)+\mathbf{J}_a. \tag{7.57b}$$

The medium is characterized by $\bar{\boldsymbol{\mu}}$, $\bar{\boldsymbol{\epsilon}}$, $\bar{\boldsymbol{\xi}}$, and $\bar{\boldsymbol{\zeta}}$. With respect to source b, we use constitutive relations characterized by $\bar{\boldsymbol{\mu}}^c$, $\bar{\boldsymbol{\epsilon}}^c$, $\bar{\boldsymbol{\xi}}^c$, and $\bar{\boldsymbol{\zeta}}^c$ such that

$$\bar{\boldsymbol{\mu}}^c=\bar{\boldsymbol{\mu}}^t, \tag{7.58a}$$

$$\bar{\boldsymbol{\epsilon}}^c=\bar{\boldsymbol{\epsilon}}^t, \tag{7.58b}$$

$$\bar{\boldsymbol{\xi}}^c=-\bar{\boldsymbol{\zeta}}^t, \tag{7.58c}$$

$$\bar{\boldsymbol{\zeta}}^c=-\bar{\boldsymbol{\xi}}^t, \tag{7.58d}$$

and call this medium the complementary medium. Maxwell's equations for source b in the complementary medium become

$$-\nabla\times\mathbf{E}_b^c=-i\omega\left(\bar{\boldsymbol{\mu}}^c\cdot\mathbf{H}_b^c+\bar{\zeta}^c\cdot\mathbf{E}_b^c\right)+\mathbf{M}_b, \tag{7.59a}$$

$$\nabla\times\mathbf{H}_b^c=-i\omega\left(\bar{\boldsymbol{\epsilon}}^c\cdot\mathbf{E}_b^c+\bar{\xi}^c\cdot\mathbf{H}_b^c\right)+\mathbf{J}_b, \tag{7.59b}$$

where $\mathbf{E}_b^c$ and $\mathbf{H}_b^c$ denote the fields produced by $\mathbf{J}_b$ and $\mathbf{M}_b$ in the complementary medium. If we define a new reaction

$$\langle b,a\rangle^c=\int\int\int_V dV(\mathbf{J}_b\cdot\mathbf{E}_a^c-\mathbf{M}_b\cdot\mathbf{H}_a^c),$$

we find from (7.57) and (7.59) that

$$\langle a,b\rangle=\langle b,a\rangle^c. \tag{7.60}$$

This result may be called the *modified reciprocity theorem*, which states that the reaction $\langle a,b\rangle$ of source a caused by source b in a bianisotropic medium is equal to the reaction $\langle b,a\rangle^c$ of source b caused by source a in the complementary medium. The medium is reciprocal if the complementary medium is identical to its original medium.

7.3 QUANTIZATION OF ELECTROMAGNETIC WAVES

In microscopic physics, quantum electrodynamics has become a well-established discipline. In studying the interactions of electromagnetic waves with material media, a full quantum theory is too complicated to carry out; therefore semiclassical approaches are usually taken. We can either treat eletromagnetic waves classically or treat material media classically. In this section, we quantize electromagnetic waves with material media characterized by constitutive relations. The Heisenberg representation will be used in the process of quantization. First we give a brief review of this representation.[29,71]

The state of a physical system is represented by a state vector, which can be viewed either as a column matrix called a *ket* and denoted by $|\psi\rangle$ or as a row matrix called a *bra* and denoted by $\langle\psi|$. The bra $\langle\psi|$ is the complex conjugate and transpose of ket $|\psi\rangle$. Physical observables are represented by hermitian operators. An operator can be viewed as a square matrix. Any measurement of a physical observable A yields a statistical expectation value for the observable. The average value of a series of measurements made on an ensemble of systems characterized by the same state vector $|\psi\rangle$ is given by

$$\langle A\rangle=\langle\psi|A|\psi\rangle. \tag{7.61}$$

This average value, or expectation value, is a real scalar number.

Each hermitian operator possesses a set of eigenvectors with associated eigenvalues. The eigenvectors are eigenstates of the corresponding observable. The result of a single measurement for an observable A on a system described by the state vector $|\psi\rangle$ yields an eigenvalue λ_n of the operator A and sends the system into the corresponding eigenstate $|\lambda_n\rangle$. The probability of obtaining this eigenvalue λ_n and resulting in this eigenstate $|\lambda_n\rangle$ is given by $|\langle\lambda_n|\psi\rangle|^2$. As time progresses, the eigenstates of the operator evolve as the operator evolves with time, and the state of the system will no longer be an eigenstate of the operator. The evolution of the operator with time is determined by the equations of the motion. In classical electromagnetic theory the equations of motion for the electromagnetic field vectors are Maxwell's equations. In quantum theory the field vectors are treated as operators and are governed by Maxwell's equations.

7.3a Uncertainty Principle

Since operators do not necessarily commute, commutation relations for noncommuting operators must be postulated. The physical interpretation of the commutation relations leads to the uncertainty principle, which states, in essence, that any measurement made on a physical system perturbs the system, no matter how slight the perturbation may be. Thus reality is forever beyond reach; only statistical results show up. This perturbation, induced by a measurement on one observable, may or may not affect the true values of measurements on other observables. When two observables interfere with each other, they are noncommuting operators. When two observables do not interfere with each other, they are simultaneously measurable and the commutation relation is zero. Commutation relations for electromagnetic fields **D** and **B** are postulated as

$$\begin{aligned}\left[D_i(\mathbf{r},t),B_j(\mathbf{r}',t)\right]&=D_i(\mathbf{r},t)B_j(\mathbf{r}',t)-B_j(\mathbf{r}',t)D_i(\mathbf{r},t)\\&=-i\hbar\varepsilon_{ijk}\frac{\partial}{\partial x_k}\delta(\mathbf{r}-\mathbf{r}')\end{aligned} \tag{7.62}$$

where $\hbar$ is Planck's constant divided by 2π and equals 1.05×10^{-34} J-s. The factor $\hbar$ signifies that quantum effects are important whenever $\hbar$ is not numerically negligible. The classical limit is obtained when we let $\hbar\rightarrow0$ and treat operators as classical variables.

In vacuum the commutation relations for other field components follow directly from the constitutive relations for vacuum:

$$[E_i(\mathbf{r},t), B_j(\mathbf{r}',t)] = -i\hbar\varepsilon_{ijk}\frac{1}{\epsilon_0}\frac{\partial}{\partial x_k}\delta(\mathbf{r}-\mathbf{r}'), \tag{7.63a}$$

$$[D_i(\mathbf{r},t), H_j(\mathbf{r}',t)] = -i\hbar\varepsilon_{ijk}\frac{1}{\mu_0}\frac{\partial}{\partial x_k}\delta(\mathbf{r}-\mathbf{r}'), \tag{7.63b}$$

$$[E_i(\mathbf{r},t), H_j(\mathbf{r}',t)] = -i\hbar c^2\varepsilon_{ijk}\frac{\partial}{\partial x_k}\delta(\mathbf{r}-\mathbf{r}'). \tag{7.63c}$$

The commutation relations state that the perpendicular components of the electric and magnetic fields interfere with each other, whereas parallel components are simultaneously measurable. For instance, to measure an electric field, we may use a test charge and observe its motion along the electric field lines. But when a charge moves, it constitutes a current. The current produces a magnetic field perpendicular to the electric field. Thus the test charge used in the measurement of electric fields interferes with a simultaneous measurement on magnetic fields.

Recall that the magnetic field **B** is expressible as the curl of a vector potential **A**:

$$\mathbf{B} = \nabla \times \mathbf{A}. \tag{7.64}$$

In terms of the vector potential **A**, the commutation relation can be written as

$$[A_i(\mathbf{r},t), D_j(\mathbf{r}',t)] = -i\hbar\delta_{ij}\delta(\mathbf{r}-\mathbf{r}'). \tag{7.65}$$

Note that, although all commutation relations are written for equal times, we can also deduce and postulate commutation relations for unequal times (Heitler[47]).

In our description of a quantized system, the operators evolve with time, as do their associated eigenvectors. The eigenvectors can be viewed as forming base vectors describing a system state vector that is not varying with time. This is known as the Heisenberg picture. The Heisenberg picture is different from the Schrödinger picture, in which the system state vectors are functions of time but the operators and their eigenstates are stationary. In the Schrödinger picture, the equation of motion for a state vector $|\psi(t)\rangle$ is given by the Schrödinger equation. In the Heisenberg picture, the time evolution of an operator A representing a physical observable is governed by the Heisenberg equation of motion. Under the assumption that A is not explicitly dependent on time, the Heisenberg equation of motion for A is

$$i\hbar \frac{dA}{dt} = [A, H], \tag{7.66}$$

where H is the Hamiltonian of the system, which corresponds to the total energy of the system. The Hamiltonian of an electromagnetic field in a source-free region is

$$H = \int d^3x \frac{1}{2}(\mathbf{E} \cdot \mathbf{D} + \mathbf{H} \cdot \mathbf{B}). \tag{7.67}$$

To determine the equation of motion for the **D** field, we use (7.66) and find

$$\begin{aligned} i\hbar \frac{dD_i}{dt} &= [D_i, H] \\ &= \frac{1}{2}\int d^3x' \{[D_i(\mathbf{r}), E_j(\mathbf{r}')]D_j(\mathbf{r}') + E_j(\mathbf{r}')[D_i(\mathbf{r}), D_j(\mathbf{r}')] \\ &\quad + [D_i(\mathbf{r}), H_j(\mathbf{r}')]B_j(\mathbf{r}') + H_j(\mathbf{r}')[D_i(\mathbf{r}), B_j(\mathbf{r}')]\}. \end{aligned}$$

Remember that all field observables are now operators. The integral can be evaluated with the use of commutation relations 7.62 and 7.63b. The first two commutators are zero because electric field operators commute. We obtain

$$\frac{dD_i}{dt} = \varepsilon_{ijk} \frac{\partial H_j}{\partial x_k}, \tag{7.68}$$

which is Ampère's law in the absence of the source term J_i. If sources are present, the Hamiltonian in (7.67) must include an interaction term $\mathbf{J} \cdot \mathbf{A}$. In view of the commutation relation 7.65, it is clear that this extra term gives rise to the source term J_i in (7.68). Following a similar procedure, we can derive Faraday's law from the Heisenberg equation of motion for **B**.

7.3b Annihilation and Creation Operators

The eigenstates of the Hamiltonian are energy eigenstates because the Hamiltonian H is an energy operator. To facilitate discussion of the energy states of a quantized wave field, it is useful to transform the operators to **k** space:

$$\mathbf{D}(\mathbf{r})=(2\pi)^{-3/2}\int d^3k\,\mathbf{D}(\mathbf{k})e^{i\mathbf{k}\cdot\mathbf{r}}, \tag{7.69a}$$

$$\mathbf{A}(\mathbf{r})=(2\pi)^{-3/2}\int d^3k\,\mathbf{A}(\mathbf{k})e^{i\mathbf{k}\cdot\mathbf{r}}. \tag{7.69b}$$

The reality conditions for $\mathbf{D}(\mathbf{r})$ and $\mathbf{A}(\mathbf{r})$ are

$$\mathbf{D}(-\mathbf{k})=\mathbf{D}^+(\mathbf{k}), \tag{7.70a}$$

$$\mathbf{A}(-\mathbf{k})=\mathbf{A}^+(\mathbf{k}). \tag{7.70b}$$

This condition can be satisfied by the following representations:

$$D_i(\mathbf{k})=i\sqrt{\frac{\hbar k}{2\eta}}\,(a_i^+(\mathbf{k})-a_i(-\mathbf{k})), \tag{7.71a}$$

$$A_i(\mathbf{k})=\sqrt{\frac{\hbar\eta}{2k}}\,(a_i^+(\mathbf{k})+a_i(-\mathbf{k})), \tag{7.71b}$$

where $\eta=\sqrt{\mu_0/\epsilon_0}$. As we shall demonstrate in the subsequent developments, the operator $a_i(\mathbf{k})$ is an annihilation operator and the operator $a_i^+(\mathbf{k})$ is a creation operator. Upon operating on an energy eigenstate, $a_i(\mathbf{k})$ annihilates a photon corresponding to wave vector $\mathbf{k}$ and polarization D_i in the state, whereas $a_i^+(\mathbf{k})$ creates such a photon.

The Hamiltonian H can be expressed as

$$\begin{aligned} H&=\frac{1}{2(2\pi)^3}\int d^3x\,d^3k\,d^3k'\left\{\frac{1}{\epsilon_0}\mathbf{D}(\mathbf{k})\cdot\mathbf{D}(\mathbf{k}')\,e^{i(\mathbf{k}+\mathbf{k}')\cdot\mathbf{r}}-\frac{1}{\mu_0}[\mathbf{k}\times\mathbf{A}(\mathbf{k})]\right.\\ &\qquad\left.\cdot[\mathbf{k}'\times\mathbf{A}(\mathbf{k}')]\,e^{i(\mathbf{k}+\mathbf{k}')\cdot\mathbf{r}}\right\}\\ &=\frac{1}{2}\int d^3k\left\{\frac{1}{\epsilon_0}\mathbf{D}(\mathbf{k})\cdot\mathbf{D}^+(\mathbf{k})+\frac{1}{\mu_0}[\mathbf{k}\times\mathbf{A}(\mathbf{k})]\cdot[\mathbf{k}\times\mathbf{A}^+(\mathbf{k})]\right\}. \end{aligned}$$

We focus our attention on a particular photon with a prespecified $\mathbf{k}$ vector. The operators $\mathbf{a}$ and $\mathbf{a}^+$ are in the same direction as $\mathbf{D}$ and are perpendicular to $\mathbf{k}$. In terms of $\mathbf{a}(\mathbf{k})$ and $\mathbf{a}^+(\mathbf{k})$, the Hamiltonian becomes

$$H=\frac{1}{2}\int d^3k\,\hbar kc\,(a_i(\mathbf{k})a_i^+(\mathbf{k})+a_i^+(\mathbf{k})a_i(\mathbf{k})). \tag{7.72}$$

And commutation relation 7.65 becomes

$$i\frac{\hbar}{2}\left[\int d^3k\left(a_i^+(\mathbf{k})\,e^{i\mathbf{k}\cdot\mathbf{r}}+a_i(\mathbf{k})e^{-i\mathbf{k}\cdot\mathbf{r}}\right),\int d^3k'\left(a_j^+(\mathbf{k}')\,e^{i\mathbf{k}'\cdot\mathbf{r}'}-a_j(\mathbf{k}')\,e^{-i\mathbf{k}'\cdot\mathbf{r}'}\right)\right]$$

$$=-i\hbar\delta_{ij}\,\delta(\mathbf{r}-\mathbf{r}').$$

We deduce that

$$\left[a_i(\mathbf{k}),a_j^+(\mathbf{k}')\right]=\delta_{ij}\,\delta(\mathbf{k}-\mathbf{k}'). \tag{7.73}$$

For a photon with a particular **k** vector, we simply write

$$[a,a^+]=1, \tag{7.74}$$

and the Hamiltonian becomes

$$H=\frac{\hbar kc}{2}(aa^+ + a^+a)$$

$$=\hbar\omega(a^+a+\tfrac{1}{2}), \tag{7.75}$$

where we made the use of commutator 7.74 and the vacuum dispersion relation $\omega=kc$.

To obtain eigenvalues and eigenvectors for the energy operator H, we write

$$H|E\rangle=E|E\rangle, \tag{7.76}$$

where $|E\rangle$ denotes the eigenstate, and E the corresponding eigenvalue. We first show that the eigenvalue E is always nonnegative. Scalar-multiplying (7.76) by the eigenbra $\langle E|$ and using (7.75), we have

$$\hbar\omega\langle E|a^+a+\tfrac{1}{2}|E\rangle=E\langle E|E\rangle.$$

The scalar $\langle E|E\rangle$ is always nonnegative because it is a product of a column matrix $|E\rangle$ and its complex-conjugate and transpose $\langle E|$. The $\langle E|a^+a|E\rangle$ term is also nonnegative because $\langle E|a^+$ is the complex-conjugate and transpose of $a|E\rangle$. Consequently, the eigenvalue E must be nonnegative.

We next show that, if $|E\rangle$ is an eigenstate of H, so are $a|E\rangle$ and $a^+|E\rangle$. Consider $H(a^+|E\rangle)$. Using commutation relation 7.74, we find

$$[H,a^+]=\hbar\omega[a^+a,a^+]=\hbar\omega a^+.$$

Thus

$$Ha^+|E\rangle = a^+H|E\rangle + \hbar\omega a^+|E\rangle$$
$$= (E+\hbar\omega)a^+|E\rangle. \tag{7.77}$$

It is seen that $a^+|E\rangle$ is an eigenstate of H with eigenvalue $(E+\hbar\omega)$. Whenever a^+ is applied to an eigenstate of H with energy E, the state changes into another eigenstate with energy $E+\hbar\omega$. Since the net effect of this operation is to create one more photon with energy $\hbar\omega$, the operator a^+ is called a *creation operator*.

Following similar reasoning, we can show that

$$Ha|E\rangle = (E-\hbar\omega)a|E\rangle. \tag{7.78}$$

When a is applied to an eigenstate $|E\rangle$, the result is another eigenstate, $|E-\hbar\omega\rangle$, with one photon annihilated. The operator a is thus called an *annihilation operator*.

We have proved that all energy states of H possess nonnegative energy eigenvalues. Suppose that we apply the annihilation operator a to the state $|E\rangle$ n times and reach the ground state $|E_0\rangle$. Further operation of a on $|E_0\rangle$ will then yield a zero:

$$a|E_0\rangle = 0.$$

We find the energy of the ground state to be

$$E_0 = \frac{\langle E_0|H|E_0\rangle}{\langle E_0|E_0\rangle} = \tfrac{1}{2}\hbar\omega. \tag{7.79}$$

The energy spectrum is illustrated in Fig. 7.14. The separation between

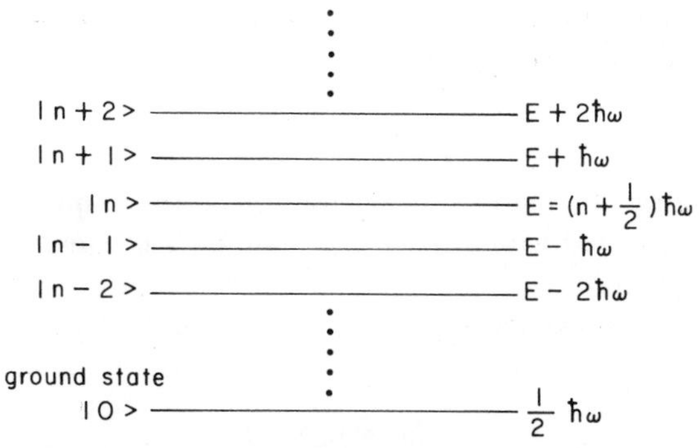

Figure 7.14 Eigenstates and the corresponding eigenvalues for the Hamiltonian operator *H*.

energy levels is $\hbar\omega$. When operated on by a, the transition is downward; when operated on by a^+, the transition is upward. The whole energy spectrum can be built up by successively applying a^+ to the ground state, which we denote by ket $|0\rangle$, 0 indicating no photon in the state. The state of n photons, $|n\rangle$, is then created by operating a^+ on $|0\rangle$ n times. The energy eigenvalue associated with $|n\rangle$ is clearly $(n+\frac{1}{2})\hbar\omega$. The energy eigenstate $|n\rangle$ can be represented by a column matrix with all elements equal to zero except the $(n+1)$th one, which is equal to unity. For instance,

$$|0\rangle = \begin{bmatrix} 1 \\ 0 \\ 0 \\ 0 \\ \vdots \end{bmatrix}, \qquad |1\rangle = \begin{bmatrix} 0 \\ 1 \\ 0 \\ 0 \\ \vdots \end{bmatrix}.$$

The Hamiltonian has been diagonalized. In matrix representation, we can write

$$H = \hbar\omega \begin{bmatrix} \frac{1}{2} & 0 & 0 & \cdots \\ 0 & \frac{3}{2} & 0 & \cdots \\ 0 & 0 & \frac{5}{2} & \cdots \\ \vdots & \vdots & \vdots & \vdots\vdots\vdots \end{bmatrix}, \tag{7.80}$$

which is a diagonal matrix. Since $H|n\rangle = \hbar\omega(a^+a+\frac{1}{2})|n\rangle = \hbar\omega(n+\frac{1}{2})|n\rangle$, we call $N = a^+a$ the number operator,

$$N|n\rangle = n|n\rangle. \tag{7.81}$$

The eigenstates of H are also eigenstates of the number operator. The eigenvalue associated with a particular state of N is equal to the number of photons in that state.

We now find an explicit representation of a and a^+ in terms of matrices. We write

$$a|n\rangle = C_n|n-1\rangle. \tag{7.82}$$

The coefficient C_n is determined from normalization. We require that the scalar product of the bra and the ket of an eigenstate be unity. We have

$$n = \langle n|a^{+}a|n\rangle = |C_n|^2.$$

Thus

$$a|n\rangle = \sqrt{n}\,|n-1\rangle. \tag{7.83}$$

The matrix elements of a can be obtained from (7.83) by noting that

$$\langle n-1|a|n\rangle = \sqrt{n}\,, \tag{7.84}$$

which is the element in the $(n-1)$th row and the nth column. All other elements are zero:

$$a = \begin{bmatrix} 0 & 1 & 0 & 0 & \cdots \\ 0 & 0 & \sqrt{2} & 0 & \cdots \\ 0 & 0 & 0 & \sqrt{3} & \cdots \\ \vdots & \vdots & \vdots & \vdots & \vdots\,\vdots\,\vdots \end{bmatrix}. \tag{7.85}$$

Following similar reasoning, we let $a^{+}|n\rangle = C_n'|n+1\rangle$ and find

$$|C_n'|^2 = \langle n|aa^{+}|n\rangle = \langle n|a^{+}a+1|n\rangle = n+1.$$

Thus

$$a^{+}|n\rangle = \sqrt{n+1}\,|n+1\rangle. \tag{7.86}$$

Forming the product

$$\langle n+1|a^{+}|n\rangle = \sqrt{n+1}\,, \tag{7.87}$$

we see that the matrix representation of a^{+} takes the following form:

$$a^{+} = \begin{bmatrix} 0 & 0 & 0 & 0 & \cdots \\ 1 & 0 & 0 & 0 & \cdots \\ 0 & \sqrt{2} & 0 & 0 & \cdots \\ 0 & 0 & \sqrt{3} & 0 & \cdots \\ \vdots & \vdots & \vdots & \vdots & \vdots\,\vdots\,\vdots \end{bmatrix}. \tag{7.88}$$

Obviously, the matrix representation for a^+ is the transpose of that for a. Its time evolution follows the Heisenberg equation of motion as **D** and **A** do, as is seen from (7.71). From matrix multiplication, we see that operating a on the state $|n\rangle$ will move the unit element in the nth position to the $(n-1)$th position and form the state $|n-1\rangle$ multiplied by $\sqrt{n}$. Similarly, operating a^+ on $|n-1\rangle$ results in the state $|n\rangle$ multiplied by $\sqrt{n}$. Operating a^+a on $|n\rangle$ will result in the same state and give the photon number n.

We have discussed energy eigenstates and their associated eigenvalues for the Hamiltonian operator H, and we have seen what results the annihilation operator a and the creation operator a^+ have when operating on the eigenstates. It is natural to ask, "What are the eigenstates and eigenvalues for the annihilation and the creation operators themselves?" First, we shall prove that a^+ has no nonzero eigenstates. Denote the eigenstates of a^+ as $|e\rangle$ and let the eigenvalues be λ. In view of (7.88), we have

$$\lambda \begin{bmatrix} e_1 \\ e_2 \\ e_3 \\ e_4 \\ \vdots \end{bmatrix} = \begin{bmatrix} 0 & 0 & 0 & 0 & \cdots \\ 1 & 0 & 0 & 0 & \cdots \\ 0 & \sqrt{2} & 0 & 0 & \cdots \\ 0 & 0 & \sqrt{3} & 0 & \cdots \\ \vdots & \vdots & \vdots & \vdots & \vdots\vdots\vdots \end{bmatrix} \begin{bmatrix} e_1 \\ e_2 \\ e_3 \\ e_4 \\ \vdots \end{bmatrix}$$

We see that, if $\lambda=0$, then $e_1=e_2=\cdots=0$ and $|e\rangle=|0\rangle$. If $\lambda\neq0$, then $e_1=0, e_2=(1/\lambda)e_1=0,\ldots,e_n=(1/\lambda)(n-1)^{1/2}e_{n-1}=0,\ldots,$ and the eigenstate is identically zero.

The same procedure can be used to find eigenstates for a. We write, in view of (7.85),

$$\lambda \begin{bmatrix} e_1 \\ e_2 \\ e_3 \\ e_4 \\ \vdots \end{bmatrix} = \begin{bmatrix} 0 & 1 & 0 & 0 & \cdots \\ 0 & 0 & \sqrt{2} & 0 & \cdots \\ 0 & 0 & 0 & \sqrt{3} & \cdots \\ \vdots & \vdots & \vdots & \vdots & \vdots\vdots\vdots \end{bmatrix} \begin{bmatrix} e_1 \\ e_2 \\ e_3 \\ e_4 \\ \vdots \end{bmatrix}. \tag{7.89}$$

We see that $e_2=\lambda e_1$, $e_3=(\lambda/\sqrt{2})e_2,\ldots,e_n=(\lambda/\sqrt{n-1})e_{n-1},\ldots$. In terms of the energy eigenstates, we obtain

$$|e\rangle=e_1\left(|0\rangle+\lambda|1\rangle+\frac{\lambda^2}{\sqrt{2!}}|2\rangle+\cdots+\frac{\lambda^n}{\sqrt{n!}}|n\rangle+\cdots\right). \tag{7.90}$$

Imposing the normalization condition $\langle e|e\rangle=1$ yields

$$|e_1|^2\left(1+\lambda^2+\frac{\lambda^4}{2!}+\cdots+\frac{\lambda^{2n}}{n!}+\cdots\right)=1. \tag{7.91}$$

Thus $e_1=e^{-\lambda^2/2}$ and (7.90) becomes

$$|e\rangle=e^{-\lambda^2/2}\sum_{n=0}^{\infty}\frac{\lambda^n}{\sqrt{n!}}|n\rangle. \tag{7.92}$$

We note that the expectation value for the photon number operator a^+a is determined from

$$\begin{aligned}\bar{n}=\langle e|a^+a|e\rangle&=e^{-\lambda^2}\sum_{n=0}^{\infty}n\frac{\lambda^{2n}}{n!}\\&=\lambda^2.\end{aligned} \tag{7.93}$$

Consequently, the eigenvalue λ is equal to the square root of the photon number expectation value. Equation 7.92 becomes

$$|e\rangle=e^{-\bar{n}/2}\sum_{n=0}^{\infty}\frac{\bar{n}^{n/2}}{\sqrt{n!}}|n\rangle. \tag{7.94}$$

This represents the eigenstate of the annihilation operator a. It is also called the *coherent state*. The probability of finding n photons in the coherent state is

$$|\langle n|e\rangle|^2=\frac{n^{\bar{n}}e^{-\bar{n}}}{n!}.$$

This is the Poisson distribution obtained by assuming complete randomness of photon detection. We appreciate that in the energy state representation the precise photon number is given, whereas in the coherent state representation the photon number obeys Poisson probability distribution.

7.3c Wave Quantization in Bianisotropic Media

With the use of the annihilation and creation operators a and a^+, we have diagonalized the Hamiltonian for an electromagnetic field in vacuum. The quantized fields have been discussed in terms of the energy states. We now generalize the procedure to carry out wave quantization in bianisotropic media. The commutation relations for **D** and **B** are the same as postulated in (7.62), and those for other field operators are derived from (7.62) by using the constitutive relations. Instead of (7.71), the annihilation and creation operators are introduced by

$$D_j(\mathbf{k}) = \frac{i}{\alpha_j}\sqrt{\frac{\hbar}{2}}\,\left[a_j^+(\mathbf{k}) - a_j(-\mathbf{k})\right], \tag{7.95a}$$

$$A_j(\mathbf{k}) = \alpha_j\sqrt{\frac{\hbar}{2}}\,\left[a_j^+(\mathbf{k}) + a_j(-\mathbf{k})\right], \tag{7.95b}$$

where α_j is a constant to be determined. In view of the reality condition 7.70, we must have $\alpha_j(-\mathbf{k}) = \alpha_j^*(\mathbf{k})$. Substituting (7.95) in (7.65), we find that the commutation relations for a and a^+ are identical to (7.73) and (7.74). Using a and a^+, we shall diagonalize the Hamiltonian by properly choosing α_j.

As an example of bianisotropic media we consider a uniaxial medium moving along the direction of its optic axis. We let the optic axis be along the $\hat{z}$ direction. The constitutive relation is given by (2.38). Before transforming the Hamiltonian to **k** space, we choose for each **k** vector a *kDB* system as defined in Chapter 3. In the *kDB* system, $D_3 = B_3 = 0$ and the constitutive relation is given by (3.73). The Hamiltonian becomes

$$\begin{aligned} H &= \frac{1}{2}\int d^3k\left\{\mathbf{D}^+(\mathbf{k})\cdot\mathbf{E}(\mathbf{k}) + \mathbf{B}^+(\mathbf{k})\cdot\mathbf{H}(\mathbf{k})\right\} \\ &= \frac{1}{2}\int d^3k\left\{\kappa D_1^+(\mathbf{k})D_1(\mathbf{k}) + (\kappa\cos^2\theta + \kappa_z\sin^2\theta)D_2^+(\mathbf{k})D_2(\mathbf{k})\right. \\ &\quad + k^2\nu A_2^+(\mathbf{k})A_2(\mathbf{k}) + k^2(\nu\cos^2\theta + \nu_z\sin^2\theta)A_1^+(\mathbf{k})A_1(\mathbf{k}) \\ &\quad - ik\chi\cos\theta\big(D_1^+(\mathbf{k})A_1(\mathbf{k}) - A_1^+(\mathbf{k})D_1(\mathbf{k}) \\ &\quad \left. + D_2^+(\mathbf{k})A_2(\mathbf{k}) - A_2^+(\mathbf{k})D_2(\mathbf{k})\big)\right\}, \end{aligned} \tag{7.96}$$

where θ is the angle between the **k** vector and the z axis. To express the

Hamiltonian in terms of the annihilation and the creation operators, we note that

$$\int d^3k\,k\{D_j^+(\mathbf{k})A_j(\mathbf{k})-A_j^+(\mathbf{k})D_j(\mathbf{k})\}$$
$$=i\hbar\int d^3k\,k\{a_j(\mathbf{k})a_j^+(\mathbf{k})+a_j^+(\mathbf{k})a_j(\mathbf{k})\}. \qquad (7.97)$$

Introducing (7.95) and (7.97) in (7.96), we obtain

$$H=\frac{\hbar}{4}\int d^3k\Bigg\{\left(\frac{\kappa}{\alpha_1^2}+k^2\alpha_1^2(\nu\cos^2\theta+\nu_z\sin^2\theta)+2k\chi\cos\theta\right)$$
$$\cdot(a_1(\mathbf{k})a_1^+(\mathbf{k})+a_1^+(\mathbf{k})a_1(\mathbf{k}))+\left(\frac{1}{\alpha_2^2}(\kappa\cos^2\theta+\kappa_z\sin^2\theta)\right.$$
$$\left.+k^2\alpha_2^2+2k\chi\cos\theta\right)(a_2(\mathbf{k})a_2^+(\mathbf{k})+a_2^+(\mathbf{k})a_2(\mathbf{k}))$$
$$+\left(k^2\alpha_1^2(\nu\cos^2\theta+\nu_z\sin^2\theta)-\frac{\kappa}{\alpha_1^2}\right)(a_1(\mathbf{k})a_1(-\mathbf{k})+a_1^+(-\mathbf{k})a_1^+(\mathbf{k}))$$
$$+\left(k^2\nu\alpha_2^2-\frac{1}{\alpha_2^2}(\cos^2\theta+\kappa_z\sin^2\theta)\right)(a_2(\mathbf{k})a_2(-\mathbf{k})+a_2^+(-\mathbf{k})a_2^+(\mathbf{k}))\Bigg\}$$

We see that the last two terms can be made to vanish by choosing

$$\alpha_1^4=\frac{\kappa}{k^2(\nu\cos^2\theta+\nu_z\sin^2\theta)}, \qquad (7.98a)$$

$$\alpha_2^4=\frac{\kappa\cos^2\theta+\kappa_z\sin^2\theta}{k^2\nu}. \qquad (7.98b)$$

The Hamiltonian H is seen to be diagonalized. It can be written as the sum of two Hamiltonians, each corresponding to a characteristic wave in the moving medium:

$$H=H_m+H_e, \qquad (7.99a)$$

$$H_m=\frac{\hbar}{2}\int d^3k\left(\frac{\kappa}{\alpha_1^2}+k\chi\cos\theta\right)(a_1(\mathbf{k})a_1^+(\mathbf{k})+a_1^+(\mathbf{k})a_1(\mathbf{k})), \qquad (7.99b)$$

$$H_e=\frac{\hbar}{2}\int d^3k\left(\frac{\kappa\cos^2\theta+\kappa_z\sin^2\theta}{\alpha_2^2}+k\chi\cos\theta\right)(a_2(\mathbf{k})a_2^+(\mathbf{k})+a_2^+(\mathbf{k})a_2(\mathbf{k})). \qquad (7.99c)$$

In the case of a stationary uniaxial dielectric medium, $\chi=0$. We see that the photons associated with H_m are the ordinary photons and those associated with H_e are the extraordinary photons. The Hamiltonians in (7.99) can be expressed in terms of the number operators. When the Hamiltonian operates on an energy state, the result is the total photon energy in the state. The photon energies of the two types of photons corresponding to H_m and H_e are as follows:

$$E_m = \hbar\left(\frac{\kappa}{\alpha_1^2} + \kappa\chi\cos\theta\right), \tag{7.100a}$$

$$E_e = \hbar\left(\frac{\kappa\cos^2\theta + \kappa_z\sin^2\theta}{\alpha_2^2} + k\chi\cos\theta\right). \tag{7.100b}$$

The photon energies can be negative when the medium velocity is sufficiently high. They correspond to classical slow waves in moving media which can also possess negative energy[52].

In general, the energy of a photon with a specified wave vector **k** is equal to $\hbar$ times the angular frequency. Classically, the angular frequency is related to the **k** vector by dispersion relations. The derivation of the dispersion relation is facilitated by the use of the *kDB* system. Since each characteristic wave has a particular dispersion relation, we expect that for each characteristic wave there will be a corresponding photon after quantization is carried out.

7.4 FOUR-DIMENSIONAL NOTATION

7.4a Contravariant and Covariant Vectors

One can imagine a four-dimensional (4D) space composed of coordinates formed by time and three-dimensional space. Space-time coordinates of a physical event possess properties of a vector in the 4D space. Let us denote the four components of an event by

$$x^0 = ct, \qquad x^1 = x, \qquad x^2 = y, \qquad x^3 = z, \tag{7.101}$$

where we use the superscript 0 to denote the time component, and 1, 2, 3 to denote the space components. A popular convention in special relativity is to write the time component with the subscript 4 and designate it as imaginary; this should be carefully distinguished from the imaginary notations used in quantum theory and in wave theory. The notation that we use does not require an imaginary signature, but we must distinguish between superscripts and subscripts. This notation is readily generalized when general relativity is considered.

The transformation of the space-time coordinate vector from one observer to another is given by Lorentz transformations. Under the Lorentz transformation,

$$x^2+y^2+z^2-c^2t^2=x'^2+y'^2+z'^2-c^2t'^2 \tag{7.102}$$

is an invariant quantity independent of velocity. The square root of (7.102) expresses the magnitude of a 4D vector and in effect defines the transformation. As numerical values of other physical quantities change from one frame to another, this number stays unchanged in all frames. We note that in 4D space the magnitude of a vector can now be imaginary as well as real. A 4D vector is called a *spacelike* vector, a *null* vector, or a *timelike* vector, according to whether its magnitude is real, zero, or imaginary, respectively. The space-time coordinates of two physical events (ct_1,x_1,y_1,z_1) and (ct_2,x_2,y_2,z_2) form a four-vector with magnitude squared

$$(X_1-X_2)^2=(x_1-x_2)^2+(y_1-y_2)^2+(z_1-z_2)^2-c^2(t_1-t_2)^2,$$

which is a Lorentz invariant. When the vector is timelike, we can always find a moving observer such that, in his rest frame, the two events occur at the same location but at different times. He therefore observes the two events in person at times t_1' and t_2':

$$-c^2(t_1'-t_2')^2=(X_1-X_2)^2.$$

When the vector is a null vector, an observer has to move at velocity c in order to see the two events in person. When the vector is spacelike, there exists an observer in whose frame the two events occur at different locations but simultaneously in time. The 4D space with coordinates governed by the Lorentz transformation laws is called *Minkowski space*.

To visualize the fourth dimension, imagine that this 4D space is spanned by four unit base vectors,

$$\hat{e}_\alpha=(\hat{e}_0,\hat{e}_1,\hat{e}_2,\hat{e}_3). \tag{7.103}$$

Any 4D vector X can thus be expressed in terms of $\hat{e}_\alpha$:

$$X=x^\alpha\hat{e}_\alpha=x^0\hat{e}_0+x^1\hat{e}_1+x^2\hat{e}_2+x^3\hat{e}_3. \tag{7.104}$$

In (7.104) we used the Einstein summation convention: the repeated Greek index α implied summation from 0 to 3. We shall use Greek letters to indicate 0–3, and Roman letters to denote 1–3. The square of X is

$$X^2 = \hat{e}_\alpha \cdot \hat{e}_\beta x^\alpha x^\beta. \tag{7.105}$$

In view of (7.102),

$$X^2 = x^2 + y^2 + z^2 - c^2 t^2. \tag{7.106}$$

Thus we must have

$$\hat{e}_\alpha \cdot \hat{e}_\beta = 0 \qquad \text{for } \alpha \neq \beta, \tag{7.107a}$$

$$\hat{e}_i \cdot \hat{e}_i = 1 \qquad \text{for } i = 1, 2, 3, \tag{7.107b}$$

$$\hat{e}_0 \cdot \hat{e}_0 = -1. \tag{7.107c}$$

By (7.107a), all four base vectors are orthogonal to one another; by (7.107b), the three space base vectors have unit magnitude as usual; and, by (7.107c), the zeroth (or the fourth) base vector describing the fourth dimension possesses a magnitude squared of -1. It follows that the length of the zeroth base vector is imaginary. These four base vectors may be called *contravariant base* vectors which span a contravariant 4D space. A vector expressed in terms of the contravariant base vectors is called a *contravariant vector.*

We can define a set of covariant base vectors $\hat{e}^0$, $\hat{e}^1$, $\hat{e}^2$, and $\hat{e}^3$ such that

$$\hat{e}^0 = -\hat{e}_0, \tag{7.108a}$$

$$\hat{e}^i = \hat{e}_i, \qquad i = 1, 2, 3. \tag{7.108b}$$

The scalar product of $\hat{e}^0$ with $\hat{e}_0$ gives unity. The vector $\hat{e}^0$ also has a magnitude squared of -1. These four base vectors e^α may be called *covariant base* vectors which describe a covariant 4D space. A vector expressed in terms of the covariant base vectors is called a *covariant vector*. We write

$$X = x_\alpha \hat{e}^\alpha, \tag{7.109}$$

Components of X in the new base are now denoted as x_α, which, in view of (7.108), are related to x^α by

$$x^0 = -x_0, \tag{7.110a}$$

$$x^i = x_i. \tag{7.110b}$$

The contravariant components of X are denoted by superscripts, and its covariant components by subscripts. We define

$$\eta_{\alpha\beta} = \hat{e}_\alpha \hat{e}_\beta = \begin{bmatrix} -1 & 0 & 0 & 0 \\ 0 & 1 & 0 & 0 \\ 0 & 0 & 1 & 0 \\ 0 & 0 & 0 & 1 \end{bmatrix} \tag{7.111a}$$

and

$$\eta^{\alpha\beta} = \hat{e}^\alpha \hat{e}^\beta = \begin{bmatrix} -1 & 0 & 0 & 0 \\ 0 & 1 & 0 & 0 \\ 0 & 0 & 1 & 0 \\ 0 & 0 & 0 & 1 \end{bmatrix} \tag{7.111b}$$

to express the transformation between the two sets of base vectors and the contravariant and covariant components of a vector:

$$\hat{e}^\alpha = \eta^{\alpha\beta} \hat{e}_\beta, \qquad \hat{e}_\alpha = \eta_{\alpha\beta} \hat{e}^\beta, \tag{7.112a}$$

$$x^\alpha = \eta^{\alpha\beta} x_\beta, \qquad x_\alpha = \eta_{\alpha\beta} x^\beta. \tag{7.112b}$$

Equation 7.112 is equivalent to (7.108) and (7.110). The scalar product of two vectors is defined to be the summation over the contravariant components of one vector and the corresponding covariant components of another. Thus the magnitude squared of x^α is

$$X^2 = x^\alpha x_\alpha = \eta_{\alpha\beta} x^\alpha x^\beta \tag{7.113}$$

in view of (7.112b).

We summarize the rules for using the indices notation as follows:

1. When the index of a vector is raised or lowered, the zeroth component of the vector changes sign, while the other components remain unchanged.
2. When an index is denoted by a Greek letter, it ranges from 0 to 3. When an index is denoted by a Roman letter, it ranges from 1 to 3.
3. When an index is repeated on the same side of an equation, a summation over the index is implied. Summation is always carried out over a contravariant index and a covariant index.

4. Free (nonrepeated) indices on one side of an equation must be balanced by the same indices on the other side of the equation.

5. Contravariant components of a vector are denoted by superscripts; covariant components, by subscripts. The notation for the base vectors is just the opposite; subscripts denote contravariant base vectors, and superscripts denote covariant base vectors.

We have discussed the transformation between contravariant and covariant representations. We shall now consider the transformation of a contravariant or a covariant vector from one frame of reference to another. When two frames are in relative uniform motion, the transformation is determined by the Lorentz transformation (2.1). We can express the space-time coordinates of a physical event either by a contravariant or a covariant vector. The transformation from an unprimed frame to a primed frame is

$$x'^{\alpha} = P^{\alpha}{}_{\beta} x^{\beta} \tag{7.114a}$$

or

$$x'_{\alpha} = Q_{\alpha}{}^{\beta} x_{\beta}. \tag{7.114b}$$

We can view $P^{\alpha}{}_{\beta}$ as a matrix, denoted as $\overline{\mathbf{P}}_4$, operating on column matrix x^{β} and giving column matrix x'^{α}. A similar view applies to $Q_{\alpha}{}^{\beta}$. In view of (2.1), the transformation matrices $\overline{\mathbf{P}}_4$ and $\overline{\mathbf{Q}}_4$ are as follows:

$$\overline{\mathbf{P}}_4 = \begin{bmatrix} \gamma & -\gamma\beta_x & -\gamma\beta_y & -\gamma\beta_z \\ -\gamma\beta_x & 1+(\gamma-1)\beta_x^2/\beta^2 & (\gamma-1)\beta_x\beta_y/\beta^2 & (\gamma-1)\beta_x\beta_z/\beta^2 \\ -\gamma\beta_y & (\gamma-1)\beta_y\beta_x/\beta^2 & 1+(\gamma-1)\beta_y^2/\beta^2 & (\gamma-1)\beta_y\beta_z/\beta^2 \\ -\gamma\beta_z & (\gamma-1)\beta_z\beta_x/\beta^2 & (\gamma-1)\beta_z\beta_y/\beta^2 & 1+(\gamma-1)\beta_z^2/\beta^2 \end{bmatrix}, \tag{7.115a}$$

$$\overline{\mathbf{Q}}_4 = \begin{bmatrix} \gamma & \gamma\beta_x & \gamma\beta_y & \gamma\beta_z \\ \gamma\beta_x & 1+(\gamma-1)\beta_x^2/\beta^2 & (\gamma-1)\beta_x\beta_y/\beta^2 & (\gamma-1)\beta_x\beta_z/\beta^2 \\ \gamma\beta_y & (\gamma-1)\beta_y\beta_x/\beta^2 & 1+(\gamma-1)\beta_y^2/\beta^2 & (\gamma-1)\beta_y\beta_z/\beta^2 \\ \gamma\beta_z & (\gamma-1)\beta_z\beta_x/\beta^2 & (\gamma-1)\beta_z\beta_y/\beta^2 & 1+(\gamma-1)\beta_z^2/\beta^2 \end{bmatrix}, \tag{7.115b}$$

A few properties of $\bar{\mathbf{P}}_4$ and $\bar{\mathbf{Q}}_4$ follow from (7.114). The magnitude squared of x^α is a Lorentz invariant. From

$$x'^{\mu}x'_{\mu} = P^{\mu}{}_{\alpha}Q_{\mu}{}^{\beta}x^{\alpha}x_{\beta},$$

we learn that

$$P^{\mu}{}_{\alpha}Q_{\mu}{}^{\beta} = \delta_{\alpha}^{\beta}. \tag{7.116}$$

The summation is on μ. In matrix form, we have

$$\bar{\mathbf{P}}_4^t \cdot \bar{\mathbf{Q}}_4 = \bar{\mathbf{I}}_4,$$

where $\bar{\mathbf{I}}_4$ is the 4×4 unit matrix. Obviously, $\bar{\mathbf{P}}_4^t$ is the inverse of $\bar{\mathbf{Q}}_4$:

$$\bar{\mathbf{P}}_4^t = \bar{\mathbf{Q}}_4^{-1}.$$

The inverse of $\bar{\mathbf{P}}_4$ is then the transpose of $\bar{\mathbf{Q}}_4$:

$$\bar{\mathbf{P}}_4^{-1} = \bar{\mathbf{Q}}_4^t.$$

With these last two relations, the inverse transformation from primed to unprimed coordinate is as follows:

$$x^{\alpha} = Q_{\beta}{}^{\alpha}x'^{\beta}, \tag{7.117a}$$

$$x_{\alpha} = P^{\beta}{}_{\alpha}x'_{\beta} \tag{7.117b}$$

When (7.117) is compared with (7.114), it can be verified that the two are equivalent by using (7.116).

Conventionally, a contravariant vector is defined as one that transforms with the transformation matrix $P^{\alpha}{}_{\beta}$ as (7.114a); a covariant vector transforms with the matrix $Q_{\alpha}{}^{\beta}$ as (7.114b). Extending the definition, an nth-rank contravariant tensor transforms from one Lorentz frame to another by using the transformation matrix $P^{\alpha}{}_{\beta}$ n times. An nth-rank covariant tensor transforms from one Lorentz frame to another by using $Q_{\alpha}{}^{\beta}$ n times. The scalar product of an nth-rank contravariant tensor with an nth-rank covariant tensor is Lorentz-invariant.

For instance, the space-time derivatives $(\partial/\partial ct, \nabla)$ form a covariant four-vector because, according to (2.12), the transformation is like (7.114b) and (7.117b). If we denote the derivatives of a scalar function $\chi(x)$ by

$\chi_{,\alpha}=(\partial\chi/\partial ct, \nabla\chi)$, (2.12) becomes $\chi'_{,\alpha}=Q_\alpha{}^\beta\chi_{,\beta}$. By (2.13), the charge current density, written as

$$J^\alpha=(c\rho,\mathbf{J}), \tag{7.118}$$

is seen to be a contravariant vector because it transforms as (7.114a) and (7.117a). The space-time derivative of J^α,

$$J^\alpha{}_{,\alpha}=\frac{\partial\rho}{\partial t}+\nabla\cdot\mathbf{J},$$

becomes a scalar. The charge current conservation law states that

$$J^\alpha{}_{,\alpha}=0. \tag{7.119}$$

From (7.111), we can show that $\eta_{\alpha\beta}$ is a second-rank covariant tensor and $\eta^{\alpha\beta}$ is a second-rank contravariant tensor. They are known as metric tensors.

The transformation matrices $P^\alpha{}_\beta$ and $Q_\alpha{}^\beta$ govern pure Lorentz transformations. A pure Lorentz transformation satisfies two assumptions: (*a*) the coordinate axes of S and S' are parallel, and (*b*) the origins of the two coordinate systems coincide at $t=0$. A Lorentz transformation (LT) that satisfies (*b*) but does not satisfy (*a*) is called a homogeneous LT (HLT). An HLT is a combination of a pure LT plus a spatial rotation. Mathematically, the whole class of HLTs satisfies the postulates of a group and is called an HLT group. It is important to note that the group multiplication of two pure LTs results, not in a pure LT, but in a pure LT plus a spatial rotation. The matrices $P^\alpha{}_\beta$ and $Q_\alpha{}^\beta$ in (7.114) can be used to represent the HLT group. Although as represented in (7.115) they appear to be symmetrical, symmetry is not a general property of all elements in the HLT group. For instance, the transformation matrix for a spatial rotation is not symmetrical. When assumption (*b*) is violated, the LT is inhomogeneous. An inhomogeneous LT can be made homogeneous by rechoosing the space-time origins. The whole class of inhomogeneous LTs also forms a group called an inhomogeneous Lorentz group or simply the Poincaré group. The HLT group is a subgroup of the Poincaré group because the identity element is there. Any element in a Poincaré group can be joined to the identity element continuously by successive LTs. For completeness, we mention the LTs involving inversions. This group of LTs is called an improper Lorentz group in which an element cannot continuously join to the identity. In our treatment, we confine ourselves to pure LTs in the HLT group.

7.4b Maxwell's Equations in Tensor Form

To write Maxwell's equations in compact tensor form, we define a field tensor $F_{\alpha\beta}$ and an excitation tensor $G_{\alpha\beta}$. Explicitly in terms of matrix representation, we have

$$F_{\alpha\beta}=\begin{bmatrix} 0 & E_x & E_y & E_z \\ -E_x & 0 & -cB_z & cB_y \\ -E_y & cB_z & 0 & -cB_x \\ -E_z & -cB_y & cB_x & 0 \end{bmatrix}, \tag{7.120}$$

$$G_{\alpha\beta}=\begin{bmatrix} 0 & cD_x & cD_y & cD_z \\ -cD_x & 0 & -H_z & H_y \\ -cD_y & H_z & 0 & -H_x \\ -cD_z & -H_y & H_x & 0 \end{bmatrix}. \tag{7.121}$$

Both tensors are skew-symmetric:

$$F_{\alpha\beta}=-F_{\beta\alpha},$$
$$G_{\alpha\beta}=-G_{\beta\alpha}. \tag{7.122}$$

They are second-rank covariant tensors and transform as

$$F'_{\mu\nu}=Q_\mu{}^\alpha Q_\nu{}^\beta F_{\alpha\beta}, \tag{7.123a}$$

$$G'_{\mu\nu}=Q_\mu{}^\alpha Q_\nu{}^\beta G_{\alpha\beta}. \tag{7.123b}$$

Using matrix notation, we write

$$\overline{\mathbf{F}}'=\overline{\mathbf{Q}}_4\cdot\overline{\mathbf{F}}\cdot\overline{\mathbf{Q}}_4^t,$$

$$\overline{\mathbf{G}}'=\overline{\mathbf{Q}}_4\cdot\overline{\mathbf{G}}\cdot\overline{\mathbf{Q}}_4^t.$$

It is straightforward to show that the result of this transformation is identical to the Lorentz transformation (2.16) obtained with three-dimensional notation.

The three-dimensional field vectors are related to the field tensor and the excitation tensor in the following manner:

$$E_i = F_{0i},$$

$$cB_i = -\tfrac{1}{2}\varepsilon_{ijk}F_{jk},$$

$$cD_i = G_{0i},$$

$$H_i = -\tfrac{1}{2}\varepsilon_{ijk}G_{jk}.$$

The contravariant tensors corresponding to $F_{\alpha\beta}$ and $G_{\alpha\beta}$ can be obtained by using the metric tensor $\eta^{\alpha\beta}$. For instance,

$$G^{\mu\nu} = \eta^{\mu\alpha}\eta^{\nu\beta}G_{\alpha\beta} = \begin{bmatrix} 0 & -cD_x & -cD_y & -cD_z \\ cD_x & 0 & -H_z & H_y \\ cD_y & H_z & 0 & -H_x \\ cD_z & -H_y & H_x & 0 \end{bmatrix}. \tag{7.124}$$

This result can be obtained by the convention of raising and lowering indices. The contravariant components G^{0i} are negatives of their corresponding covariant components because an index 0 is raised.

In tensor notation, Maxwell's equations read as

$$F_{\alpha\beta,\gamma} + F_{\beta\gamma,\alpha} + F_{\gamma\alpha,\beta} = 0, \tag{7.125a}$$

$$G^{\alpha\mu}{}_{,\alpha} = J^{\mu}. \tag{7.125b}$$

We demonstrate that (7.125) is equivalent to the full set of Maxwell's equations. If none of the three indices α,β,γ in (7.125a) is zero, we have Gauss' law of magnetism:

$$\nabla \cdot \mathbf{B} = 0.$$

If one of α, β, γ is zero, (7.125a) gives Faraday's law:

$$\nabla \times \mathbf{E} + \frac{\partial \mathbf{B}}{\partial t} = 0.$$

If $\mu=0$ in (7.125b), we obtain Gauss' law of electricity:

$$\nabla\cdot\mathbf{D}=\rho.$$

For $\mu\neq 0$, (7.125b) gives

$$\nabla\times\mathbf{H}-\frac{\partial\mathbf{D}}{\partial t}=\mathbf{J},$$

which is Ampère's law with Maxwell's displacement current term $\partial\mathbf{D}/\partial t$.

The conventional exercise of expressing field vectors in terms of vector and scalar potentials is observed from (7.125a). It is quite easy to show that (7.125a) is satisfied if

$$F_{\alpha\beta}=A_{\alpha,\beta}-A_{\beta,\alpha}, \tag{7.126}$$

where A_α is a covariant four-vector, its 0th contravariant component is the scalar potential ϕ, and its space components are the vector potential $\mathbf{A}$ times c:

$$A_\alpha=\begin{bmatrix}-\phi\\ c\mathbf{A}\end{bmatrix},\qquad A^\alpha=\begin{bmatrix}\phi\\ c\mathbf{A}\end{bmatrix}. \tag{7.127}$$

Writing (7.126) in three-dimensional notation, we have the familiar expressions

$$\mathbf{E}=-\frac{\partial\mathbf{A}}{\partial t}-\nabla\phi,$$

$$\mathbf{B}=\nabla\times\mathbf{A}.$$

We can make a gauge transformation from A_α to A'_α so that

$$A_\alpha=A'_\alpha+\chi_{,\alpha}, \tag{7.128}$$

where χ is any scalar function of space-time. By introducing (7.128) in (7.125a), we have

$$F_{\alpha\beta}=A_{\alpha,\beta}-A_{\beta,\alpha}=A'_{\alpha,\beta}-A'_{\beta,\alpha}.$$

This shows that both A_α and A'_α give rise to the same field tensor. This arbitrariness is fixed by the gauge condition. The Lorentz gauge is

$$A^\mu{}_{,\mu}=0, \tag{7.129}$$

which takes the same form as the continuity equation for charge current densities in (7.119). The Lorentz gauge is relativistically covariant.

7.4c Constitutive Relations in Tensor Form

The constitutive relations in tensor notation provide a relation for the excitation tensor $G^{\alpha\beta}$ and the field tensor $F_{\alpha\beta}$. We write

$$G^{\alpha\beta}=\tfrac{1}{2}C^{\alpha\beta\rho\sigma}F_{\rho\sigma}. \tag{7.130}$$

We call the fourth-rank tensor $C^{\alpha\beta\rho\sigma}$ the constitutive tensor. Because of the skew-symmetric properties of $F_{\rho\sigma}$ and $G^{\alpha\beta}$, we see that

$$\begin{aligned} C^{\alpha\beta\rho\sigma} &= -C^{\beta\alpha\rho\sigma} \\ &= -C^{\alpha\beta\sigma\rho} \\ &= C^{\beta\alpha\sigma\rho}. \end{aligned} \tag{7.131}$$

The constitutive tensor is skew-symmetric with respect to the first pair, as well as the second pair, of indices. In general, a fourth-rank tensor in a four-dimensional space possesses 256 elements. Because of this skew symmetry, the first pair of indices has 6 independent elements and so does the second pair, giving rise to a total of 36 independent elements. Thus the 6×6 constitutive matrix $\overline{\mathbf{C}}$ in (1.16) is a faithful representation of the constitutive tensor.

We shall establish relations between the tensor elements of $C^{\alpha\beta\rho\sigma}$ and the matrix elements of $\overline{\mathbf{C}}$ in (1.16). In view of (7.131), we have

$$C^{0i0j}=-p_{ij}, \tag{7.132a}$$

$$C^{ijkl}=\varepsilon_{ijm}\varepsilon_{kln}q_{mn}, \tag{7.132b}$$

$$C^{0kij}=\varepsilon_{ijm}l_{km}, \tag{7.132c}$$

$$C^{ij0k}=-\varepsilon_{ijn}m_{nk}. \tag{7.132d}$$

In Chapter 1 we demonstrated that symmetric conditions exist for lossless media which limit the independent constitutive parameters to 21. In terms of the constitutive tensor, (1.20) corresponds to

$$C^{\alpha\beta\rho\sigma}=(C^{\rho\sigma\alpha\beta})^*. \tag{7.133}$$

At this point it is appropriate to mention the work of Tischer and Hess[103] on a covariant description of a conducting medium. Ohm's law relates conduction current to electric field by conductivity, which can be isotropic as well as anisotropic. For moving anisotropic conducting media, Tischer and Hess introduced a new three-dimensional vector, together with the conducting current, to form a four-dimensional skew-symmetric tensor just like $F_{\alpha\beta}$ and $G_{\alpha\beta}$. They thus obtained a covariant description of what we may call bianisotropic conducting media. The implications of the new vector have not been explored.

Other covariant descriptions of moving isotropic media exist, such as

$$(\delta_\rho^\alpha + u^\alpha u_\rho)J_c^\rho = -\sigma u_\beta F^{\alpha\beta}$$

for Ohm's law and

$$c\mu G_{\lambda\mu} = F_{\lambda\mu} + (n^2-1)(F_{\mu\sigma}u^\sigma u_\lambda - F_{\lambda\sigma}u^\sigma u_\mu)$$

for isotropic nonconducting media. The velocity four-vector $u^\alpha = \gamma(1,\boldsymbol{\beta})$. Thus the manifestly convariant descriptions explicitly display the velocity dependence. When reference is made to the rest frame of the medium, $u^\alpha = (1,0,0,0)$ and the two equations yield $\mathbf{J} = \sigma\mathbf{E}$, and $\mathbf{D} = \epsilon\mathbf{E}$ and $\mathbf{H} = \mathbf{B}/\mu$.

7.5 LAGRANGIAN FORMULATION OF MACROSCOPIC ELECTROMAGNETIC THEORY

7.5a Action Integral

Starting from a postulated Lagrangian density, the variational principle provides an elegant and systematic way of deriving the equations of motion and the conservation laws of a physical system. In the case of macroscopic electromagnetic fields the Lagrangian density is postulated as

$$L\left[x_\alpha, A_\alpha(x_\mu), A_{\alpha,\beta}(x_\mu)\right] = \tfrac{1}{4}G^{\alpha\beta}F_{\beta\alpha} + J^\alpha A_\alpha$$

$$= \tfrac{1}{8}C^{\alpha\beta\sigma\rho}(A_{\alpha,\beta} - A_{\beta,\alpha})(A_{\rho,\sigma} - A_{\sigma,\rho}) + J^\alpha A_\alpha. \qquad (7.134)$$

The Lagrangian density $L(x, A_\alpha, A_{\alpha,\beta})$ is a function of the space-time coordinates x_α, the potential functions A_α, and the space-time derivatives of the potential functions $A_{\alpha,\beta}$. The potential functions A_α are also called state functions. The charge current four-vector J_α is an externally given state function.

The variational principle applies to an action integral I, defined by integration of the Lagrangian density over a four-dimensional space R:

$$I = \int_R d^4x\, L\left[x_\alpha, A_\alpha(x_\mu), A_{\alpha,\beta}(x_\mu)\right]. \qquad (7.135)$$

The variation of the action integral is caused either by a variation of the state functions A_α or a variation of the domain of integration R which induces variations on the space-time dependent state function A_α, and on the externally given J_α because both are space-time dependent.

7.5b Hamilton's Principle and Maxwell's Equations

In Hamilton's principle, the domain of integration R is not varied. The state functions A_α inside the domain R are varied by an arbitrary and infinitesimal amount δA_α:

$$A'_\alpha(x_\mu) = A_\alpha(x_\mu) + \delta A_\alpha(x_\mu), \tag{7.136}$$

where A'_α are new state functions. The state functions on the boundary of R are not varied, where $\delta A_\alpha = 0$. The principle requires that the action integral be stationary under such variations:

$$\delta I = \int_R d^4x \left[L'(x_\alpha, A'_\alpha, A'_{\alpha,\beta}) - L(x_\alpha, A_\alpha, A_{\alpha,\beta}) \right] = 0. \tag{7.137}$$

The new Lagrangian density $L'(x_\alpha, A'_\alpha, A'_{\alpha,\beta})$ is related to the old Lagrangian density L as follows:

$$L'(x_\alpha, A'_\alpha, A'_{\alpha,\beta}) = L(x_\alpha, A_\alpha, A_{\alpha,\beta}) + \frac{\partial L}{\partial A_\alpha}\delta A_\alpha + \frac{\partial L}{\partial A_{\alpha,\beta}}\delta A_{\alpha,\beta}$$

$$= L(x_\alpha, A_\alpha, A_{\alpha,\beta}) + \frac{\partial L}{\partial A_\alpha}\delta A_\alpha - \frac{d}{dx^\beta}\left(\frac{\partial L}{\partial A_{\alpha,\beta}}\right)\delta A_\alpha + \frac{d}{dx^\beta}\left(\frac{\partial L}{\partial A_{\alpha,\beta}}\delta A_\alpha\right).$$

Substituting in (7.137) yields

$$\int_R d^4x \left\{ \left[\frac{\partial L}{\partial A_\alpha} - \frac{d}{dx^\beta}\left(\frac{\partial L}{\partial A_{\alpha,\beta}}\right)\right]\delta A_\alpha + \frac{d}{dx^\beta}\left(\frac{\partial L}{\partial A_{\alpha,\beta}}\delta A_\alpha\right)\right\} = 0. \tag{7.138}$$

The last term vanishes after integration because of the condition that δA_α vanishes on the boundary of R. Since δI vanishes for all variations of the state function δA_α, we have

$$\frac{\partial L}{\partial A_\alpha} - \frac{d}{dx^\beta}\left(\frac{\partial L}{\partial A_{\alpha,\beta}}\right) = 0, \tag{7.139}$$

which is known as the *Euler-Lagrangian equation*.

Maxwell's equation

$$F_{\alpha\beta,\gamma} + F_{\beta\gamma,\alpha} + F_{\gamma\alpha,\beta} = 0$$

is a direct consequence of the definition

$$F_{\alpha\beta}=A_{\alpha,\beta}-A_{\beta,\alpha}.$$

The other Maxwell's equation,

$$G^{\beta\alpha}{}_{,\beta}=J^{\alpha},$$

is derived from the Euler-Lagrangian equation by noting that $\partial L/\partial A_{\alpha}=J^{\alpha}$ and $\partial L/\partial A_{\alpha,\beta}=-G^{\alpha\beta}$.

7.5c Noether's Theorem and Energy Momentum Tensors

In Noether's theorem the domain of integration R is rendered an infinitesimal transformation, which induces variations on the state functions, both A_{α} and J_{α}. Suppose that the domain R is mapped onto a new domain R' such that

$$x'_{\alpha}=x_{\alpha}+\delta x_{\alpha}.$$

This mapping transplants a state function from x_{α} to x'_{α}:

$$A'_{\alpha}(x')=A_{\alpha}(x)+\delta A_{\alpha}. \tag{7.140}$$

Note that δA_{α} gives the difference between the new state function $A'_{\alpha}(x')$ at the new location x'_{α} and the old state function $A_{\alpha}(x)$ at the old location x_{α} before it is transplanted. Usually the state functions are not explicit functions of space-time coordinates. Consider an infinitesimal translation of the domain R:

$$x'_{\alpha}=x_{\alpha}+\varepsilon_{\alpha}. \tag{7.141}$$

From the active viewpoint of a coordinate transformation, all state functions are transplanted by an infinitesimal amount. The new and the old state functions are equal in magnitude, and there is no change in the orientation. From the passive viewpoint of a coordinate transformation, the coordinate axes are translated by an infinitesimal amount and the state functions are left unchanged. We have

$$\delta A_{\alpha}=0. \tag{7.142}$$

Next, consider an infinitesimal rotation of the coordinate axes:

$$x'_{\alpha}=x_{\alpha}-\omega_{\alpha}{}^{\beta}x_{\beta}.$$

The induced variations on the state functions are

$$\delta A_\alpha = \omega_\alpha{}^\beta A_\beta. \tag{7.143}$$

Thus, the state functions are rotated by an infinitesimal amount equal to the change of the component projections on the new and the old coordinate axes.

The variation denoted by δA_α is rather awkward because from the active viewpoint it compares two state functions at different locations. We define instead

$$\begin{aligned}\bar{\delta} A_\alpha &= A'_\alpha(x') - A_\alpha(x') \\ &\approx A'_\alpha(x) - A_\alpha(x)\end{aligned} \tag{7.144}$$

to denote the difference between the new and the old state functions at the same location. The second equality is valid up to first order, as is shown in the following:

$$\begin{aligned}\bar{\delta} A_\alpha &= A'_\alpha(x') - A_\alpha(x') = A'_\alpha(x) - A_\alpha(x) + (A'_{\alpha,\beta} - A_{\alpha,\beta})\delta x^\beta + \cdots \\ &\approx A'_\alpha(x) - A_\alpha(x).\end{aligned} \tag{7.145}$$

As a matter of fact, all first-order quantities with infinitesimal space-time separation can be shown to be equal. The relation between δA_α and $\bar{\delta} A_\alpha$ is easily established:

$$\begin{aligned}\bar{\delta} A_\alpha &= A'_\alpha(x') - A_\alpha(x') = A'_\alpha(x') - A_\alpha(x) - [A_\alpha(x') - A_\alpha(x)] \\ &= \delta A_\alpha - A_{\alpha,\beta}\,\delta x^\beta.\end{aligned} \tag{7.146}$$

Note that, when the domain variation is a translation, $\delta A_\alpha = 0$ because the new state function is the same as the old one. But $\bar{\delta} A_\alpha \neq 0$ because the new state function at x'_α is transplanted from a neighboring location and is certainly different from the old state function at x'_α, which has been transplanted to another location.

Under the infinitesimal domain variations, variations are induced on both the state functions A_α, $A_{\alpha,\beta}$ and the externally given J_α. The new Lagrangian density is related to the old one by

$$\begin{aligned}L'(x'_\alpha, A'_\alpha, A'_{\alpha,\beta}) &= L + L_{,\alpha}\delta x^\alpha + \frac{\partial L}{\partial A_\alpha}\delta A_\alpha + \frac{\partial L}{\partial J_\alpha}\delta J_\alpha + \frac{\partial L}{\partial A_{\alpha,\beta}}\delta A_{\alpha,\beta} \\ &= L + \frac{dL}{dx^\rho}\delta x^\rho + \frac{\partial L}{\partial A_\alpha}\bar{\delta} A_\alpha + \frac{\partial L}{\partial A_{\alpha,\beta}}\bar{\delta} A_{\alpha,\beta} + \frac{\partial L}{\partial J_\alpha}\bar{\delta} J_\alpha,\end{aligned}$$

where

$$\frac{dL}{dx^{\rho}} = L_{,\rho} + \frac{\partial L}{\partial A_{\alpha}} A_{\alpha,\rho} + \frac{\partial L}{\partial J_{\alpha}} J_{\alpha,\rho} + \frac{\partial L}{\partial A_{\alpha,\beta}} A_{\alpha,\beta\rho}.$$

The new domain of integration R' is related to the old domain of integration by the Jacobian, which is

$$\det\left|\frac{\partial x'^{\alpha}}{\partial x^{\beta}}\right| = 1 + (\delta x^{\alpha})_{,\alpha}. \tag{7.147}$$

The variation of the action integral under this domain variation becomes, to first order,

$$= \int_R d^4x \left\{ (1 + (\delta x^{\alpha})_{,\alpha}) \left(L + \frac{dL}{dx^{\rho}} \delta x^{\rho} + \frac{\partial L}{\partial A_{\alpha}} \bar{\delta} A_{\alpha} + \frac{\partial L}{\partial J_{\alpha}} \bar{\delta} J_{\alpha} + \frac{\partial L}{\partial A_{\alpha,\beta}} \bar{\delta} A_{\alpha,\beta} \right) - L \right\}$$

$$= \int_R d^4x \left\{ \frac{d}{dx^{\rho}} \left[L\delta x^{\rho} + \frac{\partial L}{\partial A_{\alpha,\rho}} \bar{\delta} A_{\alpha} \right] + \left[\frac{\partial L}{\partial A_{\alpha}} - \frac{d}{dx^{\beta}} \left(\frac{\partial L}{\partial A_{\alpha,\beta}} \right) \right] \bar{\delta} A_{\alpha} + \frac{\partial L}{\partial J_{\alpha}} \bar{\delta} J_{\alpha} \right\}.$$

Note that, although this variation of the action integral is derived from induced variations because of infinitesimal domain changes, the result is fairly general and includes Hamilton's principle as a special case. Let us keep the domain unchanged; then we have $\delta x^{\rho} = 0$ and $\bar{\delta} A_{\alpha} = \delta A_{\alpha}$. The externally given J_{α} is also not varied: $\delta J_{\alpha} = 0$. This result is seen to reduce to (7.138) in the preceding section.

Noether's theorem requires that the action integral be stationary, and the Euler-Lagrangian equations be satisfied for the state functions A_{α} under the domain variations. As a result, we obtain

$$\frac{d}{dx^{\rho}} \left(L\delta x^{\rho} + \frac{\partial L}{\partial A_{\alpha,\rho}} \bar{\delta} A_{\alpha} \right) + \frac{\partial L}{\partial J_{\alpha}} \bar{\delta} J_{\alpha} = 0. \tag{7.148}$$

This equation gives all of the conservation laws. We shall first consider the case of translation, which yields energy momentum tensors for macroscopic electromagnetic fields. The case of four-dimensional rotation is then considered, and the angular momentum conservation laws are derived.

Under the infinitesimal translation (7.141), we have

$$\bar{\delta}A_\alpha = -A_{\alpha,\beta}\varepsilon^\beta, \tag{7.149a}$$

$$\bar{\delta}J_\alpha = -J_{\alpha,\beta}\varepsilon^\beta. \tag{7.149b}$$

Equation 7.148 gives

$$\varepsilon^\beta\left[\frac{d}{dx^\rho}\left(\tfrac{1}{4}F^{\mu\nu}G_{\mu\nu}\delta^\rho_\beta - J^\alpha A_\alpha\delta^\rho_\beta - G^{\alpha\rho}A_{\alpha,\beta}\right) - A^\alpha J_{\alpha,\beta}\right] = 0.$$

This equation can be expressed in terms of field variables alone. After some manipulation and by use of the fact that $G^{\alpha\rho}A_{\beta,\alpha\rho} = 0$, we eliminate the potentials and obtain

$$T^{\alpha\beta}{}_{,\alpha} = -f^\beta, \tag{7.150}$$

where

$$T^{\alpha\beta} = -\tfrac{1}{4}\eta^{\alpha\beta}F_{\rho\sigma}G^{\rho\sigma} + G^{\rho\alpha}F_\rho{}^\beta \tag{7.151}$$

is the four-dimensional energy momentum tensor, and

$$f^\beta = J_\alpha F^{\alpha\beta}. \tag{7.152}$$

In three-dimensional vector notation, we find that

$$f^\beta = \begin{bmatrix} \mathbf{J}\cdot\mathbf{E} \\ c\rho\mathbf{E} + \mathbf{J}\times c\mathbf{B} \end{bmatrix}, \tag{7.153}$$

$$T^{\alpha\beta} = \begin{bmatrix} cW & c^2\mathbf{G} \\ \mathbf{S} & \overline{\mathbf{T}} \end{bmatrix}, \tag{7.154}$$

where

Electromagnetic energy density, $W=\frac{1}{2}(\mathbf{D}\cdot\mathbf{E}+\mathbf{B}\cdot\mathbf{H})$,

Power flow density, $\mathbf{S}=\mathbf{E}\times\mathbf{H}$,

Momentum density, $\mathbf{G}=\mathbf{D}\times\mathbf{B}$,

Maxwell stress tensor, $\overline{\mathbf{T}}=\frac{1}{2}(\mathbf{D}\cdot\mathbf{E}+\mathbf{B}\cdot\mathbf{H})\overline{\mathbf{I}}-\mathbf{DE}-\mathbf{BH}$.

Conservation law 7.152, written in vector notation, takes the form

$$\nabla\cdot\mathbf{S}+\frac{\partial W}{\partial t}=-\mathbf{J}\cdot\mathbf{E}, \tag{7.155a}$$

$$\nabla\cdot\overline{\mathbf{T}}+\frac{\partial\mathbf{G}}{\partial t}=-(\rho\mathbf{E}+\mathbf{J}\times\mathbf{B}). \tag{7.155b}$$

The first equation is Poynting's theorem. The second equation was derived also in Problem 1.2. Under an infinitesimal rotation, we have

$$\begin{aligned}\bar{\delta}A_\alpha&=\delta A_\alpha-A_{\alpha,\beta}\delta x^\beta\\&=\omega_\alpha{}^\beta A_\beta-A_{\alpha,\beta}\omega_\rho{}^\beta x^\rho,\end{aligned} \tag{7.156a}$$

$$\bar{\delta}J_\alpha=\omega_\alpha{}^\beta J_\beta-J_{\alpha,\beta}\omega_\rho{}^\beta x^\rho. \tag{7.156b}$$

After some manipulations (7.148) gives

$$\omega_{\alpha\beta}\left[\frac{d}{dx^\rho}(T^{\rho\alpha}x^\beta+G^{\alpha\rho}A^\beta+G^{\beta\rho}A^\alpha)+J_\rho F^{\rho\alpha}x^\beta+\eta^{\alpha\beta}J^\rho A_\rho\right]=0.$$

Interchanging α and β, we have

$$\omega_{\beta\alpha}\left[\frac{d}{dx^\rho}(T^{\rho\beta}x^\alpha+G^{\beta\rho}A^\alpha+G^{\alpha\rho}A^\beta)+J_\rho F^{\rho\beta}x^\alpha+\eta^{\beta\alpha}J^\rho A_\rho\right]=0.$$

We add the two equations above and note that $\omega_{\beta\alpha}=-\omega_{\alpha\beta}$, obtaining

$$\frac{d}{dx^\rho}M^{\rho\alpha\beta}=x^\alpha J_\rho F^{\rho\beta}-x^\beta J_\rho F^{\rho\alpha}, \tag{7.157}$$

where

$$M^{\rho\alpha\beta}=T^{\rho\alpha}x^\beta-T^{\rho\beta}x^\alpha \tag{7.158}$$

is the four-dimensional angular momentum tensor for electromagnetic fields.

PROBLEMS

7.1. Calculate the diffraction pattern produced by a plane wave normally incident on a rectangular slit.

7.2. Determine the field radiated by an aperture antenna made of a rectangular waveguide. Apply the equivalence principle by assuming that the field at the opening is equal to the TE_{10} mode inside the waveguide.

7.3. Using the equivalence principle, calculate the field radiated from the open end of a coaxial line. Assume that the field inside the waveguide is TEM and that the radius of the coax is very small compared with a wavelength.

7.4. Consider a plane wave polarized in the $\hat{x}$ direction and incident upon a sphere of radius a. The radius of the sphere is much larger than a wavelength. Assume that the following approximate values hold for the scattered fields $\mathbf{E}_s$ and $\mathbf{H}_s$ on the surface of the sphere: (a) in the shadow region $\mathbf{E}_s \approx -\mathbf{E}_i$ and $\mathbf{H}_s \approx -\mathbf{H}_i$, and ($b$) in the illuminated region $\hat{r}\times\mathbf{E}_s \approx -\hat{r}\times\mathbf{E}_i$ and $\hat{r}\times\mathbf{H}_s \approx \hat{r}\times\mathbf{H}_i$, where $\mathbf{E}_i$ and $\mathbf{H}_i$ are the fields of the incident wave. This assumed source distribution is a result of the physical optics approximation. Solve for the scattered radiation fields and calculate the echo area, defined by

$$A_e = \lim_{r\to\infty}\left(\frac{4\pi r^2 P_s}{P_i}\right),$$

where P_s is the back-scattered power density, and P_i is the incident power density of the plane wave.

7.5. Consider a plane wave normally incident on a conducting plate with area A. Assume that area A is very large as compared with square wavelength. Using the equivalence principle, approximate the source on the plate and show that the echo area is $A_e \approx k^2A^2/\pi$.

7.6. Calculate the diffracted field caused by a plane wave normally incident on a semiinfinite conducting plane. Find field solutions in both the Fresnel zone and the Fraunhofer zone.

7.7. Calculate the field radiated from an open parallel-plate waveguide. Assume that the field inside is the TE_1 mode.

7.8. Consider a conducting plate of length a in the $\hat{x}$ direction and width b

in the $\hat{y}$ direction. A plane wave

$$E_i = \hat{x}E_0 e^{ik_y^i y + ik_z^i z}$$

is incident obliquely upon the plate. Solve for the scattered field by two approximate source distributions.

(*a*) Use the source distribution for the scattered field, given by

$$\mathbf{J}_s = -\hat{n}\times(\mathbf{H}-\mathbf{H}_i) \qquad \text{and} \qquad \mathbf{M}_s = \hat{n}\times(\mathbf{E}-\mathbf{E}_i),$$

where **E** and **H** represent the total field at the scatterer. These are impressed currents which radiate in the presence of the conductor. By the image theorem, the source responsible for the scattered field in the absence of the scatterer is $2\mathbf{M}_s = -2\hat{n}\times\mathbf{E}_i$, since the total field **E** at the conducting plate is zero and $\mathbf{J}_s$ does not radiate in front of perfect conductors.

(*b*) According to the image theorem, the tangential components of **H** at the conducting plate are twice those from the same source in unbounded space and the tangential component of **E** is zero. The source distribution which gives rise to the scattered field is

$$\mathbf{J}_s = 2\hat{n}\times\mathbf{H}_i.$$

Compare the results obtained with the two approaches. The first approach is referred to as the induction theorem; the second approach is the physical optics approximation.

7.9. Consider a plane wave normally incident upon a rectangular aperture in a conducting screen occupying the x-y plane. The incident wave is polarized in the $\hat{x}$ direction. Replace the aperture with equivalent electric current source and find the electric field in the radiation zone for the entire half space $z>0$. Next consider the problem with the aperture and screen interchanged. Let the incident field be polarized in the $-\hat{y}$ direction. Construct equivalent magnetic current sources and find the scattered magnetic field in the radiation zone for the entire half-space $z>0$. Show that the solutions to the two problems are related by Babinet's principle. If in the second problem we construct electric current sources instead of magnetic current sources, will the solutions obey Babinet's principle?

7.10. By the image theorem, a vertical monopole antenna on a conducting plane is equivalent to a dipole with the conductor removed. In radio broadcasting stations, the Earth is used as the conducting plane. Calculate the power, the gain, and the radiation pattern arising from a monopole on a conducting plane.

7.11. Using the image theorem, find the field inside a parallel-plate waveguide caused by an electric dipole antenna. The dipole is placed in the middle between the two plates, and its axis is parallel to the plates. Repeat the calculations when its axis is perpendicular to the plates.

7.12. Find multiple images for a dipole source in a rectangular metallic waveguide.

7.13. Find a stationary formula for cutoff frequencies in a cylindrical metallic waveguide similar to the one for resonator cavities as given by (7.43).

7.14. Derive a stationary formula for the resonant frequency of a resonator in terms of magnetic fields **H**. Show that it is

$$k^2 = \frac{\iiint dV(\nabla \times \mathbf{H})^2}{\iiint dV H^2}$$

and that **H** need not satisfy any boundary conditions. Repeat the example for a circular cavity as given in the text by assuming approximate magnetic field distributions and compare with the exact value in (7.44).

7.15. Show that, if the inside of a waveguide is filled with more than one isotropic medium, the stationary formula for the cutoff frequency in terms of the **E** field is

$$\omega_c^2 = \frac{\iint dS(1/\mu)(\nabla \times \mathbf{E})^2 + 2\oint dl\,\hat{n}\cdot[(1/\mu)(\nabla \times \mathbf{E}) \times \mathbf{E}]}{\iint dS \epsilon E^2},$$

where $\hat{n}$ is the outward-pointing unit vector normal to the waveguide walls.

7.16. Use the stationary formula as derived in Problem 7.15 to solve for the cutoff frequency of a rectangular waveguide filled with two different dielectric media. Let the waveguide dimension be a along $\hat{x}$ and b along $\hat{y}$. The dielectric ϵ_1 fills the space from $x=0$ to $x=a/2$, and the dielectric ϵ_2 fills the space from $x=a/2$ to $x=a$. Assuming an electric field of $\mathbf{E}=\hat{y}\sin(\pi x/a)$, show that the cutoff frequency is

$$\omega_c \approx \frac{\pi}{a\sqrt{\mu(\epsilon_1+\epsilon_2)/2}}.$$

7.17. Using stationary formula 7.52, show that the echo area for back scattering resulting from a linearly polarized field

$$\mathbf{E}^i = \hat{z}E_0 e^{ikx}$$

is

$$A_e = \lim_{r\to\infty} 4\pi r^2 \left|\frac{E_s}{E^i}\right|^2 = \pi \left| \frac{\eta}{\lambda} \frac{\left(\oiint J_z^a e^{ikx}\, dS\right)^2}{\oiint \mathbf{E}^a \cdot \mathbf{J}^a dS} \right|^2.$$

7.18. A plane wave $\mathbf{E}^i = \hat{z}E_0 e^{ikx}$ is normally incident upon a conducting wire of length $L = \lambda/2$ along the z axis. Assume that the current on the wire is $I^a = I_0 \cos kz$. Using the stationary formula in Problem 7.17 and the fact that $\langle a,a\rangle = -73 I_0^2$ (see Problem 6.6), show that the echo area for back scattering is $A_e \approx 0.86\lambda^2$.

7.19. Let Z_a be the input impedance of an aperture antenna made from a plane conductor, and Z_e be the input impedance of its complementary antenna, made from the metal removed from the plane conductor in order to form the aperture.[54] Show that $Z_a Z_e = \eta^2/4$.

7.20. Which of the following media are reciprocal? For the nonreciprocal ones, what are their complementary media?

(*a*) A biaxial medium.
(*b*) A moving biaxial medium.
(*c*) A biisotropic medium with a real χ.
(*d*) A biisotropic medium with an imaginary χ.
(*e*) A ferrite in a dc magnetic field.

7.21. What are the expectation values for the field operator $\mathbf{D}$ with a given energy state $|n\rangle$? What are the expectation values for $\mathbf{D}^2$?

7.22. Let A, B, and C all be Hermitian operators, and let $[A,B] = iC$. Define the mean-square deviation by $(\Delta A)^2 = \langle\psi|(A - \langle A\rangle)^2|\psi\rangle$ and $(\Delta B)^2 = \langle\psi|(B - \langle B\rangle)^2|\psi\rangle$ with respect to the state function $|\psi\rangle$, where $\langle A\rangle = \langle\psi|A|\psi\rangle$ and $\langle B\rangle = \langle\psi|B|\psi\rangle$ are expectation values of A and B. Using the Schwartz inequality $\langle\phi|\phi\rangle\langle x|x\rangle \geqslant |\langle\phi|x\rangle|^2$, show that

$$(\Delta A)^2(\Delta B)^2 \geqslant \tfrac{1}{4}|\langle\psi|C|\psi\rangle|^2.$$

When $C = \hbar$, the commutation relation for A and B implies the uncertainty relation $\Delta A\, \Delta B \geqslant \hbar/2$.

7.23. The coherent state is also a minimum uncertainty state. Show that with the coherent state $(\Delta A)\cdot(\Delta D)=\hbar/2$.

7.24. Given $[a,a^+]=1$, compute $[a,(a^+)^n],[a^+,a^n]$, and $[a,e^{a^+}]$.

7.25. Prove that, if $[a,a^+]=1$, then $e^{xa}e^{xa^+}=e^{xa^+}e^{xa}e^{x^2}$.

7.26. Show that $e^{-xa^+}f(a,a^+)e^{xa^+}=f(a+x,a^+)$, where x is a number.

7.27. What are the commutation relations for field vectors at different space-time points? Interpret these relations, and use them to derive Maxwell's equations from the postulated Hamiltonian[47].

7.28. The Hamiltonian of a harmonic oscillator is given by

$$H=\frac{1}{2m}(p^2+m^2\omega^2x^2).$$

Define $a=(m\omega x+ip)/\sqrt{2m\hbar\omega}$, and $a^+=(m\omega x-ip)/\sqrt{2m\hbar\omega}$. Given $[x,p]=i\hbar$, what are the Hamiltonian and the commutation relation for a and a^+? Show that

$$(\Delta p)^2=\frac{m\hbar\omega}{2}(2n+1) \quad \text{and} \quad (\Delta x)^2=\frac{\hbar}{2m\omega}(2n+1)$$

by using the energy states of the Hamiltonian.

7.29. Show that the unitary transformation U which transforms from the Heisenberg picture to the Schrödinger picture is governed by the differential equation

$$i\hbar\frac{dU}{dt}=H_sU,$$

where H_s is the Hamiltonian in the Schrödinger picture.

7.30. The Schrödinger equation is

$$i\hbar\frac{d}{dt}|\psi_s\rangle=H_s|\psi_s\rangle.$$

Write a Hamiltonian in the Schrödinger picture as

$$H_s=H_{0s}+H_{is},$$

where H_{0s} is time-independent and H_{is} is called the interaction Hamiltonian. Transform to the interaction picture by a unitary transformation V such that $H_{ii}=VH_{is}V^+$ and

$$i\hbar\frac{d}{dt}|\psi_i\rangle=H_{ii}|\psi_i\rangle,$$

where H_{ii} is the interaction Hamiltonian in the interaction picture. Show that the differential equation is equivalent to the following integral equation:

$$|\psi_i(t)\rangle = |\psi_i(0)\rangle + \frac{1}{ih}\int_0^t d\tau\, H_{ii}(\tau)|\psi_i(\tau)\rangle.$$

For first-order perturbation, $|\psi_i(t)\rangle$ in the integrand is approximated by $|\psi_i(0)\rangle$. Thus

$$|\psi_i(t)\rangle = 1 + \frac{1}{i\hbar}\int_0^t d\tau\, H_{ii}(\tau)|\psi_i(0)\rangle.$$

For second-order perturbation, this solution is substituted in the integral equation, and we have

$$|\psi_i(t)\rangle = \left\{1 + \frac{1}{ih}\int_0^t d\tau\, H_{ii}(\tau)\left[1 + \frac{1}{ih}\int_0^\tau d\tau'\, H_{ii}(\tau')\right]\right\}|\psi_i(0)\rangle.$$

The process can be continued. The transition from state $|\psi_i(0)\rangle$ to state $|\psi_i(t)\rangle$ is governed by a scattering operator S, $|\psi_i(t)\rangle = S|\psi_i(0)\rangle$. Show that the transition amplitude between an initial state $|E_i\rangle$ and a final state $|E_f\rangle$ under the first-order perturbation theory is

$$S_{fi} = \langle E_f|S|E_i\rangle = -2ie^{i(E_f - E_i)t/2\hbar}\langle E_f|H_{ii}|E_i\rangle \frac{\sin[(E_f - E_i)t/2\hbar]}{E_f - E_i},$$

where $|E_f\rangle$ and $|E_i\rangle$ are the energy eigenvectors of the unperturbed Hamiltonian H_{0s}.

7.31. Show that the homogeneous Lorentz transformations (LTs) form a group by demonstrating that (*a*) two successive LTs are equivalent to another LT (closure), (*b*) associative law applies to successive LTs, (*c*) the identity of the group is no transformation at all, and (*d*) for each LT there exists an inverse LT.

7.32. Show that pure Lorentz transformations violate the closure postulate of a group.

References

1. Abramowitz, M., and I. A. Stegun. *Handbook of Mathematical Functions*, Dover Publications, New York, 1965.
2. Anderson, J. L. *Principles of Relativity Physics*, Academic Press, New York, 1967.
3. Anderson, J. L., and J. W. Ryon. "Electromagnetic Radiation in Accelerated System," *Phys. Rev.*, **181**, 1765–1775 (1969).
4. Astrov, D. N. "The Magnetoelectric Effect in Antiferromagnetics," *Zh. Eksp. Teor. Fiz.*, **38**, 984–985 (1960).
5. Baños, A., Jr. *Dipole Radiation in the Presence of a Conducting Half-Space*, Pergamon Press, New York, 1966.
6. Barut, A. O. *Electrodynamics and Classical Theory of Fields and Particles*, Macmillan, New York, 1964.
7. Beckmann, P., and A. Spizzichino. *The Scattering of Electromagnetic Waves from Rough Surfaces*, Pergamon Press, New York, 1963.
8. Bergmann, P. G. *Introduction to the Theory of Relativity*, Prentice-Hall, Englewood Cliffs, N. J., 1942.
9. Birss, R. R., "Macroscopic Symmerty in Space-Time," *Rep. Progr. Phys.*, **26** 307–310 (1963).
10. Bjorken, J. D., and S. D. Drell. *Relativistic Quantum Mechanics*, McGraw-Hill, New York, 1964.
11. Bjorken, J. D., and S. D. Drell. *Relativistic Quantum Fields*, McGraw-Hill, New York, 1965.
12. Born, M., and E. Wolf. *Principles of Optics*, Pergamon Press, New York, 1970.
13. Boyd, G. D., and J. P. Gordon. "Confocal Multimode Resonator for Millimeter through Optical Wavelength Masers," *Bell System Tech. J.*, **40**, 489–508 (1961).
14. Brekhovskikh, L. M. *Waves in Layered Media*, Academic Press, New York, 1960.
15. Brevik, I. "Electromagnetic Energy Momentum Tensor within Material Media. 1: Minkowski's Tensor," *Kgl. Dan. Vidensk. Selsk. Mat. Fys. Medd.*, **37**, No. 11, 1–51 (1970).
16. Brevik, I. "Electromagnetic Energy Momentum Tensor within Material Media. 2: Discussion of Various Tensor Forms," *Kgl. Dan. Vidensk. Selsk. Mat. Fys. Medd.*, **37**, No. 13, 1–70 (1970).
17. Brevik, I., and B. Lautrap. "Quantum Electrodynamics in Material Media," *K. Dan. Vidensk. Selsk. Mat. Fys. Medd.*, **38**, No. 1, 3–36 (1970).
18. Carniglia, C. D., and L. Mandel. "Quantization of Evanescent Electromagnetic Waves," *Phys. Rev.* D, **3**, No. 2, 280–296 (1971).
19. Čerenkov, P. A. "Visible Radiation Produced by Electrons Moving in a Medium with Velocities Exceeding That of Light," *Phys. Rev.*, **52**, 378–379 (1937)
20. Chawla, B. R., and H. Unz. *Electromagnetic Waves in Moving Magneto-Plasmas*, University Press, Lawrence, Kan., 1971.
21. Cheng, D. K., and J. A. Kong. "Covariant Descriptions of Bianisotropic Media," *Proc. IEEE*, **56**, 248–251 (1968).

22. Collin, R. E. *Field Theory of Guided Waves*, McGraw-Hill, New York, 1960.

23. Compton, R. T., Jr. "The Time-Dependent Green's Function for Electromagnetic Waves in Moving Simple Media," *J. Math. Phys.*, **7**, 2145–2152 (1966).

24. Costen, R. C., and D. Adamson. "Three-Dimensional Derivation of the Electrodynamic Jump Conditions and Momentum-Energy Laws at a Moving Boundary," *Proc. IEEE*, **53**, 1181–1196 (1965).

25. Daly, P., and H. Gruenberg. "Energy Relations for Plane Waves Reflected from Moving Media," *J. Appl. Phys.*, **38**, 4486–4489 (1967).

26. Deringin, L. N., "The Reflection of a Longitudinally Polarized Plane Wave from a Surface of Rectangular Corrugations," *Radio Tekhnol.*, **15**, 9–16 (1960).

27. DeVries, H. "Rotatory Power and Other Properties of Certain Liquid Crystals," *Acta Crystallogs.*, **4**, 219–226 (1951).

28. DeGroot, S. R., and L. G. Suttorp, *Foundations of Electrodynamics*, North-Holland, Amsterdam, 1972.

29. Dirac, P. A. M. *Principles of Quantum Mechanics*, 4th ed., Oxford University Press, London, 1958.

30. Dirac, P. A. M. *Lectures on Quantum Field Theory,* Yeshiva University, New York, 1966.

31. Du, L. J., and R. T. Compton, Jr. "Cutoff Phenomena for Guided Waves in Moving Media," *IEEE Trans. Microwave Theory Tech.*, **MTT-14**, No. 8, 358–363 (1966).

32. Dzyaloshinskii, I. E. "On the Magnetoelectrical Effect in Antiferromagnets," *Sov. Phys.-JETP*, **10**, 628–629 (1960).

33. Einstein, A., et al. *Principle of Relativity*, Dover Publications, New York, 1923.

34. Fano, R. M., L. J. Chu, and R. B. Adler. *Electromagnetic Fields, Energy, and Forces*, Wiley, New York, and M.I.T. Press, Cambridge, Mass., 1960.

35. Felsen, L. B., and N. Marcuwitz. *Radiation and Scattering of Electromagnetic Waves*, Prentice-Hall, Englewood Cliffs, N. J., 1973.

36. Feynmann, R. P., and A. R. Hibbs. *Quantum Mechanics and Path Integrals*, McGraw-Hill, New York, 1965.

37. Frank, I., and Ig. Tamm. "Coherent Visible Radiation of Fast Electrons Passing through Matter," *Compt. Rend. (Dokl.)*, **14**, No. 3, 109–114 (1937).

38. Fuchs, R. "Wave Propagation in a Magnetoelectric Medium," *Phil. Mag.*, **11**, 647 (1965).

39. Fulton, R. L. "Macroscopic Quantum Electrodynamics," *J. Chem. Phys.*, **50**, 3355–3377 (1969).

40. Ginzburg, V. L. *Propagation of Electromagnetic Waves in Plasma*, Addison-Wesley, Reading, Mass., 1964.

41. Goldstein, H. *Classical Mechanics*, Addison-Wesley, Reading, Mass., 1950.

42. Goodman, J. W. *Introduction to Fourier Optics*, McGraw-Hill, New York, 1968.

43. Goos, Von F., and H. Hänchen, "Ein neuer und fundamentaler Versuch zur Totalreflexion," *Ann. der Phys.* **6**, 333–346 (1947).

44. Harrington, R. F. *Time-Harmonic Electromagnetic Fields*, McGraw-Hill, New York, 1961.

45. Harrington, R. F., and A. T. Villeneuve. "Reciprocity Relationships for Gyrotropic Media," *IRE Trans. Microwave Theory Tech.*, **MTT-6**, No. 3, 308–310 (1958).

46. Haus, H. A. "Steady-State Quantum Analysis of Linear Systems," *Proc. IEEE*, **58**, 1599–1611, 1970.

47. Heitler, W. *The Quantum Theory of Radiation*, 3rd ed., Oxford University Press, London, 1966.

48. Hill, E. L. "Hamilton's Principle and the Conservation Theorems of Mathematical Physics," *Rev. Mod. Phys.*, **23**, 253–260 (1951).

49. Indenbom, V. L. "Irreducible Representations of the Magnetic Groups and Allowance for Magnetic Symmetry," *Sov. Phys.-Crystallogr.*, **5**, 493 (1960).

50. Jackson, J. D. *Classical Electrodynamics*, Wiley, New York, 1962.

51. Jauch, J. M., and K. M. Watson. "Phenomenological Quantum Electrodynamics," *Phys. Rev.*, **74**, 950–957 (1948).

52. Jauch, J. M., and K. M. Watson. "Phenomenological Quantum Electrodynamics. Part II: Interaction of the Field with Charges," *Phys. Rev.*, **75**, 1485 – 1493 (1948).

53. Jauch, J. M., and K. M. Watson. "Phenomenological Quantum Electrodynamics. Part III: Dispersion," *Phys. Rev.*, **75**, 1249–1261 (1949).

54. Jordan, E. C., and K. G. Balmain, *Electromagnetic Waves and Radiating Systems*, Prentice-Hall, Englewood Cliffs, N. J., 1968.

55. Kapany, N. S., and J. J. Burke. *Optical Waveguides*, Academic Press, New York, 1972.

56. King, R. W. P., and C. W. H. Harrison, Jr. *Antennas and Waves*, M.I.T. Press, Cambridge, Mass., 1969.

57. Klauder, J. R., and E. C. G. Sudarshan. *Fundamentals of Quantum Optics*, Benjamin, New York, 1968.

58. Kong, J. A., and D. K. Cheng. "Guided Waves in Moving Anisotropic Media," *IEEE Trans. Microwave Theory Tech.*, **MTT-16**, 99–103 (1968).

59. Kong, J. A. "Quantization of Electromagnetic Radiation Fields in Moving Uniaxial Media," *J. Appl. Phys.*, **41**, 554–559 (1970).

60. Kong, J. A., "Interaction of Accoustic Waves with Moving Media," *J. Acoust. Soc. Am.*, **48**,236–241 (1970).

61. Kong, J. A. "Reflection and Transmission of Electromagnetic Waves by Stratified Moving Media," *Can. J. Phys.*, **49**, 2786–2797 (1971).

62. Kong, J. A. "Electromagnetic Fields Due to Dipole Antennas over Stratified Anisotropic Media," *Geophys.*, **37**, 985–996 (1972).

63. Kong, J. A. "Theorems of Bianisotropic Media," *Proc. IEEE*, **60**, 1036–1046 (1972).

64. Kong, J. A. "Optics of Bianisotropic Media," *J. Opt. Soc. Am.* **64**, 1304–1308 (1974).

65. Landau, L. D., and E. M. Lifshitz. *Electrodynamics of Continuous Media*, Addison-Wesley, Reading, Mass., 1960.

66. Landau, L. D., and E. M. Lifshitz. *Mechanics* 2nd ed., Addison-Wesley, Reading, Mass., 1960.

67. Landau, L. D., and E. M. Lifshitz. *Classical Theory of Fields*, 3rd ed., Addison-Wesley, Reading, Mass., 1971.

68. Landau, L. D., and E. M. Lifshitz. *Quantum Mechanics: Non-Relativistic Theory*, Addison-Wesley,Reading, Mass., 1958.

69. Lee, K. S. H., and C. H. Papas. "Electromagnetic Radiation in the Presence of Moving Simple Media," *J. Math. Phys.*, **5**, 1668–1672 (1964).

70. Lee, S. W., and Y. T. Lo. "Reflection and Transmission of Electromagnetic Waves by a Moving Uniaxially Anisotropic Medium," *J. Appl. Phys.*, **38**, 870–875 (1967).

71. Louisell, W. *Radiation and Noise in Quantum Electronics*, McGraw-Hill, New York, 1964.

72. Marcuse, D. *Engineering Quantum Electrodynamics*, Van Nostrand-Reinhold, New York, 1970.

73. Marcuse, D. *Light Transmission Optics*, Van Nostrand-Reinhold, New York, 1972.

74. Mason, W. P. *Crystal Physics and Interaction Processes*, Academic Press, New York, 1966.

75. Maxwell, J. C. *A Treatise on Electricity and Magnetism*, Dover Publications, New York, 1954.

76. McKenzie, J. F. "Electromagnetic Waves in Uniformly Moving Media," *Proc. Phys. Soc.*, **91**, 532–536 (1967).

77. Mergelyan, O. S. "A Point Charge in a Gyrotropic Dielectric," *Sov. Phys.-Tech. Phys.*, **12**, 594–597 (1967).

78. Møller, C. *The Theory of Relativity*, Oxford University Press, London, 1966.

79. Morse, P. M., and H. Feshbach. *Methods of Theoretical Physics*, McGraw-Hill, New York, 1953.

80. O'Dell, T. H. "Magnetoelectrics—A New Class of Materials," *Electron. Power*, **11**, 266–267 (1965).

81. O'Dell, T. H. *The Electrodynamics of Magneto-Electric Media*, Vol. 11: *Selected Topics in Solid State Physics*, E. P. Wohlforth (Ed.), North-Holland, Amsterdam, 1970.

82. Panofsky, W. K. H., and M. Phillips. *Classical Electricity and Magnetism*, 2d ed., Addison-Wesley, Reading, Mass., 1962.

83. Penfield, P., Jr., and H. A. Haus. *Electrodynamics of Moving Media*, M.I.T. Press, Cambridge, Mass., 1967.

84. Rado, G. T. "Observation and Possible Mechanisms of Magnetoelectric Effects in a Ferromagnet," *Phys. Rev. Lett.*, **13**, 335 (1964).

85. Ramo, S., J. R. Whimery, and T. Van Duzer. *Fields and Waves in Communication Electronics*, Pergamon Press, New York, 1970.

86. Rao, B. R., and T. T. Wu. "On the Applicability of Image Theory in Anisotropic Media," *IEEE Trans. Antennas Propagation*, **AP-13**, No. 5, 814–815 (1965).

87. Roetgen, W. C. "Ueber die durch Bewegung eines in homogenenen electrischen Felde befindlichen Dielectricums hervorgerufene electrodynamische Kraft," *Ann. Physik.*, **35**, 264–270 (1888).

88. Rohrlich, F. *Classical Charged Particles*, Addison-Wesley, Reading, Mass., 1965.

89. Rumsey, V. H. "Reaction Concept in Electromagnetic Theory," *Phys. Rev.*, **94**, 1483–1491; **95**, 1706 (1954).

90. Schiff, L. I. *Quantum Mechanics*, McGraw-Hill, New York, 1968.

91. V. T. Schlomka, "Das Ohmsche Gesetz bei Bewegten Körpern," *Ann. Phys.*, **6**, 246–252 (1951).

92. Schmutzer, E. von. "Zur Relativistischer Elektrodynamik in Beliebigen Medien," *Ann. Phys. (Leipzig)*, **6**, 171–180 (1956).

93. Shiozawa, T., and N. Kumagai, "Total Reflection at the Interface between Relatively Moving Media," *Proc. IEEE*, **55** 1243–1244 (1967).

94. Silver, S. *Microwave Antenna Theory and Design*, Dover Publications, New York, 1949.

95. Sommerfeld, A. *Partial Differential Equations*, Academic Press, New York, 1962.

96. Sommerfeld, A. *Electrodynamics*, Academic Press, New York, 1949.

97. Sommerfeld, A. *Optics*, Academic Press, New York 1949.

98. Stratton, J. A. *Electromagnetic Theory*, McGraw-Hill, New York, 1941.

99. Synge, J. L. *Relativity: The Special Theory*, North-Holland, Amsterdam, 1965.

100. Tai, C. T. "A Study of Electrodynamics of Moving Media," *Proc. IEEE*, **52**, 685–689 (1964).

101. Tai, C. T. *Dyadic Green's Function in Electromagnetic Theory*, Intext Publishers, New York, 1971.

102. Tellegen, B. D. H. "The Gyrator, a New Electric Network Element," *Philips Res. Rep.*, **3**, 81–101 (1948).

103. Tischer, M., von, and S. Hess. "Die Materialgleichungen in Beliebigen Medien," *Ann. Phys. (Leipzig)*, **3**, No. 3–4, 113–121 (1959).

104. Tolman, R. C. *Relativity, Thermodynamics, and Cosmology*, Oxford University Press, London, 1966.

105. Tsang, L., J. A. Kong, and G. Simmons, "Interference Patterns of a Horizontal Electric Dipole over Layered Dielectric Media," *J. Geophys. Res.*, **78**, 3287–3300 (1973).

106. Tyras, G. *Radiation and Propagation of Electromagnetic Waves*, Academic Press, New York, 1969.

107. Van Bladel, J., *Electromagnetic Fields*, McGraw-Hill, New York, 1964.

108. Wait, J. R. (Ed.). *Electromagnetic Probing in Geophysics*, Golem Press, Boulder, Colo., 1971.

109. Wait, J. R. *Electromagnetic Waves in Stratified Media*, 2nd ed., Pergamon Press, New York, 1970.

110. Watson, G. N. *A Treatise on the Theory of Bessel Functions*, 2nd ed., Macmillan, New York, 1944.

111. Wilson, H. A. "On the Electric Effect of Rotating a Dielectric in a Magnetic Field," *Phil. Trans. Roy. Soc. London*, **204A**, 121–137 (1905).

112. Yariv, A., *Quantum Electronics*, Wiley, New York, 1967.

113. Yeh, C. "Propagation along Moving Dielectric Waveguides," *J. Opt. Soc. Am.*, **58**, 767–770 (1968).

114. Zernike, F., and J. E. Midwinter, *Applied Nonlinear Optics*, Wiley-Interscience, New York, 1973.

115. Ziman, J. M., *Elements of Advanced Quantum Theory*, Cambridge University Press, London 1969.

Index